American Geographers,
1784–1812

American Geographers, 1784–1812

A Bio-Bibliographical Guide

Ben A. Smith and James W. Vining

PRAEGER

Westport, Connecticut
London

Library of Congress Cataloging-in-Publication Data

Smith, Ben A., 1936–
American geographers, 1784–1812 : a bio-bibliographical guide / Ben A. Smith and James W. Vining
p. cm.
Includes bibliographical references (p.).
ISBN 0–313–32336–4 (alk. paper)
1. Geography—United States—Bio-bibliography. 2. Geographers—United States—Biography. I. Vining, James W. II. Title.
Z6004.B6 S65 2003
016.910′92′273—dc21 2002029897

British Library Cataloguing in Publication Data is available.

Library of Congress Catalog Card Number: 2002029897
ISBN: 0–313–32336–4

First published in 2003

Praeger Publishers, 88 Post Road West, Westport, CT 06881
An imprint of Greenwood Publishing Group, Inc.
www.praeger.com

Printed in the United States of America

The paper used in this book complies with the Permanent Paper Standard issued by the National Information Standards Organization (Z39.48–1984).

10 9 8 7 6 5 4 3 2 1

This volume is dedicated to two women who spent many hours in dusty archives helping us uncover who were the first Americans to engage in geographic types of activities, our wives:

Sylvia Sizemore Smith
and
Peggy Ann Vining

Contents

Preface

Since we are geographers with an interest in American history, perhaps it was inevitable that we would develop an interest in the history of geography. Many years ago, before we knew each other, we each began building a collection of geography books published before the Civil War. Eventually, each of us would own one of the country's largest personal collections of early American geographies. The research potential of these collections became apparent to us long after we began assembling them. By the 1980s, we were giving papers based on early American geographies in our collections. In 1988, we found out about each other's interests and decided to join our efforts in an attempt to make some significant contributions to the history of American geography. Since then, we have given a number of joint papers and co-authored a number of publications.

Almost two decades before we began our joint efforts, Antonelli (1970) divided early American geographers into the "post-Revolutionary geographers" and the "second generation of American geographers." The time periods of publications by the first group was given as 1784 to 1812 and that of the second group as 1812 to 1845. Inspired by, but slightly modifying, his assertions, we began speaking of the "first generation of American geographers" (including writings published between 1784 and 1812) and the "second generation of American geographers" (publications between 1812 and 1860). Our focus, in this work, is on the first generation.

Geographical writers in those days, of course, had much less information at their disposal than the geographers of today. In 1812, the states of the United States numbered only 18, and the population of the country was less then eight million. Some geographers of our time have criticized the early geographers for the inaccuracies in their work, but we consider that unfair.

We presented a paper on "How Twentieth Century Geographers Have Viewed Their Pre-Civil War Predecessors." In it we identified 13 modern generalizations about the early geographers, and we determined that many of them were unreasonable or unfounded. (Vining and Smith, 1998). We were impressed by much of the work of the early American geographers.

The material you are now reading identifies people who were engaged during the period 1784 to 1812 with activities that were instrumental in forming a body of knowledge that grew to be the foundation of the American geography profession. "The year 1859 is sometimes viewed as the end of the period during which a person might command a universal knowledge of many content areas"; thereafter sufficient literature and other material existed to allow greater specialization in professions such as geography, than was previously possible (Smith and Vining, 1989).

As this publication originally began to develop, we set out to identify geography books published up to 1812. Our supposition was that the majority of those who contributed to geography would have published books, pamphlets, or maps. We checked the Dissertation Abstracts International and other references and found one dissertation which indicated an investigation of early American geographers during the period (Sahli 1941). The limited number of early American geographers included in the Sahli publication prompted us to reconsider who should be included as contributors to the body of knowledge that comprises American geography. We ultimately decided to add other people who made contributions as surveyors, military geographers, cartographers, engravers, and authors of geographic books that were not intended to be textbooks. Some of the people who are included in this volume were educated as clergy, several of them published geography textbooks, others were military engineers and engravers, while others explored and surveyed the new United States.

The information contained in this text may be of use to educators, engineers, geographers, historians, and planners. The identification of those who were early American geographers is complicated by the fact that, other than for military purposes, geographers were not formally trained (Harley, Petchenik, and Towner 1978). During the early stage of our investigation we found numerous people who worked at activities directly related to geography. A part of the problem is that most early geographers were involved in a variety of activities and were not professionally trained as geographers.

The investigation then turned to the Evans files and the National Union Catalog Pre-1956 Imprints (American Library Association 1978). The imprints we found encouraged us to look further. By that time we were beginning to ask ourselves a few questions regarding who should be included in our list:

(1) Can we appropriately refer to anyone as a geographer who has produced geography materials?

Virtually none of those we identified were trained to be geographers. Many of those we identified wrote materials other than geography books.

Many contributors to geographic knowledge were engravers and publishers.
Many of the books written as geographies are textbooks, some intended for primary audiences.
Some people published only one geography, nothing else.
Some publications consisted of a map or journal that provided geographic information.
Some individuals worked as surveyors, leaving behind little besides their name.

(2) Can we call these people Americans?
Virtually all of them were British citizens prior to 1783.
At least one of them was an officer in the British army.

It was decided to make this work inclusive and identify as many individuals as possible who produced materials or engaged in activities that we might today attribute to the work of a geographer or one who was intent on spreading geographic knowledge. During the search of literature we found that no one has done an intensive investigation of early American geographers. We decided to examine imprints held in two of the better collections within the United States: The Special Collections of the Library of Congress and the American Antiquarian Society. The Library of Congress produced 19 imprints and numerous maps. The American Antiquarian Society produced 27 imprints. Only 3 imprints by the same author were found in either collection. That, along with a geography text not in either collection, that was found by David Kromm (Professor of Geography at Kansas State University) encouraged us to feel that we had not identified all those who might be appropriately included in the first generation of American geographers.

Cartographers and engineers of the Revolutionary War are presented in a text that provides considerable detail about those engaged, on opposing sides, during the American Revolution (Harley, et al. 1978). Some of the Americans noted in that are not herein addressed. Three of those identified as geographers of the United States are included: Robert Erskine, Simeon DeWitt, and Thomas Hutchins.

Harley, Petchenik, and Towner (1978) provide a chart entitled, "Table 2:2. Regimental Attachment of American and British Mapmakers in the Revolution." That chart lists Headquarters mapmakers, mapmaking corps, engineers and artillery, line regiments and cavalry, and civilian and others under Numbers of Cartographers, American, and British. The British had more cartographers than did the Americans (p.77).

Other useful publications were those of Carter and Wilson:

Carter, Clarence E. (Ed.). (1934). The Territorial Papers of the United States. Washington: Government Printing Office.

Volumes

II-III. (1803). Territory Northwest of the River Ohio,

1787-1803.
V. Territory of Mississippi, 1798-1817.
VII. Territory of Indiana, 1800-1810.
Wilson, George R. (1919). Early Indiana Trails and Surveys. Indiana Historical Society Publications, 6, 347-457.

THE AMERICAN PHILOSOPHICAL SOCIETY

Of particular note is the American Philosophical Society. Guthorn (1984) reveals that members of the American Philosophical Society, the first scientific society in America, recognized the need for and organized a survey of the American coast. The Society was founded in 1769 with Benjamin Franklin as its first president. It was the outgrowth of two earlier organizations: The American Philosophical Society held at Philadelphia for Promoting Useful Knowledge, was founded in 1727 by Franklin; and the American Philosophical Society, formed in 1742. Meisel (1929) provides a list of officers and publications by the Society that covers the period 1769-1880.

The Society was important because of the variety of thinkers who contributed to its prestige and the extent to which it published and promoted matters regarding geography in America. According to its bylaws, "The members of the Society shall be classed into one or more of the following committees: 1. Geography, Mathematics, Natural Philosophy and Astronomy; 2. Medicine and Anatomy; 3. Natural History and Chemistry; 4. Trade and Commerce; 5. Mechanics and Architecture; and 6. Husbandry and American Improvements" (Barton, 1813, p.353). Further support for its importance to the new nation is provided by the fact that the Pennsylvania General Assembly of 1782-1783 granted to the Society a sum of four hundred dollars. Barton (1813) went on to list the following members: John Ewing, William Smith, Thomas Jefferson, James Hutchinson, Jonathan Williams, Samuel Magaw, David Rittenhouse, James Anderson, David Stewart Erskine, Don Joseph J. Ferrer, Mr. Adet, John Lukens, O. Biddle. Several of the members are among those included in this text.

AFRICAN AMERICANS

Only one African American, who was a prominent contributor to the knowledge of American geography, has been identified during the time,

Benjamin Banneker. We suspect that there must have been other African Americans involved in one or another aspect of geography during the period (Harley, Petchenik & Towner 1978).

NATIVE AMERICANS

Though the authors of this manuscript have little information regarding who they were, Native Americans supplied many explorers and surveyors with information of territory unknown to European Americans. Among those to benefit from such information were Lewis and Clark who were led by Sacajawea (Rowland 1989; Schwartz & Ehrenberg 1980, p. 221).

SPANISH AMERICA

Spanish America was being mapped and explored as the new United States of America was being born. Juan Perez, a sea pilot, mapped the present area of the Golden Gate, depicting the islands of Yerba Buena and Alcatraz in 1775. Also in 1775 Juan Bautista de Anza mapped the land route from Sonora to Alta California (Schwartz & Ehrenberg 1980). Since Spanish America was not a part of the United States between the years 1784 and 1812, those geographers and others are not included in this text.

WOMEN

Women, as a rule did not participate in geographic activities. Susanna Rowson was an exception. She opened a school for girls and wrote geography textbooks, as well as other books.

METHOD

The United States citizenship of those included in these pages is verified where possible, including a brief biography, though some of those included could not be absolutely verified. As individuals were identified, their citizenship was checked in a variety of biographies. Although varification was possible for many of those listed in this text, additional information regarding those included will no doubt surface. Additionally, there have likely been some individuals and works that are still unknown. At this time, this text is as complete and inclusive as we can determine.

SYMBOLS AND ABBREVIATIONS

As above = [indicates that this publication is identical to the previous, with only information that is differently noted]
... = continuation of the title
(c. or C. & date) = circa or near the date presented, e.g., (c.1790)
ed. = editor
col. = colored
corr. = corrected
eng. = engraved
fold. = folded
front. = frontispiece
illus. = illustrated
incl. = including
nd = no date
no. = number
np = no publisher
Other = publications containing additional information about the person
pl. = plate
pt. = part
rev. = revised
tbl. = table
tr. = translator
V. & v. = volume
= unnumbered pages

For bibliographic style the Publication Manual of the American Psychological Association, Fourth Edition, was followed (1994). Only initials of author's first and middle names are provided. Titles of books are underlined and the first letter of the title, the first word following a colon, and proper names are capitalized. Titles of articles in journals and books are not underlined and only proper names and the first word following a colon are capitalized. Titles of journals are underlined and capatilized. Examples follow:

Books

Hopkins, J.G.E. (Ed.). (1964). Concise dictionary of American biography. New York: Charles Scribner's Sons.

Journals

Barton, B.S. (1791). An account of the most effectual means of preventing the deleterious consequences of the bite of the Crotalus Horridus, or rattle-snake. Transactions of the American Philosophical Society, 3, pp.100-115.

American Geographers, 1784-1812

ABERNETHIE, THOMAS (C.1780)

Thomas Walker and Abernethie collaborated to produce one of the first road maps in the United States, 1787. The two men were natives of Scotland who settled near Charleston about 1783. Abernethie, shortly after arriving in Charleston, opened a shop for engraving and copper plate printing (Ristow 1985, p. 157). Two of his maps appeared in David Ramsay's History of the Revolution in South Carolina, 1785.

(1787). Specimen, on an intended traveling map of the roads of the state of South Carolina, from actual survey. By Thomas Walker and Thomas Abernethie. (np).

ABERT, JOHN J., MAJ. (1788-1863)

Guthron (1984) reports that Abert was recruited by Ferdinand Hassler to assist in surveying the American coast from Maine to Georgia. Abert lived in Mount Holly, New Jersey in 1816 and was a member of the American Topographical Engineers. Since the survey was authorized during the period recognized as that including the First Generation of American Geographers and it is likely that Abert was already active in geographic matters, indicated by his membership in the Engineers, he is included here. That survey was authorized by Congress on February 10, 1807, but it was not executed until 1816 though considerable planning took place between the two dates (p.17).

Brown (1897) confirms the above. Abert was born in Shepherdstown, Virginia and died in Washington, DC. He was the son of John Abert who came to America from France with Count Rochambeau in 1780. The younger

Abert was educated as a soldier, graduating from West Point in 1811. He was employed by the government war department in Washington where he also studied law. His topographical bureau work brought him the brevetted rank of lieutenant-colonel. He worked from 1829 through 1861 with and commanded, from 1838-1861 the Corps of Topographical Engineers. Abert thereafter retired. He was a member of numerous American scientific and historical associations and of the Geographical Society of Paris, France. He was among the founders of the National Institute of Science which came to be the Smithsonian Institute (v.1, p.13).

White (1897) reports that as the head of the Corps of Engineers Abert was responsible for geographical and other information concerning the continent which its officers collected and published. Abert married Ellen Matlack Stretch, the granddaughter of Col. Timothy Matlack, the Revolutionary War patriot (V.IV. p.380).

(1846). Expedition by Lieut. Abert on the upper Arkansas. W. [American politics] (np).

(1857). Map of the Patuxent and St.Mary's Rivers Maryland. From surveys by J. J. Abert Topl. Engrs. and Major J. Kearney Topl. Engrs. In 1824. Compiled in the Bureau of Topl. Engrs. By Order of the Sec. Of War In Accordance with a Resolution of the Senate of the United States. Drawn by Dennis Callahan. Lithograph by Sarony, Major & Knapp. Washington & New York: Bureau of Topl. Engrs.

(1845). Message from the Pres. U.S... Communicating a report of an expedition led by Lieut. Abert on the Upper Arkansas and through the country of the Comanche Indians in the fall of the year 1845. (np). Voy. & T.T. 8mo V.2.

(1843). Reply of the Col. Abert and Mr. Markoe to the Hon. Mr. Tappan, of the United States Senate. Washington: W.Q. Force, printer. 18p. 23.5cm.

(1838). Report from J. J. Abert, in reference to the canal to connect the Chesapeake and Ohio Canal with the city of Baltimore. Washington: Printed by Gales and Seaton. 42p. 23cm.

(1852). Report from the Topographical Bureau, relative to the improvement of the St. Clair Flats. 3 pamphlets. Engineering, 3(1).

(1846). Report in favor of the improvement of the navigation of the Hudson River. Albany: (np).

(1874). Report in reference to the canal to connect the Chesapeake and Ohio Canal with the city of Baltimore, by Colonel J.J. Abert... Reprinted for the use of the Engineer Department, U.S.A. Washington, DC: Government Printing Office. 40p. folding map. 32cm.

(1831). Report of Col. John J. Abert and Col. James Kearney, of the United States Topographical Engineers, upon an examination of the Chesapeake and Ohio Canal from Washington City to the "Point of Rocks;" made by order of the Secretary of War, at the request of the President and directors of the Chesapeake and Ohio Canal Company. Printed by order of the stockholders of the Chesapeake and Ohio Canal Company. Washington: Printed by Gales and Seaton. 24p. 25cm.

(1839). Statement in reply to an order of the House of Delegates, 9th February, 1839, calling for certain information relative to surveys made in the summer of 1838, for the route of the Maryland Canal. [Annapolis?]: (np). Reply to the clerk of the House of delegates. 39p. 22cm.

Other

(1893). "Abert, John J., Maj." American Soc. Civil Eng. Proc. V. 19, pp.88-92.

ADAMS, JAMES (PRINTER) (C.1785)

Ristow (1985) identifies Adams as a publisher (p.59). The National Union Catalog Pre-1956 Imprints (The American Library Association 1976). Mansell (1986) listed two publications by a James Adams Junior.

(1784). The discovery, settlement, and present of state of Kentucke [sic]. By John Filson. Engraved by Henry D. Pursell. Wilmington, DE: James Adams. Printed by T. Rook of Philadelphia.

(1823 &1827). The Portland directory & register; containing the names, professions and residence, of the heads of families & persons of business... Accompanied with a new and correct plan of the town. Portland: J. Adams, Jr..

(1795). A synopsis of geography with the use of terrestrial globe; intended for the benefit of youth, especially that of the students in the public grammar schools in Wilmington. Wilmington, DE: Author.

ALLEN, IRA (1751-1814)

Ristow (1985) reveals that Allen was Vermont's surveyor general from 1779-1787. He was the youngest brother of Colonel Ethan Allen, famous as the commander of the Green Mountain Boys during the Revolutionary War. Surveys by Allen and others were the basis on which William Blodget compiled his Topographical Map of the State of Vermont (p. 88-89).

White (1897) goes on to note that Allen was a soldier, statesman born in Cornwall, Litchfield Co., Connecticut and died in Philadelphia, Pennsylvania. He had a common-school education and took up surveying as his occupation. Allen was married to Jerusha, the daughter of Gen. Roger Enos, of Simsbury, Connecticut. He had two sons and one daughter.

He, his cousin, Remember Baker, and five others explored the country lying near the lower Winooski and Onion rivers. They built Fort Frederick near the falls of the Winooski, and cut a road through the forest to Castleton. Their efforts, however did not flourish because the fort was isolated, bringing it to abandonment. His support of the American cause and as arbiter between rivals such as Vermont and New Hampshire brought him fame, but little fortune. He was one of seven commissioners who settled the boundary dispute with New York. In 1795 he sailed to Britain to secure support for the construction of a canal between Lake Champlain and the St. Lawrence River. While on the trip he was also in pursuit of arms for the Vermont militia. That resulted in a purchase from the French, transportation aboard the Olive Branch, and capture by the British which culminated in imprisonment that lasted until 1801. That along with other misfortunes depleted his fortune leaving him in poverty at his demise in 1814 (V.IV. pp.29-30).

(1806). A concise summary of the second volume of the Olive Branch, a book containing an account of Governor Chittenden's giving written instructions to General Ira Allen in 1795, to purchase military stores in Europe for the militia of Vermont...To which is subjoined, General Allen's circular letter, on the subject of a ship canal of commerce, and the advantage of British America in preserving peace between Great Britain and the United States. Philadelphia: Printed and sold by Thomas T. Stiles.

(1807). As above. Printed for the author.

(1809). Copies of letters to the governor of Vermont & address to the legislature thereof, respecting a conspiracy against the author & respecting a ship canal from Lake Champlain to the river St. Lawrence, with the opinion of the attorney general of the case of the Olive Branch... Philadelphia: author. 61p.

(1810). As above. By Ira Allen, claimant of said cargo. 32p. 21cm.

(1811). As above. Printed by John Binns - for the author. 61p. 21.5cm

(c.1800). Extracts from the First volume of the particulars of the capture of the ship Olive Branch. 8mo 1-152p.

(1798). A map of the state of Vermont drawn under the direction of Ira Allen Esq. late Surveyor Genl. of the said state for his history thereof. London: Engraved by I. Palmer.

(1798). The natural and political history of the state of Vermont, one of the United States of America. To which is added, an appendix containing answers to sundry queries, addressed to the author. London: J. W. Myers. pp.vii, 281, xviii. 21cm.

Other

(1939). Calendar of Ira Allen papers in the Wilbur Library, University of Vermont. Prepared by the historical records survey, Division of Professional and Service Projects, Work Projects Administrations. Montpelier, VT: The Historical Records Survey.

ALLEN, JOEL (1755-1825)

Ristow (1985) notes that Allen was born in Farmington, Connecticut. He established an engraving shop in Middletown (p. 88). No other biographic information regarding Allen nor any publications have been located (The American Library Association, 1974; Johnson and Malone 1928). He was the engraver of William Blodget's map:

(1791). A new and correct map of Connecticut one of the United States of North America, from actual survey-humbly dedicated by permission to His Excellency Samuel Huntington Esquire Governor and Commander in

Chief of said state, by his most humble servant William Blodget. Middletown: author. Joel Allen engraver.

ANDERSON, D. (C.1790)

Winterbotham's atlas contains two maps engraved by Anderson (Ristow 1985, p.153). Publications and biographical information about Anderson have not been located.

Other

Winterbotham, W. (1796). The American atlas. Philadelphia: John Reid.

ARMSTRONG, JOHN (1758-1843)

Armstrong was a soldier, born in Carlisle, Pennsylvania who died in Red Hook, New York. He was the son of a brigadier-general in the Continental Army. It appears that young Armstrong was an aid to General Gates during the encampment of the American army at Newburg which led to the "Newburg Letters." In those Armstrong alleged wrongs of the soldiers and called for redress. He was elected U. S. Senator in 1800 and served until 1810. The biography offers indicates that he was an enemy of Monroe (White 1897, V.1, p.48). Armstrong did publish several manuscripts, but none were found in The National Union Catalog Pre-1956 Imprints (The American Library Association 1976). What we know of the person is that Robert Erskine names him on one of the "rough draughts [sic] of Surveys" with which he is credited.

118 From Morristown towards Sussex C. H. by J. Armstrong A. B. & C.

BALDWIN, JEDUTHAN (1732-1788)

Baldwin was a "Captain Assistant Engineer" in the Continental Army. Though he appears to have had no formal training, his work indicated an understanding of planning, cartography, and fortification that was representative of "officers trained in European military academies" (Harley et al. 1978, p.6).

Other

Pell, Stephen H. P. (Ed.). (1777). Baldwin's diary. Extracts from the diary of Colonel Jeduthan Baldwin, Chief Engineer of the Northern Army, July 6th, 1776 to July 5th, 1777. (np).

BANKER, EVART (C.1780)

Banker published the first map of New Jersey in 1784 (Schwartz and Ehrenberg 1980, p.205). No other biographic information regarding Banker nor any publications have been located (The American Library Association, 1974; Johnson and Malone 1928).

BANKER, GERARD (C.1770)

Possibly related to Evart Banker, this Banker is known only as a collaborator of Bernard Ratzer and co-producer of the following map. No other biographic information regarding Banker nor any publications have been located (The American Library Association, 1974; Johnson and Malone 1928).

(1777). The province of New Jersey, divided into east and west, commonly called The Jerseys. This map drawn by Bernard Ratzer and By Gerard Banker. (np). 80.1x60.2cm.

BANNEKER, BENJAMIN (1731-1806)

Banneker was a farmer, astronomer, mathematician and surveyor who worked as an assistant to Andrew Ellicott to lay out the boundaries of the District of Columbia (Bedini 1972). He was born near Baltimore and was taught how to read by his grandmother, an English woman. He attended, for several winters, a school that was open to blacks and whites. That is where he developed his love of mathematics and science. Much of his knowledge was self taught.

Banneker made all the astronomical and tide calculations and weather predictions for yearly almanacs that he published between 1791 and 1796. A copy of the almanac was sent to Thomas Jefferson who sent a copy to the Royal Academy of Sciences in Paris. That became evidence of the abilities of black people, supporting calls for the abolition of slavery. Bedini (1972) provides numerous references for study of Banneker.

(1791). Benjamin Banneker's Pennsylvania, Delaware, Maryland and Virginia almanack [sic] and ephemeris, for the year of our Lord, 1792; Being bissextile, or leap-year, and the sixteenth year of American independence, which commenced July 4, 1776. Containing, the motions of the Sun and Moon, the true places and aspects of the planets, the rising and setting of the Sun, and the rising setting and southing, place and age of the Moon, &c. The lunation, conjunction, eclipses, judgment of the weather, festivals and other remarkable days; Days for holding Supreme and Circuit Courts of the United States, as also the usual courts in Pennsylvania, Delaware, Maryland, and Virginia.- Also, several useful tables and valuable receipts.- Various selections from the Commonplace Book of the Kentucky philosopher, an American sage; with interesting and entertaining essays, in prose and verse-the whole comprising a greater, more pleasing, and useful variety, than any work of the kind and price in North-America. Baltimore: Printed and sold wholesale and retail, by William Goddard and James Angell, at their printing-office, in Market Street.- Sold, also by Mr. Joseph Cruckshank, printer, in Market-Street, and Mr. Daniel Humphrey's, printer in South-Front Street, Philadelphia-and by Messrs. Hanson and Bond, printers in Alexandria.

(1791). Benjamin Bannider's [sic] Pennsylvania, Delaware, Maryland and Virginia almanack and ephemeris for 1792. Baltimore: William Goddard and James Angell.

(1791). Banneker's almanack for 1792. Philadelphia: Printed for William Young.

(1792). Banneker's almanack, and ephemeris for the year of our Lord, 1793; Being the first after bissextile or leap-year: Containing the motions of the Sun and Moon; the true places and aspects of the planets; the rising and setting of the Sun; rising, setting, and southing of the moon: the lunations, conjunctions, eclipses; and the rising, setting, and southing of the planets and noted fixed stars. Philadelphia: Printed and sold by Joseph Curckshank.

(1792). Benjamin Banneker's Pennsylvania, Delaware, Maryland and Virginia almanack and ephemeries, for the year of our Lord, 1793; being the first after bissextile, or leap-year, and the seventeenth year of American independence, which commenced July 4, 1776. Containing, the motions of the Sun and Moon, the true places and aspects of the planets, the rising and setting of the Sun, and the rising setting and southing, place and age of the Moon, &c. The lunations, conjunction, eclipses, judgment of the weather, festivals and other remarkable days; Days for holding Supreme and Circuit Courts of the United States, as also the usual courts in Pennsylvania, Delaware, Maryland, and Virginia. Also, several useful tables and valuable receipts. Various selections from the Commonplace Book of the Kentucky philosopher, an American sage; with interesting and entertaining essays, in prose and verse-the whole comprising a greater, more pleasing, and useful variety, than any work of the kind and price in North-America. Baltimore: Printed and sold wholesale and retail, by William Goddard and James Angell, at their printing-office, in Market Street.

(1793). Banneker's almanack, and ephemeris for the year of our Lord, 1794; Being the second after bissextile or leap-year: Containing the motions of the Sun and Moon; the true places and aspects of the planets; the rising and setting of the Sun; Rising, setting, and southing of the Moon: the lunations, conjunctions, eclipses; and the rising, setting, and southing of the planets and noted fixed stars. Philadelphia: Printed and sold by Joseph Curckshank.

(1793). As above- Baltimore: Printed and sold, wholesale and retail by James Angell.

(1793). As above- Philadelphia: Printed for William Young, bookseller.

(1793). As above- Petersburg: Printed for William Prentis.

(1794). Bannaker's [sic] New Jersey, Pennsylvania, Delaware, Maryland and Virginia almanack and ephemeries, for the year of our Lord, 1795; being the third after leap-year, and the nine-teenth year of American independence, and the seventh of our Federal Government-which may the Governor of the world prosper. Containing, the motions of the Sun and Moon, the true places and aspects of the eight planets, the rising and setting of the Sun, and the rising setting and southing, node, age and latitude of the Moon, &c. Also the lunations, conjunction, eclipses, remarkable days; judgment of the weather; length of days and nights; the time that courts are held in New-Jersey, Pennsylvania, Delaware, Maryland, and Virginia; post-roads, with a variety of instructive and entertaining matter in prose and verse. Wilmington: Printed by S. & J. Adams.

(1794). Bannaker's [sic] Wilmington almanac, or ephemeris, for the year of our Lord 1795; being the third after leap-year, and the nineteenth year of American independence, and the seventh of our Federal Government- Which may the Governor of the world prosper! Containing, the motions of the Sun and Moon, the true places and aspects of the eight planets, the rising and setting of the Sun, and the rising setting and southing, node, age and latitude of eclipses; remarkable days and nights; the time that courts are held in New-Jersey, Pennsylvania, Delaware, Maryland, and Virginia; post-roads, with a variety of instructive and entertaining matter in prose and verse. Wilmington: Printed by S. & J. Adams for Frederick Carig.

(1794). Banneker's almanac, for the year 1795: Being the third after leap year. Containing, (Besides every thing necessary in an almanac,) an account of the yellow fever, lately prevalent in Philadelphia; with the number of those who died, from the first of August till the ninth of November, 1793. Philadelphia: Printed for William Young, bookseller.

(1794). Benjamin Bannaker's [sic] Pennsylvania, Delaware, Maryland and Virginia almanack and ephemeris for the year of our Lord 1795; Being the third after leap year. Philadelphia: Printed for William Gibbons.

(1794). As above- Philadelphia: Printed for Jacob Johnson & Co.

(1794). Bannaker's [sic] Wilmington almanack and ephemeris for the year of our Lord 1795; Being the third after leap year. Wilmington: Printed by S. & J. Adams.

(1794). As above- Wilmington: S. & J. Adams for W. C. Smyth.

(1794). Benjamin Bannaker's [sic] Pennsylvania, Delaware, Maryland and Virginia almanack and ephemeris for the year of 1795; Being the third after leap year. Wilmington: S. & J. Adams.

(1794). Bannaker's [sic] New Jersey, Pennsylvania, Delaware, Maryland and Virginia almanack and ephemeries, for 1795. Baltimore: S. & J. Adams.

(1794). New Jersey & Pennsylvania almanack and ephemeries, for the year of our Lord, 1795; Being the third after leap-year, and the twentieth year of American independence, after the fourth of July. Containing, besides the usual requisites of an almanac, a variety of entertaining matter, in prose and verse. To which is added, an account of the yellow fever, in Philadelphia. The astronomical calculations by Benjamin Banneker, an African. Trenton: Printed and sold, wholesale and retail by Matthias Day.

(1794). Bannaker's [sic] New Jersey, Pennsylvania, Delaware, Maryland and Virginia almanack and ephemeries, for the year of our Lord, 1795; Being the third after leap-year. Wilmington: Printed by S. & J. Adams.

(1794). The Pennsylvania, Delaware, Maryland and Virginia almanack and ephemeries, for the year of our Lord, 1795. Baltimore: Printed by James Angell for Fisher and Coale.

(1794). The Pennsylvania, Delaware, Maryland and Virginia almanack and ephemeries, for the year of our Lord, 1795; Being the third after leap-year. Wilmington: Printed by S. & J. Adams.

(1794). Benjamin Bannaker's [sic] Pennsylvania, Delaware, Maryland and Virginia almanack and ephemeries, for the year of our Lord, 1795; Being the third after leap year. Baltimore: Printed and sold by John Fisher.

(1795). Bannaker's [sic] Maryland, Pennsylvania, Delaware, Virginia, Kentucky and North Carolina almanack and ephemeries, for the year of our Lord, 1796; being bissextile, or leap-year; the twentieth year of American independence and eighth year of the Federal Government. Baltimore: Printed for Philip Edwards, James Keddie, and Thomas, Andrews and Butler.

(1796). Bannaker's [sic] Virginia and North Carolina almanack and ephemeries, for the year of our Lord, 1797; being first after bissextile, or leap-year; The twenty-first year of American independence and ninth year of the Federal Government. Baltimore: Printed for Philip Edwards, James Keddie, and Thomas, Andrews and Butler.

(1796). Bannaker's [sic] Virginia, Pennsylvania, Delaware, Maryland and Kentucky almanack and ephemeries, for the year of our Lord, 1797; being first after bissextile, or leap-year; the twenty-first year of American independence and ninth year of the Federal Government. Baltimore: Printed by Christopher Jackson for George Keatinge's bookstore.

(1796). Bannaker's [sic] Virginia, Pennsylvania, Delaware, Maryland and Kentucky almanack and ephemeries, for the year of our Lord, 1797; being first after bissextile, or leap-year; the twenty-first year of American independence and ninth year of the Federal Government. Richmond: Printed by Samuel Pleasants.

(1796). Bannaker's [sic] Maryland and Virginia almanack and ephemeries, for the year of our Lord, 1797; being first after bissextile, or leap-year; the twenty-first year of American independence and ninth year of the Federal Government. Baltimore: Printed by Christopher Jackson for George Keatinge's bookstore.

BARKER, ELIHU (C.1790)

Ristow (1985) reveals that Barker's map was by far the most accurate early map of Kentucky. That recognition probably places him among the First Generation of American Geographers. Barker died shortly after publishing his map (p.135).

(c.1795). Kentucky, reduced from Elihu Barker's large map. W. Barker sculp. (np). 22x49cm. Scale about 22 American miles to an inch.

(1794). A map of Kentucky from actual survey by Elihu Barker. Philadelphia: Mathew Carey. 17 3/8 x 38 11/16 in.

BARKER, WILLIAM (C.1785)

The map of the Seven Ranges of Townships includes the geographer's line that would establish the northern boundary of the townships (Schwartz and Ehrenberg 1980). The actual survey was made by Thomas Hutchins to implement the Ordinance of 1785, establishing a standard of measurement for future land division (p.209).

The American Library Association (1974) identified a few William Barkers, but none of them appear to have been the person noted above.

(1812). Carey's American atlas. Philadelphia: Mathew Carey. Fig.9-5 engraved by Barker.

(1794). A general atlas for the Present War. (np): Mathew Carey.

(c.1787). Plat of the Seven Ranges of townships being part of the territory of the United States N. W. of the river Ohio which by a late act of Congress are directed to be sold...W. Barker sculp... Surveyed in conformity to an ordinance of Congress of May 20th, 1785. Under the direction of Thomas Hutchins late Geographer to the United States. Philadelphia: Matthew Carey. 24.2 x 13.6 in. (over all 26.9 x 18.7 in.).

BARTON, BENJAMIN SMITH (1766-1815)

Barton (1813) notes several people engaged in establishing the boundary line between Virginia and Pennsylvania: "This boundary-line was begun in May, 1785, by Messrs. Rittenhouse, Ellicott, Porter, and Nevill; assisted by the present Dr. Benjamin Smith Barton, then a youth about nineteen years of age, whose medical and other scientific acquirements rendered him, even at that early period of life, a useful associate of the commissioners" (pp.306-307).

Barton was born in Lancaster, Pennsylvania and died in Philadelphia. He was a physician and a naturalist. He was educated at Edinburgh, in London, and in Gottengen. Barton practiced medicine in Philadelphia and he taught at the University of Pennsylvania (Hopkins, 1964, p.52)

(1814). Archaeologicae Americanae telluris collectanea et specimena; or, collections with specimens for a series of memoirs on certain extinct animals and vegetables of North America, together with facts and conjectures relative to the ancient condition of the lands and waters of the continent. Philadelphia: (np), 505p. 1pl.

(1799). Fragments of the natural history of Pennsylvania... Part I, Philadelphia: Printed for the author by Way & Groff, 42p. folio.

(1805). Letter to M. Lacepe'de, of Paris, on the natural history of North America. London: (In Philadelphia Magazine, 22, ed. by Tilloch, pp.97-103 & 204-211).

(1804). Memorandums concerning the earthquakes of (North) America.

Philadelphia Medical and Physical Journal, 1, pt. 1, art. 17, pp.60-67.
(1797). New views of the origin of the tribes and nations of America. Philadelphia: Printed for the author by J. Bioren, 199p.
(1799). Notes, number 1, on the falls of Niagara. Proceedings of The American Philosophical Society, Philadelphia, 22, appx. p.284.
(1806). Notice of the botany of the White Mountains in... Philadelphia Medical and Physical Journal, Supplement no. 1, art. 9, pp.57-8.
(1800). Sketch of a geographical view of the trees and shrubs of N[orth] America. Proceedings of the American Philosophical Society, Philadelphia, 22, appx. p.194.
(1809). Specimen of a geographical view of the trees and shrubs, and many of the herbaceous plants of North America, between the latitudes of seventy-one and twenty-five... Philadelphia: (np), 36p. pl. 4°.

BARTRAM, A. (PRINTER) (C.1805)

Like many authors who contributed to the geographic knowledge in the new United States of America, nothing is now known about the life of Bartram. No entries or references were found in The American Library Association (1974) or Johnson and Malone (1928).

(1805). Sketch, of the geography and present state of the united territories of North America; to which is added, a list of the several nations and tribes of Indians in Canada and the United States & c. & c. Philadelphia: author.

BARTRAM, WILLIAM (1739-1823)

Bartram was principally a botanist who traveled widely and in great detail, described the flora of those places which he visited. That contribution places him among those citizens of the new United States of America who contributed to the knowledge of the country's geography. He, on February 9th, 1739, was born in Kingsessing, Pennsylvania and he died there on July 22, 1823. He was a son of the noted botanist, John Bartram and his wife Ann Mendenhall. He traveled Pennsylvania, North and South Carolina, Georgia, and other states. Bartram's observations included the customs and manners of the Creek and Cherokee Indian (Herringshaw 1898) .

He spent the years 1773-1777 exploring, with the support of Dr. John Fothergill, the southeastern portion of the United States. In return for Fothergill's support, Bartram was to provide seeds, specimen, and drawings. In 1782 he was elected to the position of Professor of Botany at the University of Pennsylvania, but due to his health declined the appointment.

Although he did not claim to be a geographer, his published travels describing the flora and fauna of numerous states. His observations may have the potential of providing current scientists with a benchmark for restoration of some damaged environments. Bartram was respected and corresponded with naturalists abroad, and he was honored with membership in foreign learned societies. In 1786 he became a member of the American Philosophical Society (Brown 1897, V.1, p.216-217; Johnson & Malone 1928, V.1, pp.28-

29; Wilson & Fiske 1889, VI, p.189).
(1804). Account of the species, hybrids, and other varieties of the vine of North America. The Medical Repository, New York City, 7(1), art. 2, pp.12-19.
(1805). Conjectures relative to the site of Bristol, in Pennsylvania. Philadelphia Medical and Physical Journal, 1, pt. 2, art. 21, pp.131-133.
(nd). Portions of the mss. of travels thru the Carolina, Georgia and E. & W. Florida, 1773-1777. (np), manuscript.
(1793). William Bartram's Reisen durch Nord- und Sud-Karolina, Georgien, Ost- und West-Florida, das gebiet der Tscherokesen, Krihks und Tschaktahs, nebst umstandlichen nachrichten von den einwohnern, dem boden und den naturprodukten dieser wenig bekannten grossen lander. Aus dem englischen. Mit erlauternden ammerkungen von E. A. W. Zimmerman... Berlin: in der Vossischen buchhandlung, 502p., 7 plates, 20cm. [Accents omitted]
(1794). Reizen door Noord- en Zuid-Carolina, Georgia, Oosten West-Florida; de landen der Cherokees, der Muscogulges, of het Creek bondgenootschap en het land der Chactaws... Uit het Engelsch vertaald, door J. D. Pasteur... Haarlem: F. Bohn, 695p., 3v. in 1, map, 8o. [Accents omitted]
(1791). Travels. Philadelphia: James & Johnson, 8o.
(1792). As above - London: for J. Johnson.
(1928). As above -The travels of William Bartram. New York: Macy-Masius, 414p., 20cm.
(1928). As above - New York: Dover Publications.
(1940). As above - ... edited by Mark Van Doren; with an introduction by John Livingston Lowes. New York: Facsimile library, exclusive distributors: Barnes & Noble, Ind., 19.5cm.
(1947). As above - New York: Dover Publications.
(1955). As above - illustrated, map, 21cm.
(1943). Travels in Georgia and Florida, 1773-1774. A report to Dr. John Fothergill. Annotated by Francis Harper. Transactions of the American Philosophical Society, 33, pt.2, 121-242.
(c.1790). Travels, through North and South-Carolina, Georgia, and Florida. Philadelphia: James and Johnson, 1p., broadside.
(1791). Travels through North and South Carolina, Georgia, East and West Florida, the Cherokee Country, the extensive territories of the Muscogulges, or Creek Confederacy, and the country of the Chactaws; containing an account of the soil and natural productions of those regions, together with observations on the manners of the Indians. Embellished with copper-plates. Philadelphia: Printed by James & Johnson, 556p., 8 plates, folded map, 20.5cm.
(1792). As above - reprint.
(1793). As above - Dublin: J. Moore, W. Jones, R. M'Allister, and J. Rice.
(1794). As above - The 2d ed. in London. Philadelphia: Printed by James and

Johnson, 1791; London: Reprinted for J. Johnson.
(1959). As above - Reproduction of 1791 edition. Louisville: Lost Cause Press.
(1798, 99, 1800). Voyage dans les Parties sud de l'Amerique septentrionale; Savior: les Carolines septentri onale et meridionale, la Georgie, les Florides orientale et occidentale, le pays des Cherokees, le vaste territoires des Muscogulges ou de la confederation Creek, et le pays des Chactaws... Paris: for Carteret et Brosson, map, 21cm. [Accents omitted]

Other

Squier, E. G. (Ed.). (1851). Observations on the Creek & Cherokee Indians. New York: (np), 81p., illustrations.

BAUMAN, SEBASTIAN (C.1782)

Bauman is responsible for the first map printed in America that represents the Battle of Yorktown (Schwartz and Ehrenberg 1980). It was dedicated to George Washington and published in Philadelphia in 1782. Ristow (1985) identifies Bauman as a surveyor (p.44).

(1782). To His Excellency Gen. Washington... this plan of the investment of York and Gloucester has been surveyed and laid down and is most humbly dedicated... This plan was taken between the 22nd and 28th of October 1781. Philadelphia: (np). 25 1/8 x 17 1/2 inches. Scale 400 yards to 1 inch.
(1785). As above. 48.5 x 63cm.
(nd). ... Map of the siege of Yorktown. (np). 2p.
(1881). Plan of the siege of Yorktown, 1781. From a survey... three days after the surrender of Cornwallis. Reproduced by the Photo Electrotype Engraving Company... from the original, in the possession of the New England Historic and Genealogical Society of Boston,... Boston: (np). 31x23 inches. Scale 425 yards to 1 inch.
(1825). Plan of Yorktown in Virginia and adjacent country. Exhibiting the operations of the American, French & English armies during the siege of that place in October 1781. Surveyed from the 22nd to the 28th Octr. Drawn by Jn. F. Renault. With a Crow-pen and presented to the Marquis De La Fayette. Engraved by B. Tanner. Philadelphia: (np). 4815 x 73.7 cm.

BEASLEY, A. M. (C.1804)

According to Guthron (1984), Beasley produced a map in 1804-1809 that was published by Blunt. Although no biographical information is available about the person, he appears to have been an engraver who contributed to the knowledge of geography about the United States.

(1804). The Bay of Chesapeake from its entrance to Baltimore. Compiled for Edmund M. Blunt, engraved by A. M. Beasley. In American Coast Pilot: Containing directions for the principal harbors, capes, and headlands on the coast of North and part of South America...

(1804). Charleston Harbour. Engraved by A. Beasley. Newburyport, MA: Edmund M. Blunt.

BELKNAP, JEREMY, D.D. (1744-1798)

Belknap's contributions to the geographic knowledge of the United States appear to have been mainly through publications. On June 4, 1744 he was born in Boston and died there on June 20, 1798. He attended Mr. Lovell's school prior to entry into Harvard at the age of fifteen years. After graduating with a degree in theology he taught school at Milton, Massachusetts and then at Portsmouth and Greenland, New Hampshire. Belknap also studied for the ministry and in 1766 he moved to Dover, New Hampshire and became pastor of the Congregational church there.

Soon after the beginning of hostilities at the beginning of the American Revolution, Belknap was appointed chaplain to the New Hampshire troops at Cambridge, but his health prevented him accepting the appointment. His literary reputation came mainly from his History of New Hampshire, remarkable for its research, impartiality, literary merit, and its treatment of the natural history of New Hampshire. His American Biography contained sketches of the lives of some famous early explorers and colonial leaders. Belknap formulated a plan for the creation of an Antiquarian Society that came to fruition as the Massachusetts Historical Society in 1794, the first of its kind in the United States. He was elected corresponding secretary of the society and endeavored to promote the formation of similar institutions in other states (Brown 1897, V.1, p.257; Johnson & Malone 1928, V.1, p.147; Tucker 1990).

Ristow (1985) mentions Belknap as the recipient of a note from Ebeling regarding "six maps" (p.176). He was engaged with the production of the maps.

(1792). Circumnavigation of Africa, by the ancients. Boston: (np). Note. Appended to Columbian discourse: Oct. 23, 1792.

(1737). A copy of the plan returned by His Majesty's Commissioners for settling the boundarys [sic] between the Province of New Hampshire and the Massachusetts Bay, along with the said Comrs: Judgment of 2. Sept. 1737. In New Hampshire, Boundary Line; Mass. The case of His Majesty's Province in N. H. Manuscript, 19 3/4 x 11 1/4 in., Scale: 5 miles to 1 inch.

(1784). Description of the White Mountains in New Hampshire. Transactions of the American Philosophical Society, 2, no.3, p. 42-49.

(1792). A discourse, intended to commemorate the discovery of America by Christopher Columbus; delivered at the request of the historical society in Massachusetts, on the 23d day of October, 1792, being the completion of the third century since that memorable event. To which are added, four dissertations, connected with various parts of the discourse, viz. 1. On the circumnavigation of Africa by the ancients. 2. An examination of the pretensions of Martin Behaim to a discovery of America prior to that of Columbus, with a chronological detail of all the discoveries made in the 15th century. 3. On the question, whether the honey-bee is a native of America? 4. On the colour [sic] of the native Americans and the recent population of this continent. Boston: Printed at the Apollo Press by Belknap and Hall, 132p., 20.5cm.

Advertisements in back - American Biography; or an historical account of those persons who have been distinguished in America as adventurers, statesmen, philosophers, divines, warriors, authors, and other remarkable characters... by the above author

The history of New-Hampshire. In three volumes, 8vo. Vol.I. Comprehending the events of one complete century, from the discovery of the river Pascataqua in 1614. Vol.II. Comprehending he events of seventy-five years, from 1715 to 1790, illustrated by a map of the state. V.III. Containing a geographical description of the state, with sketches of its natural history, productions, improvements, and present state of society and manners, laws and government. by the above author.

(1795). The discovery and description of the islands called the Marquesas in the South Pacific ocean, with a farther account of the seven adjacent islands, discovered first by Capt. Joseph Ingraham... Boston: Massachusetts Historical Society. Collections, 4, pp.238-246. Note. For account of Ingraham's discovery see v.2, pp.20-24.

(1795). Gentlemen. I am informed that Winterbotham's "Historical, geographical, commercial and philosophical view of the United States of America:... has been reprinted in this country. [Charging plagiarism.] Boston: (np), Boradside.

(1784). The history of New Hampshire. V.I. Comprehending the events of one complete century from the discovery of the river Pascataqua. Philadelphia: Printed for the author by R. Aitken, 361p., 21.5cm.

(1792). As above - Boston: Printed for the author by Belknap and Young, 3v. (V.2: folded map), 20.5cm.

(1812). As above - Dover, NH: Printed for O. Crosby and J. Varney, by J. Mann and J. K. Remick, 22.5cm.

(1812-13). As above - 2d ed., Boston and Dover: (np).

(1813). As above - 2d ed., with large additions and improvements, published from the author's last manuscript. Boston: Bradford and Read.

(1831). As above - ... By John Farmer... Dover, NH: S. C. Stevens and Ela & Wadleigh, 512p., 23cm.

(1862). As above - Dover, NH: G. Wadleigh, 26cm.

(1876). Journal of a tour to the White Mountains in July, 1784. Printed from the original manuscript, with a prefatory note by the editor [Charles Deane]. Boston: Massachusetts Historical Society, 21p., folded map, 25cm.

(1950). Journey to Dartmouth in 1774. Edited by Edward C. Lathem from the manuscript journal. Hanover, NH: Dartmouth Publications, 25p., map, 17cm.

(1725). Letter from Jeremy Belknap: 1725. References to topographical and historical description of Boston. Boston: Mass. Historical Society. Coll., 4.

(1791). A new map of New Hampshire. Dover: (np), Manuscript, 14 1/2 x 9 3/4 in., Scale: 12 miles to 1 inch.

(1914). A rough sketch map of the portion of the town of Boston affected by the fire of July 30, 1794. In Massachusetts Historical Society. Proceedings. 47, Boston: p. 293.

(1877). Sketch plan of the part of Boston between Washington street and Fort Hill, showing the New Tontine Crescent, and the region overrun by the fire of July 30, 1794. In Collections of the Massachusetts Historical Society. 43, p.350, Boston: Mass. Hist. Soc.

(1877). Sketch plan of the part of Boston overrun by the fire of April, 1787. In Collections of the Massachusetts Historical Society. 42, p.470, Boston: Mass. Hist. Soc.

Other

Tucker, Louis Leonard. (1990). Clio's consort: Jeremy Belknap and the founding of the Massachusetts Historical Society. Boston: The Society, 149p., illustrated, 24cm. Tucher, St. George.

(1847). Life of Dr. Belknap, with selected letters. [published by his grand daughter (Brown 1897, V.1, p.257)]

Bever, John (c.1800)

Bever was one of the deputy surveyors appointed under the Land Act of 1800 to work under the direction of Rufus Putnam (Pattison 1957, p.207). Beyond that we have little information regarding John Bever. Perhaps the will noted in Mansell (1969, V.51, p.580) is that of this John:

Other

(1837). [last will and testament] Of Columbiana Co., Ohio... 7 leave, printed on one side.

BIGGS, ZACCHEUS (C.1800)

Biggs was one of the deputy surveyors appointed under the Land Act of 1800 to work under the direction of Rufus Putnam (Pattison 1957, p.207). Although no biographical information or publications, beyond the reference by Pattison, were found, surveyors most certainly contributed to the geographic knowledge of the United States.

BINGHAM, CALEB (1757-1817)

Among other things, Bingham was the author of geography textbooks. He was born on April 4, 1757 in Salisbury, Connecticut, and he died in Boston on April 6, 1818 (the National Cyclopedia, 1898). He attended a district school prior to college which he completed at Dartmouth in 1782. Following that he became the master of Moor's Indian Charity School which was maintained as a part of Dartmouth. He moved to Boston in 1784 and kept a private school established for the instruction of young women. In 1785 he published his first book, an English grammar, the second one, published in the United States. Bingham established the first children's library in Boston (Bingham Library for Youth). After teaching for seven years, poor health caused him to quit and he became a book seller.

His book store was a favorite gathering place for teachers where education was the topic of conversation. Bingham was an advocate of free primary schools. He was also a Jeffersonian Republican who on several occasions unsuccessfully ran for the state senate. His textbooks were his biggest contribution. The Geographical Catechism had 22 editions and sold over 100,000 copies. This was originally published in 1795 and was the only geography book used in Boston. He also published the first set of copy-slips (similar to contemporary work books) printed for teachers of writing published in the United States. Bingham was a conservative Congregationalist, fond of music, and he was a member of the choir of the churches he attended (Johnson & Malone 1928, v. 1, pp. 273-274; Hopkins 1964, p.75).

(1795). An astronomical and geographical catechism for the use of children. Boston: Printed and sold by S. Hall. [Examined - BS] 32p., 14.5x8.5cm.
(1796). As above - 2d ed. Boston: S. Hall, 35p., 13cm.
(1798). As above - 4th ed.
(1800). As above - 5th ed.
(1802). As above - 6th ed. Boston: Printed by David Carlisle for Caleb Bingham, 36p., 13.5cm.
(1805). As above - 8th ed.
(1806). As above - 9th ed.
(1809). As above - Pittsfield: Printed by Milo Smith & co., 14cm.
(1810). As above - Burlington: Samuel Mills.
(1813). As above - 13th ed. Boston: Printed by Lincoln & Edmands for C. Bingham.
(1815). As above - 14th ed.
(1816). As above - Hallowell: E. Goodale.
(1818). As above - 15th ed. Boston: Printed by Thomas G. Bangs, 35p., 14cm.
(1819). As above - New Haven: Sidney's press for John Babcock & son, 36p. 13.5cm.

Other

Litch, Samuel. (1814). An astronomical and geographical catechism, for the

use of common schools and private persons. Jaffrey, NH: Printed by Salmon Wilder.
Rosenfield, Lillian O. (1954). Caleb Bingham 1757-1817. Boston.

BLASKOWITZ, C. (C1784)

Blaskowitz is named by Samuel Holland as one of his deputies helping to survey New Hampshire. No other information is to be found about the man.

(1777...1780). Plan de la baie de Narraganset dans la Nouvelle Angleterre avec toutes les iles qu'elle renferme parmi lesquelles se touvent Rhode-Island et l'lle de connonicut leve par Charles Blaskowitz. Londres: author. 47.5 x 62 cm

(1777). A plan of the town of Newport in Rhode Island. London: Faden. 13 1/ 8 x 14 1/2 inches. Scale 500 ft to 1 inch

(1778). A survey of Frog's Neck and the route of the British Army to the 24th of October, 1778, under the command of His Excellency the honorable William Howe, General and Commander in Chief of His Majesty's Forces, etc., etc., Scale 1 in + 2000ft/ 29 1/4 x 16 1/2 in Faden Collections in Phillips

(1777). A topographical chart of the Bay of Narraganset in the province of New England with all the isles contained therein among which Rhode Island and Connecticut have been particularly surveyed... Engraved & Printed for Wm. Faden. Charing Cross: author. 65.2 x 95.5cm

Other

Holland, Capt. Samuel. (1784). Topographical map of the state of New Hampshire, surveyed under the direction of S. Holland, Surveyor General of Lands for the Northern District of North America; by the following gentlemen, his deputies: T. Wright, G. Sproule, J. Grant, T. Wheeler, & C. Blaskowitz. London: Printed for W. Faden, geographer to the King. In New Hampshire. Geological Survey, 1868-1878. Geology of New Hampshire, Atlas. Scale: 8 miles to an inch.

BLODGET, WILLIAM (1754-1809)

Blodget published the first map of the first state added to the original thirteen, Vermont in 1789 (Schwartz and Ehrenberg 1980). It was engraved and printed in New Haven by Amos Doolittle. Ristow (1985) observes that nothing on the map indicates the nature of the survey. Blodget may have had access to the town surveys that were deposited in the office of Ira Allen, Vermont's Surveyor General from 1779-1787 (pp. 86-88, 164).

Dr. Benjamin and Mary Satterlee Blodget were the parents of William. He was born in Stonington, Connecticut on June 8, 1754. Blodget's early education is unknown except for the fact that he appears to have had musical training. He moved with his family to Providence, Rhode Island where he met and married Ann Phillis Chace.

He enlisted in Colonel Daniel Hitchcock's Second Regiment of the Rhode Island Continental Infantry. Shortly thereafter he was named ensign and promoted to lieutenant as military secretary to General Nathanael Greene. He rose to the rank of major as an aid to Greene. Blodget served as chaplain on the frigate Deane.

After the revolution he established himself in Bennington where he operated a general store and operated an iron forge. In the January 28, 1788 issue of Hartford's American Mercury, he advertised land, maps and drafts of any part of the state. Blodget is remembered for being the first of the independent compilers and publishers of state maps (Ristow, 1985, pp. 86-88).

(1789). A topographical map of the state of Vermont; from actual survey. Most humbly dedicated to His Excellency Thomas Chittenden Esqr. Governor and Commander in Chief; the honorable the council, and the honorable representatives of said state; by their most obedient and devoted humble servant William Blodget. Bennington State of Vermont Jan. 1st. 1789. New Haven: Engrav'd and printed for William Blodget, by Amos Doolittle at his House in College-street.

(1791). A new and correct map of Connecticut one of the United States of North America, From actual survey-humbly dedicated by permission to His Excellency Samuel Huntington Esquire Governor and Commander in Chief of said state, by his most humble servant William Blodget. Middletown: author. Joel Allen, engraver.

Other

Carnac, Turner. [Blodget, William "supposed author"]. (1805). Facts and arguments respecting the great utility of an extensive plan of inland navigation in America. By a friend to national industry... Philadelphia: Printed by Wm. Duane.

BLUNT, EDMUND MARCH (1770-1862)

Blunt was a newspaper publisher, printer, and bookseller in Newburyport, Massachusetts (Guthron 1984). His contribution to geographic knowledge of the United States was that of a promoter of information about the American coast, in part attested to by the change of his shop's name from "At the sign of the Bible," to "At the sign of the Bible and quadrant" (p.9). He published Lawrence Furlong's The American Coast Pilot, which went through thirty two editions, some of which were translated into numerous European editions (1796). Guthron provides several examples of Blunt's maps.

Blunt was born in Portsmouth, New Hampshire and died in Sing Sing, New York. Brown (1897, V.1, p.333) names Blunt as author of The American Coast Pilot which enjoyed thirty two editions and was translated into many European languages. Additional information regarding Blunt is in Ristow (1985, pp.227-228, 234 & 307).

(1796). American coast pilot: Containing directions for the principal harbors, capes, and headlands on the coast of North and part of South America... with the prevailing winds, setting of the currents, &c., and the latitudes and longitudes of the principal harbors and capes; together with tide tables and variation. (np). [Brown 1897, V.1. p.333].

(1867). As above- 21st ed. New York: G. W. Blunt..

(c.1797). The American practical navigator. Newburyport, MA: Edmund Blunt.

(1819). Analysis of the authorities upon which the dangers have been inserted in Blunt's new chart of the Atlantic, or Western Ocean. To which is added, an accurate table of the variations of the magnetic needle, as observed in different parts of that sea; with the dates of the respective observations... With an exhibition of facts, showing the progress of ice islands from Greenland to Newfoundland, &c. &c. Compiled by Edmund M. Blunt... New York: Printed for William Hooker, by J. Seymour. 76p. 22cm.

(1804). The Bay of Chesapeake from its entrance to Baltimore. Compiled for Edmund M. Blunt, engraved by A. M. Beasley. In American coast pilot: Containing directions for the principal harbors, capes, and headlands on the coast of North and part of South America...

(1825). Blunt's chart of the South Atlantic Ocean. New York: (np).

(1840). As above- New York: E. & G.W. Blunt.

(1813). Blunt's New Chart of the North Eastern Coast of North America, Extending from lat. 39°30' N. Long. 74°8'W. to Lat 45°N. Long. 66° W. New York: Published as the Act Directs.

(1813). Blunt's new, and correct chart of Long Island Sound from Montauk Point to Frogs Point, including Fisher's Island Sound, and Watch Hill Reef. Surveyed in 1805 by John Cahoone. Newport: N. Fosdick; New York: E. M. Blunt.

(1881). Blunt's reisen in Nord Arabien. Gotha: (np). [Accents omitted]

(1804). Charleston harbour. Engraved by A. Beasley. Newburyport, MA: Edmund M. Blunt.

(1797). A chart of George's Bank including Cape Cod, Nantucket and the shoals lying on their coast, with directions for sailing over the same &c. Surveyed by Capt. Paul Pinkham. Compiled by Paul Pinkham by 1796, engraved by Amos Doolittle of New Haven in 1797, and printed for Edmund M. Blunt, 1797. New York: E. M. Blunt.

(c.1838). The coast of the United States of North America from New York to St. Augustine: Drawn and regulated according to the latest surveys and astronomical observations. Engraved and printed by W. Hooker. New York: E. & G. W. Blunt.

(1809). Journal of a voyage from... Edmund M. Blunt, Printer. Newburyport: (np).

(1826). Le pilote americain, contenant la description des cotes orientales de l"Amerique du Nort, depuis le fleuve Saint-Laurent jusqu au Mississippi; suivi d'une notice sur le Gulf-stream; tr. de l'anglais, par P. Magre... Paris: Imprimerie royale, se trouve chez Bachelier. [Accents omitted]

(1818). The stranger's guide to the city of New York... To which is prefixed, an historical sketch, general description, plan and extent of the city. With an appendix... By Edmund M. Blunt... London: Printed for S. Leigh, by W. Clowes. xvi,14-304p. 15.5cm.

(1815). To the Marine Society of the City of New York; This Chart, Extending from Lat. 39°30'N. Long: 74°8'W to Lat. 23°35' N. Long: 80°25'W. Is respectfully dedicated, By their obedient servant, Edmund M. Blunt. Engraved by Samuel Harrison. New York: E. M. Blunt. $102^{1/2}$x$27^{3/4}$in. Mercator projection, obliquely oriented. Scale 1:1,000,000.

(1833). Traveller's guide to and through the state of Ohio with sailing directions for Lake Erie. New York: Sold by Betts & Anstice. 28p. map. 12cm.

BOSSLER, FREDERICK (C.1807)

Virginia's first state map, engraved by Bossler, was prepared in 1807 by William Prentis and William Davis from information provided by local surveyors (Schwartz & Ehrenberg 1980, p.224). Ristow (1985) reports, "Notwithstanding its claims for accuracy, the Madison map contained a number of errors" (p.121). The reference to Madison is because it was dedicated to the General Assembly of Virginia by James Madison, William Prentis, and William Davis. Following the 1807 publication, a corrected map served as Virginia's most authoritative map until 1827. It was also the first comprehensive road map of Virginia.

(1807). A map of Virginia formed from actual surveys, and the latest as well as most accurate observations. Engraved by Frederick Bossler. Richmond: To the General Assembly of Virginia... by their Fellow Citizens, James Madison, William Prentis, William Davis, Proprietors.

BOWDITCH, NATHANIEL (1773-1838)

Ristow (1985) reports that Edmund Blunt and Bowditch collaborated on map production (p.228). White (1893) observed that Bowditch was widely respected for his knowledge of mathematics and navigation. He was engaged to correct Morris' navigation, but found that it contained so many errors that it needed to be completely rewritten. That prompted him to write his American Practical Navigator which became the standard for American navigation.

Bowditch was born in Salem, Massachusetts to Habakkuk and Mary (Ingersoll) Bowditch. At the age of ten he went to work in his father's cooper shop. He was apprenticed to a ship chandler two years later. Though he had no formal instruction, he became proficient in mathematics, acquired knowledge of navigation, and studied Latin along with Greek, Italian, Spanish, French, Portuguese, and German. From 1805 through 1807 he made a survey of Salem, Marblehead, Beverly, and Manchester, Massachusetts.

In 1799 he was elected a member of the American Academy of Arts and Sciences. Among numerous organizations to which he belonged, were The Royal Astronomical Society of London (Brown 1897, V.1, p.360-361).

At his demise, Bowditch was praised for his contributions to American and worldwide science. His death came on March 16, 1838 and he was buried in Mount Auburn Cemetery marked with a beautiful tomb topped by a statue of Bowditch (White 1893, V.VI, pp.374-375). Some Bowditch charts can be seen in Guthron (1984).

(c.1797). The American practical navigator. Newburyport, MA: Edmund Blunt.

(1806). Chart of the harbours [sic] of Marblehead, Beverly & Manchester from a survey taken in the years 1804-1806. Assisted by George Burchmore & William Ropes, 3rd. Engraved by Hooker and Fairman. (np).

(1834). As above- ...by Nath Bowditch, assisted by Geo. Burchmore & Wm. Ropes, 3d. 2d. edition. Additions and alterations by Charles M. Endicott and Joseph Perkins, Jr. [Boston?]: (np). $20^{3/8}$x$24^{3/4}$ inches. Scale:1689.6 feet to 1 inch.

(1806). Directions for sailing into the harbours of Salem, Marblehead, Beverly, and Manchester. Newburyport: Printed by Edmund M. Blunt. V.6, 30p. 21.5cm.

(c.1808). An estimate of the height, direction, velocity & magnitude of the meteor, that exploded over Weston in Conn. Dec. 14, 1807. (np).

(1809). As above- ...With methods of calculating observations made on such bodies. American Academy of Arts and Sciences, 4(3), pp.213-126.

(1910). Extracts from American practical navigator; an epitome of navigation and nautical astronomy. Washington: Government Printing Office. 92p. 26.5cm.

(1802). The improved practical navigator; containing all necessary instruction for determining the latitude by various methods, and for ascertaining the longitude by lunar observations, in a complete epitome of navigation... to which is added a number of new tables, including those requested to be used with The Nautical Almanac, with improvements and additions, and corrected from many thousand errors of former publications. Originally written and calculated Nathaniel Bowditch... revised, re-calculated, and newly arranged by Thomas Kirby... London: Printed for J. and J. Hardy , and for D. Steel. 4,vi-xvi,18-22,304p. charts. diagrs. front. plates. tables. 21.5cm.

(1809). As above- Third edition.

(nd). Laplace's mecanique celeste. (np).

(1834). Methods of computing the orbit of a comet or planet. Appendix to the third volume of the translation of the Mecanique Celeste. Boston: Hilliard, Gray, Little, and Wilkins. 761-910p. tables. 107p. 28cm.

(1825). Modern astronomy. North American Rev., April, 1825; 20, pp.309-366.

(1802). New American practical navigator. Washington: Government Printing Office. illus. maps. diagrs. charts. 23x26cm.

(c.1808). Observations of the comet of 1807. (np). 17p. From the American Academy of Arts and Sciences. Memoirs. 3.

(c.1811). On the eclipse of the sun of Sept. 17, 1811... Boston: Memoirs of the American Academy of Arts and Sciences, 3, 255-304. 22.5cm.

Other

(1838). A discourse on the life and character of Hon. Nathaniel Bowditch delivered in the church on Church Green, March 25, 1838. Boston: C. C. Little and J. Brown. 119p. 23cm.

At head of title: The varieties of human greatness. Dr. Bowditch's scientific papers; 116-117.

Bowditch, J. Ingersoll. (Ed.). (1864). Bowditch's useful tables. Fifth edition. New York: E. & G.W. Blunt.

(1868). Useful tables from Bowditch's practical navigator. A new ed., with additional tabls [sic]. Navy Department. Washington: Government Printing Office. Hydrographic Office. Publication no. 18. 181p. tables. 24.5cm.

(1881). As above. ...Bureau of Navigation, Navy Department.

(1882). As above. 220p. tables. 27.5cm.

(1885 & 1891). As above. 287p.

(1903). Useful tables from the American practical navigator, by Nathaniel Bowditch revised by G.W. Logan. Washington: Government Printing Office. (U.S. Hydrographic Office Publication no.9, pt. 2). iii,343-652p. 27cm.

(1907, 09, 10, 11, 12, 14 & 1916). As above- 621-716, 739-945p.

(1918 & 1920). As above. 739-949p.

(1925). As above. 278-289, 415-835p.

(1927). As above. 429-849p.

(1929, 31, 32, 34 & 1937). As above. 429-852p.

(1938). As above. 386p.

(1940). As above. 316-319, 387p.

BRADLEY, ABRAHAM JR. (1767-1838)

Ristow (1985) reveals that Bradley was Assistant Postmaster General of the United States. That position allowed him to acquire at least some of his maps based on information from postmasters in various parts of the country. His map of the post roads differs significantly from earlier maps in that it was not copied from other cartographic works.

Bradley was born in Connecticut. He studied law prior to settling in Pennsylvania, first at Wilkes-Barre, then in Philadelphia. He briefly practiced law before he was appointed to the Post Office Department. Ristow asserts that, "The map he compiled and published represented the first clear cartographic break from European-dominated map making and introduced a new, more distinctly American style of cartography to the United States" (pp. 70-71).

Schwartz and Ehrenberg (1980) noted that American and commercial transportation interests near the end of the eighteenth century are represented in Bradley's Map of the United States... (p.215). The authors go on to relate that Bradley and Arrowsmith, a respected British cartographer, worked together and issued revisions of their large 1804 United States maps (p.219).

(c.1800). Gazatteer of the Western Continent. (Engraved for Morse's American Gazetteer.)

Note. Found by the author in the South Caroliniana Library at the University of South Carolina, Columbia.

(1797). Map of the northern parts of the United States of America. In Morse, Jedidiah. The American gazetteer. Boston, Mass.: Presses of S. Hall, and Thomas and Andrews. 22x39cm.

(1804). Map of the southern parts of the United States of America. In Morse, Jedidiah. The American gazetteer. Charlestown, MA: author. 20x38cm.

(1796). A map of the United States, exhibiting post-roads, the situations, connexions [sic] & distances of the post-offices, stage roads, counties, & principal rivers. Philadelphia: Engraved by Francis Shallus. 3 single sheets.

Sculptors: sheet 1, I. Smither; sheets 2, 3, W. Harrison, Jr.
Contents: - Sheet 1 ($17^{1/4}$x$21^{5/8}$ in.). The first sheet, comprehending the nine northern states, with parts of Virginia and the territory north of Ohio. - [Sheet 2 (19 1/4 x 25 5/8 in.)]. Map of the United States, exhibiting the post roads, the situations, connections & distances of the post offices, stage roads, counties, ports of entry and delivery for foreign vessels and the principal rivers. [Covering the Southern states east of the Mississippi]. [Sheet 3] ($15^{1/4}$x$15^{1/4}$ in.), showing Northwestern Territory.

(1799). As above- Georgetown: (np). $33^{3/8}$x$36^{1/4}$ in. Scale: 37 miles to 1 inch. Table.

(1804). As above- Engraved by Francis Shallus. Philadelphia: F. Shallus. 131x94cm. Scale 1:1,960,000.

(1804). As above- Washington, DC: (np). 1p.

(1804). As above- Washington, DC: (np). 4 sheets.

(c.1825). As above- Engraved and printed by W. Harrison. Georgetown, DC: (np). $45^{1/2}$x$59^{3/4}$ in. Scale: 27.6 miles to 1 inch.

Note. Abraham Bradley was Assistant Post Master General from 1800 to 1829. The map appears to cover the service of 1825; it contains Lieut. Talcott's route in 1820 between St. Anthony's Falls and Council Bluff.

Other

(1877). 1731-1822, [birthday meditations by Abraham Bradley on his 92d anniversary Dec. 22d, 1822, with an introduction by his grandson Jos[eph] H. Bradley]. Chevy Chase, MD: (np). 18cm.

Bacot, Thomas W. (1989). Papers (1806-1818): Letterbook containing correspondence addressed to Postmasters General Gideon Granger and R. J. Meigs; Assistant Postmaster General Abraham Bradley, Jr.; fellow postmasters in Georgia, North Carolina, Pennsylvania, Virginia, Washington, D.C.; and to South Carolina postmasters under his jurisdiction. Outgoing letters signed chiefly by Bacot, but also by postal clerks A. Lechais and C. Tesson. Note. Provides detailed view of postal schedules, regulations, and problems; also gives insight into political unrest during the War of 1812. In University of South Caroliniana Society Program, pp.27-29.

Abram Bradley. (1859). The art of taming horses: the Rarey system fully explained. By an experienced horseman. A new ed., rev., with important additions, notes, plates, and diagrams. Buffalo, NY: Clapp, Matthews & Co.'s printing house. 115p. illustrated. plates. 17.5cm.

Note. The preceding entry may be yet another author or it may be that the name is a variation or a typographical error (The American Library Association 1976).

BRANAGAN, THOMAS (1774-1843)

Branagan is one of those people about whom no biographical information is to be found. Like many contributors to geographic knowledge in the new United States, the only information we have about him are his publications.

(1806). The flowers of literature; being an exhibition of the most interesting geographical, historical, miscellaneous and theological subjects, in miniature. To which are prefixed, preliminary addresses, to parents, teachers and pupils. Trenton, NJ: Printed by James Oram, for the author and Daniel Fenton. 324p. 15.5cm.

Intended to facilitate the improvement of youth in particular, and adults in general, whose pecuniary resources will not admit them to purchase nor relative avocations allow time to peruse voluminous productions on these important heads...

(1806) Intended to facilitate the improvement of youth in particular, and adults in general, whose pecuniary resources will not admit them to purchase nor relative avocations allow time to peruse voluminous productions on these important heads...

w = 9.5 cm H = 14.5 cm pp=324.

(1810). As above- ...A new ed. to which is added the Constitution of the United States. Philadelphia: Printed for D. Fenton; and Trenton, NJ: John Cline, printer.

(1810). As above- Philadelphia: Printed for P. Ward, by J. Rakestraw.

BRIDGES, WILLIAM (C.1807)

John Randel was, in 1807, Secretary and Chief Surveyor of the commission charged by the New York State legislature to lay out streets and roads for the extent of Manhattan Island (Ristow 1985). The result was three manuscript maps completed in 1811, A Map of the City of New York and Island of Manhattan as Laid Out by the Commissioners Appointed by the Legislature April 3d 1807. The commissioners were Governor Morris, Simeon DeWitt, and John Rutherford. The map, engraved by Maverick, was issued in six separate sheets for wall mounting, but it was published by William Bridges, an architect and surveyor of the city. Randel promptly decided to cancel his plans to publish the map (pp. 80, 249 & 316).

(1807). Map of the city of N.Y. and the island of Manhattan, as laid out by the commissioners appointed by the legislature. New York: William Bridges.

(1811). As above- ... April 3d, 1807. By Wm. Bridges, city surveyor. Engraved by P. Maverick. 63x223cm. Scale 1:9,850.

(1811). Map of the city of New York and island of Manhattan with explanatory remarks and references. New York: Printed by T. & J. Swords.

BRIGGS, ISAAC (1775-1862)

Other than what is presented by Guthron (1984), we know little about the life of Isaac Briggs. He reveals that Briggs of Maryland and Andrew Ellicott of Pennsylvania, both distinguished surveyors, were asked to "Survey the coast" from Maine to Georgia. That survey was authorized by Congress on February 10, 1807, but it was not executed until 1816 (p.17). Ristow (1985) addresses him as "Surveyor General South of the State of Tennessee" (p.145).

(1798). Isaac Brigg's almanac; for the year of our Lord, 1799; Being the third after leap-year. The astronomical part (which is much more copious than usual) is adopted [sic] to the latitude and meridian of Philadelphia: but may, without material error, be used in Pennsylvania, New Jersey, and Delaware. The selected pieces (of which there are considerable variety in prose and verse) have been chosen with care; and nothing inconsistent with the tenets of friends, is admitted into any part of this work. Baltimore: Printed by W. Pechin. 72p. 17cm.

(1823). Maryland Commissioners appointed to examine into the practicability of a canal from Baltimore to the Potomac. Baltimore: Fielding Lucas, Jun.

(1805). Message from the President of the United States, communicating information in part, on the subject of a post road, from the city of Washington to New Orleans. In pursuance of a resolution of the House of the 31st December last. 1st February, 1805. Referred to the committee of the whole House... (U.S. 8th Congress. 2d sess. House. Document) Washington City: Printed by William Sunne & Son. 10p. 22cm.

BROWN, CHARLES BROCKDEN (1771-1810)

Brown contributed to the geographic knowledge of Americans through his publications. White (1892) describes Brown as a "novelist" who was born in Philadelphia, Pennsylvania on January 17, 1771 and died there of "consumption" on February 22, 1810. He was the child of Quakers who came to the American colony aboard the same ship as William Penn. His education was guided, after his tenth year, by Robert Proud - the historian. Brown's favorite subject was geography. When he was a child he enjoyed tracing the courses of rivers and mountains on a map. White reports that, "At the age of ten there was not a geographical question broached by the family he could not answer" (V.VII, p.59-60).

Following his elementary schooling, Brown studied law under Alexander Wilson. While studying law, he also wrote verses. His publications varied broadly. "He was the first person in America to enter the field of literature as a profession." He published a series of papers in the Weekly Magazine entitled "The Man at Home." In 1799 he attempted to publish a magazine, Monthly Magazine and American Review, to which he was the principal contributor, but it did not flourish. He started another magazine, The Literary Magazine and American Register, but it lasted only five years. Brown published the first volume of his American Register in 1806. He married Elizabeth, the daughter of Dr. William Linn of New York, in 1804. It appears the marriage lasted until his demise.

(1803). An account of Louisiana, being an abstract of documents, in the offices of the Departments of State, and of the Treasury. Philadelphia: John Conrad & Co.; Baltimore: M. & J. Conrad & Co. Washington City: Rapin, Conrad, & Co. 50p. [Brown is possible editor]

(1803). As above- Philadelphia: William Duane.

(1803). An account of Louisiana, laid before Congress by direction of the President of the United States, November 14, 1803; Comprising an account of its boundaries, history, cities, towns, and settlements; of the original number and strength of the inhabitants; of its rivers, canals, mountains, minerals, and productions of soil... Providence: Printed by Heaton & Williams. 72p.

BUCKINGHAM, EBENEZER (C.1800)

Buckingham was one of the deputy surveyors appointed by the Land Act of 1800 to work under the direction of Rufus Putnam (Pattison 1957, p. 207). Although no biographic information or publications have been found, activities relating to surveying certainly place Buckingham among contributors to the geographic knowledge, among citizens of the United States.

BUELL, ABEL (1750-1825)

Buell was an engraver of maps. He was born in Killingworth, Connecticut and died in New Haven. He was a man who had numerous skills: engraver, jeweler, goldsmith, undertaker, military bugler, teacher of singing and choir leader. He was an expert engraver which led him to alter a colonial note for which he served time in prison. A special act of the Connecticut legislature pardoned him for honorable services rendered the state restoring his civil rights. His publication, The Devil's Club or Iron Cane, earned him the wrath of Puritans which brought Buell six months confinement in Symsbury Mines. He repented and was forced to carry an iron cane on the Sabbath. Buell heated the end of his Iron Cane white hot and used it to touch off the first cannon fired in the American Revolution at the Battle of Bunker Hill. Disguised as a Kickapoo Indian he was one among participants in "The Boston Tea Party" (Brown 1897, V.1, p. 482).

Schwartz and Ehrenberg (1980) provide examples of Buell's engravings, one of which was the first map of Connecticut designed, engraved and published in Connecticut (pp.177, 182). His map of the United States was the first such map exhibiting the American flag that was compiled, engraved, and printed in the United States (p. 201).

Buell's map of the United States was the first published by an American after the signing of the Treaty of Paris in 1783 (Schwartz and Ehrenberg 1980). It was advertised as the, "first map ever compiled, engraved and finished by one man and an American" (p. 205). It was based on the boundaries established by John Mitchell and Lewis Evans and it included errors. The announcement of the availability of the new map reads:

> This map is the effect of the compiler's long and unwearied application, diligence and industry, and as perfection has long been the great object of his labours [sic], and it being the first ever compiled, engraved, and finished by one man, and an American, he flatters himself, that every patriotic gentleman, and lover of geographical knowledge, will not hesitate to encourage the improvement of his own country. (Ristow 1985, p. 66)

(1784). <u>A new and correct map of the United States of North America Layd [sic] down from the latest observations and best authority agreeable to the peace of 1783</u>. New Haven, CN: author.

(1774). Part of the province of East Florida. Engraved by Paul Revere and Abel Buell. Published with Roman's <u>A concise natural history of East and West Florida</u>. New York: author. 62x221cm. [The Florida panhandle.]

(1774). Part of the province of East Florida. Engraved by Paul Revere and Abel Buell. Published with Roman's <u>A concise natural history of East and West Florida</u>. New York: author. 147x169cm. [Tampa Bay through Cape Canaveral.]

(1766). <u>Plan of the colony of Connecticut in North-America</u>. Connecticut: Lord Halifax.

BURCHMORE, GEORGE (C.1804)

Guthron (1984) lists Burchmore as an assistant to Bowditch when he surveyed Marblehead, etc. in 1804, 1805, and 1806 (p. 34). Little else is known about the man.

(1806). <u>Chart of the harbours of Marblehead, Beverly & Manchester from a survey taken in the years 1804-1806. Assisted by George Burchmore & William Ropes, 3rd.</u> Engraved by Hooker and Fairman. (np).

(1834). <u>Chart of the harbours of Marblehead, Beverly and Manchester from a survey taken in the years 1804, 5 & 6 by Nathaniel Bowditch AM, AAS, assisted by Geo. Burchmore & Wm. Ropes 3d. Engraved by Hooker and Fairman. Second edition published in 1834, with additions and alterations by Charles M. Endicott and Joseph Perkins, Jr</u>. (np).

BURGES, BARTHOLMEW (C.1740-1807)

Burges was among those who were assigned to approve of Clark's (1790) A Complete Set of Charts of the Coast of America, extending from Cape Breton on the East, into the Gulph [sic] of Mexico, with Nautical Directions (Guthorn 1984). That assignment was in regard to knowledge of the American coast, that attests to Burges' status. He advertised lectures in astronomy in Boston in 1789 and published a short treatise on the solar system. His account of the solar system included a study of the comet of 1532. In 1807 he was lost at sea while on his way from Boston to Trinidad.

(1790). A series of Indostan [sic] letters, by Bar. Burges; containing a striking account of the manners & customs of the Gentoo [sic] nations, & of the Moguls & other Mahomedan [sic] tribes in Indostan, with other polemical East India tracts... New York: Printed & published for the author by W. Ross. xxv, 168p. 16.5cm.

(1789). A short account of the solar system, and of comets in general: together with a particular account of the comet that will appear in 1789. Boston: B. Edes & Son. 16p. plates. 16.5cm.

CAHOONE, JOHN (C.1805)

Guthron (1984) noted that Cahoone was from Newport. He worked with Nicoll Fosdick to survey and produce a map of Long Island Sound in 1805 (p.64). Like many of those who contributed to knowledge of America's geography, no other publications nor a biography are to be found by or about Cahoone.

(1813). Blunt's new, and Correct Chart of Long Island Sound from Montauk Point to Frogs Point, including Fisher's Island Sound, and Watch Hill Reef. Surveyed in 1805 by John Cahoone. Newport: N. Fosdick. New York: E. M. Blunt.

CALLENDAR, JOSEPH (1751-1821)

Callendar was an engraver of maps. Osgood Carleton's 1795 map of Maine depicted towns, rivers, lakes, roads, mills, iron works, and public buildings along with the distances from Boston and each county seat and was engraved by Hill and J. Callendar (Schwartz & Ehrenberg 1980, p. 215). The engagement of several persons as contributors to the knowledge of United States geography was a direct result of legislation such as that of the Massachusetts State Legislature which resulted in a resolution in 1794 directing each town to prepare an accurate town map (p. 219).

(1802). Map of Massachusetts proper compiled from Actual surveys made by Order of the General Court... By Osgood Carleton. Engraved by Joseph Callendar and Samuel Hill.

(1802). Map of the District of Maine, drawn from the latest surveys and other best authorities. Boston: B. & J. Loring. By Osgood Carleton. Engravers Joseph Callendar and Samuel Hill.

CARLETON, OSGOOD (1742-1816)

Carleton, at the time of the publication of Clarke's (1790) A Complete Set of Charts of the Coast of America, Extending from Cape Breton on the East, into the Gulph of Mexico, with Nautical Directions... was a well-known surveyor, publisher, and leading teacher of astronomy and mathematics in Boston (Guthorn 1984). He was the person identified by the Boston Marine Society to sign the plate and proof sheets, attesting to the accuracy of each chart for the society. His 1795 map of Maine depicted towns, rivers, lakes, roads, mills, iron works, and public buildings along with the distances from Boston and each county seat (Schwartz & Ehrenberg 1980, p. 215). Carleton enlisted the support of the Massachusetts State Legislature which resulted in a resolution in 1794 that directed each town to prepare an accurate town map (p. 219).

Ristow (1985) asserts that Carleton was, "One of the most active American-born cartographers in the post-revolutionary war decades" (p. 246). Guthron (1984) provides further evidence of Carleton's stature when we note that Carleton provides, "A certificate of accuracy" for maps (pp. 7-8, 43 & 96).

(1798). An Accurate Map of the Commonwealth of Massachusetts exclusive of the District of Main compiled pursuant to an act of the General Court from actual surveys of the several towns &c. Taken by Their Order... By Osgood Carleton. Boston: Published and sold by O. Carleton and J. Norman. sold also by W. Norman. 89X119cm. Scale 1:253,440.

(1795). Map of the District of Maine, Drawn from the latest surveys and other best authorities. Engraved by John Norman. (np).

(1801). Map of Massachusetts Proper compiled from actual surveys made by order of the General Court...By Osgood Carleton. Callendar and Hill, engravers. Boston: B. & J. Loring.

(1802). As above- Drawn by G. Graham; Engraved by J. Callender and S. Hill.

(1796). Plan of the city of Boston. In Boston directory of 1796. Boston: John West.

CAREY, MATHEW (1760-1839)

Ristow (1985) reveals a number of geographic publications with which Carey was involved (pp.135,138,151 & 179). On January 28, 1760 Carey was born in Dublin, Ireland and died on September 16, 1839 in Philadelphia, Pennsylvania. He was a publisher, economist, and greatly contributed to early American geography as an editor and publisher.

Carey was dropped by his nurse when a year old and was lame from that time on. He read widely, but appears to only have had the rudiments of formal education. At an early age, seventeen years, he published an essay against duelling.

Another paper regarding the wrongs done to Irish Catholics, produced by Carey, was used in the Lords and Commons as proof of the treasonable views of Irish Catholics. That paper prompted Carey's family to send him to Paris where he was introduced to Benjamin Franklin for whom he worked for a short period. Lafayette asked him to discover the strength of the revolutionary sentiment in Ireland which prompted Carey to return to Dublin in 1783. There he established the Freeman's Journal and then the Volunteer's Journal. The paper was a year old when Carey was arrested for his published attacks on the House of Commons and the Premier. After his release, he departed by ship for America.

Once in America, Lafayette came to his economic rescue. In 1785 he issued the first issue of his Pennsylvania Herald. He followed that venture with other publications and publishing activities.

Carey volunteered along with his friend Stephen Girard and others to serve on the Committee of Health for Philadelphia during the yellow fever epidemic of 1793. In a pamphlet he described the pestilence and horrors of that epidemic. He was a charter member of the Philadelphia Society for the Promotion of National Industry and through papers published under that organizations identity presented the classic American argument in favor of the protective system. Carey also was an advocate of universal education (Johnson & Malone 1928, V.2, pp.489-491).

(1810). A Egyptiaca. Atlas of plates illustrating Egyptian antiquities. London: author.

(1795). American atlas; containing twenty maps and one chart. Philadelphia: Engraved and published by, Mathew Carey. 21 folded maps. 38cm.

(1802). American minor atlas. (see Minor American atlas).

(1800). Carey's American atlas; containing 20 maps and one chart... Philadelphia: author. 21 maps. 38x25cm.

(1809 & 1823). Carey's American atlas. Philadelphia: author. 26 folded maps.

Note. All maps except nos. 2,4,7,9,12,15,18,23-26 carry the following: Engraved for Carey's American edition of Gutherie's geography improved.

(1806). Carey's American minor atlas. Philadelphia: author. 20 maps. 24x18cm.

(1796). Carey's American pocket atlas - with concise description of each state. Philadelphia: Printed for M. Carey by Lang & Ustick. 118p. 19 folded maps. 17cm.

(1801). As above- ... containing nineteen maps... with a brief description of each state. 2d ed. Philadelphia: Printed by H. Sweitzer for M. Carey. 114p. 19 maps (1 folded). 21cm.

(1805). As above- ... containing twenty maps... with a brief description of each state, and of Louisiana, also, the census of the inhabitants of the United States, for 1801. The exports from the United States for ten years. 3d ed. greatly improved and enlarged. Philadelphia: author. 114p. 19 maps. 12°.

Note. Maps engraved by W. Barker, J. H. Seymour and A. Doolittle.

(1813). As above- 4th ed., greatly improved and enlarged. Philadelphia: author. 168p. 23 maps. 2 tables.

(1959). As above- Louisville: Lost Cause Press.
Note. Facsimile of 1801 ed. (Ed.).
(c.1795). A chart of the West Indies, from the latest marine journals and surveys. W. Barker, sculpt. Philadelphia: M. Carey. map. 27x38.5cm.
(1796). General atlas. Philadelphia: author. 45 maps (44 folded). 38cm.
Note. Most of the maps have the note "Engraved for Carey's American edition of Gutherie's Geography Improved."
(1800). As above- 49 numbered maps (part folded, part doubled). 44x38cm.
(1802). As above- 45 maps.
(1802). As above- 49 maps.
(1804 & 1811). As above- 51 maps on 49 numbers. 45x40cm.
(1814). A general atlas, being a collection of maps of the world and quarters, their principal empires, kingdoms, &c. containing fifty-eight maps and charts. Philadelphia: M. Carey and Johnson & Warner. 58 maps. 28cm.
(1794). A general atlas for the present war. Containing six maps and one chart. 1. The Seven United Provinces of Holland... Engraved by Tiebout. 2. The Austrian, French and Dutch Netherlands. By J. Scott. 3. The Empire of Germany and the thirteen cantons of Switzerland. By J. Scott. 6. Italy and Sardinia. By Tiebout. 7. The West Indies... By Barker... Philadelphia: Printed for the author. 7 maps. 39cm.
(1814). Carey's general atlas, improved and enlarged; being a collection of maps of the world and quarters; their principal empires, kingdoms, &c... Philadelphia: M. Carey. 58 maps (part doubled). 45cm.
(1815). As above- 58 maps (1 folded, 49 doubled).
(1816). As above- 57 maps. 44cm.
(1817). As above- 58 maps. 45x30cm.
Note. Some of the maps bear the note, "From the latest surveys by Samuel Lewis."
(1832). Letters on the colonization society; with a view of its probable results, under the following heads: the origin of the society; increase of the coloured [sic] population; manumissions of slaves in this country; declarations of legislatures, and other assembled bodies, in favour [sic] of the society; situation of the colonists at Monrovia and other towns. Addressed to the Hon. C. F. Mercer... Published by the Connecticut colonization society, from the 2d enl. and improved ed.... Hartford: P. B. Gleason & co. printers. 32p. 22.5cm.
(1832). As above- ...Addressed to the Hon. C. F. Mercer. Philadelphia: Young, printer. map.
(1832). As above- 3d ed. 2 maps.
(1832). As above- 4th ed. Philadelphia: Stereotyped by L. Johnson.
(1832). As above- 5th ed., greatly enl. and improved.
(1832). As above- 6th ed.
(1832 &1833). As above- 7th ed.
(1834). As above- ...9th ed. To which is prefixed the important information

collected by Joseph Jones, a coloured man, lately sent to Liberia, by the Kentucky colonization society, to ascertain the true state of the country - its productions, trade, and commerce - and the situation and prospects of the colonists.
(1835). As above- 10th ed. Philadelphia: For sale by Carey & Hart.
(1838). As above- 12th ed. Philadelphia: E. G. Dorsey, printer. 2 maps on 1 plate. 24cm.
(1838). As above - 13th ed.
(1802). Minor American atlas. Philadelphia: (np). 19 maps. 4°.
(nd). Mississippi territory and Georgia. (np). map. 15x19.5cm. Scale: 1 inch to 100 miles.
(1817). A scripture atlas: Containing all the maps necessary for understanding the geography of the countries mentioned in the Old and New Testaments: to which are added, a table of the lives of the patriarchs, and a plan of the temple of Solomon. Philadelphia: author. 13 folded maps. 1 plate. 1 table. 8°.
(1794). A short account of Algiers, containing a description of the climate of that country, of the manners and customs of the inhabitants, and of their several wars against Spain, France, England, Holland, Venice, and other powers of Europe, from the usurpation of Barbarossa and the invasion of the Emperor Charles V. to the present time; with a concise view of the origin of the rupture between Algiers and the United States... Philadelphia: Printed by J. Parker for the author. 46p. 1 folded map. 20.5cm.
(1794). As above- ... to which is added, a copious appendix, containing letters from Captains Penrose, M'Shane, and sundry other American captives, with a description of the treatment those prisoners experience. 2d ed., improved. Philadelphia: author. 50p.
(1796). Vermont, from actual survey... Engraved for M. Carey by J. H. Seymour. Philadelphia: author. map. 18x13.5cm. Scale: 10mi. to 1cm.
(1823). View of the very great natural advantages of Ireland: and of the cruel policy pursued... towards that island... Added, A sketch of the present deplorable condition of the Irish peasantry. Extracted from Vindiciae Hibernicae. Philadelphia: Carey & Lea. 24p. 8°.

Other

Lea and Febiger. (nd). Carey, Mathew, and sons, Philadelphia. (np).
Lea and Febiger. (nd). One hundred years of publishing. (np).
Moore, S. S., & Jones, T. W. (1802). Traveler's directory, or A pocket companion: shewing [sic] the course of the main road from Philadelphia to New York, and from Philadelphia to Washington... From actual survey. Philadelphia: Printed for and published by Mathew Carey.
Pennsylvania society for the promotion of manufactures and the mechanic arts. (nd). Constitution of Pennsylvania Society for the Promotion of Manufactures and the Mechanic Arts. (np).

Publicola [pseud.]. (1830). Thirteen essays on he policy of manufacturing in this country. From the New York Morning Herald. Philadelphia: Printed by Clark & Raser.

CHRISTMAS, WILLIAM (C.1808)

Christmas, according to Ristow (1985), worked on the Price-Strothers map of North Carolina. He was a surveyor and draftsman [cartographer] (p. 124). No other information was found by or about Christmas.

(1808). Actual survey of the state of North Carolina. Respectfully dedicated by the humble servants Jona. Price and John Strothers. Draftsman William Christmas. Philadelphia: printed by C. P. Harrison. 74X152cm. Scale 1:500,000.

CHURCHMAN, JOHN (1753-1805)

Churchman, likely in 1787, published the map of the Peninsula between Delaware & Chesapeake Bays with the adjacent bays and shores drawn from contemporary surveys. The date is based on advertisements in 1786 and 1787 (Ristow 1985). He was born in East Nottingham, Pennsylvania. His professional work was that of mathematician and surveyor. He surveyed the Chesapeake Bay area during the Revolutionary War, submitting a map to the American Philosophical Society in 1779. The Society responded:

> We are of the opinion that he is possessed of sufficient materials, both astronomical observations and actual surveys, to enable him to construct an accurate map, and have no doubt that he has executed his design with exactness and care, but we cannot help expressing our desire of seeing the map laid down upon a much larger scale, which would render it more serviceable for promoting the knowledge of geography. (p.118)

Churchman did not choose to redraw his map, but he dedicated it to the Society. His scholarly pursuit of scientific information took Churchman to Europe. On a return voyage from Europe he died (White 1892, V.IX p.287; Wright 1898, V.9, p.287). .

(1790). An explanation of the magnetic atlas, or variation chart, hereunto annexed; projected on a plan entirely new, by which the magnetic variation on any part of the globe may be precisely determined for any time past, present, or future: and the variation and latitude being accurately known, the longitude is of consequence truly determined. Philadelphia: Printed by James & Johnson. x[11]-46],[6]p. fold. chart. 2 fold. tab. 19.5cm.

(1782). Magnetic atlas or variation charts of the Northern Hemisphere. (np): John Churchman. 57x43cm.

(1794). The magnetic atlas, or variation charts of the whole terragueous [sic] globe; comprising a system of the variation and dip of the needle, by which the observations being truly made, the longitude may be ascertained. London: Printed for the author by Darton and Harvey. vii, [1],xxix,[31]-80p. 3 charts. 28.5x22.5cm.

(1800). As above- ...the third edition with additions. By John Churchman, Fellow of the Russian Imperial Academy. New York: Printed for the author and sold by Gaine & Ten Eyck. vii,[1],xxix,[31]-82p. maps. 27.5cm.

(1787). Map of the penninsula [sic] between Delaware & Chesopeak [sic] Bays with the said bays and shores adjacent drawn from the most accurate surveys. (np).

(1778). To the American Philosophical Society this map of the peninsula between Delaware & Chesopeak [sic] Bays with the said bays and shores adjacent drawn from the most accurate surveys is inscribed by John Churchman. Baltimore: (np). 22.25 x 17 in.

(1786). As above- 43x57cm.

(1937). As above- Washington: Reprinted, U.S. Geological Survey.

(1800). A variation chart. London: (np). 2 maps.

CLARK, GEORGE ROGERS (1752-1818)

Clark was half of the famous Lewis and Clark Expedition team, both captains in the American Army. Klemp (1976) provides a map of the tract that was prompted by Thomas Jefferson who gave orders to explore the Rocky Mountains and find a way to the Pacific coast (plate 49). The expedition took place between 1804 and 1806.

Clark was born near Monticello, Albemarle County, Virginia and died at Locust Grove, near Louisville, Kentucky. He was the son of John and Ann (Rogers) Clark. He practiced surveying and in 1771 or 1772 made a long tour through the upper Ohio Valley for the purpose of locating land for himself. He spent several years clearing and improving the land. In 1775 he was Deputy surveyor under Captain Hancock Lee to survey the lands on the Kentucky River for the Ohio Company. He was an active member and developer of the western portion of the state, Kentucky. Clark drew up a plan for the town of Louisville, Kentucky that would have "made Louisville one of the most beautiful cities on the continent" (Brown 1897, V.2, p.24; White 1898, V.1, p.82).

Application to confirm Indian grant, not confirmed by Virginia petition of George Rogers Clark. American State Papers: Public Lands, V.XXVIII (Doc. 111).

Claim for services under General George Rogers Clark against Kaskaskia and Vincennes in 1778. American State Papers: Military Affairs, V.XXI (Doc.672).

U.S. Senate. 29th Congress, 2nd Session. Legislature of Illinois, for adjustment of claims for advances to expedition under General George Rogers Clark in 1778-1779, known as Illinois campaign. (S.Doc.174). Washington: Government Printing Office, 1846-1847. (Serial Set 495).

Other

James, James A. (Ed.). (1912). George Rogers Clark Papers, 1771-1781. Springfield, IL: Illinois State Historical Library.

James, James A. (1928). The life of George Rogers Clark. Chicago: University of Chicago Press.

Thwaites, Reuben Gold. (1904). A brief history of Rocky Mountain exploration with especial reference to the expedition of Lewis and Clark. New York: Appleton & Co. 289p.

Thwaites, Reuben Gold. (Ed.). (1904-1907). Early western travels, 1748-1846; a series of annotated reprints of some of the best and rarest contemporary volumes of travel, descriptive of the aborigines and social and economic conditions in the middle and far West, during the period of early American settlement; edited, with notes, introduction, index, etc. Cleveland: The A. H. Clard Co. 32v. illus. plans. facsim. maps. tab.

CLARK, MATTHEW (1714-1798)

According to Guthorn (1984), Clark was a Boston trader and auctioneer born at Chilmark on Martha's Vinyard. His relations with pilots in the nearby waters prompted him to embark on and publish a set of charts of the coast of America. That was advertised in the Boston Gazette on July 5, 1790, with a note that it had just been published. Ristow (1985) asserts that his American Coast Pilot was the "first post-Revolution American nautical atlas" published (p. 225-6).

(1790). American coast pilot by Matthew Clark. Boston: Matthew Clark.

(1789). Chart of the coast of America, from Port Howe to the entrance of the Gulph [sic] of Mexico. Boston: (np). 18 sheets.

(1790). A complete chart of the coast of America, from Cape Breton into the Gulf of Mexico. Boston: Published and sold by Mathew Clark.

U.S. House of Representatives. 23rd Congress, 2nd Session. Contract to publish Documentary History of Revolution, by Matthew St. Clair Clarke and Peter Force. (H.doc.36). Washington: Government Printing Office, 1834-1835. (Serial Set 272).

U.S. House of Representatives. 24th Congress, 1st Session. Memorial of Matthew St. Claire Clarke, Revolutionary claims agent for Georgia. (H.doc.87). Washington: Government Printing Office, 1835-1836. (Serial Set 288).

U.S. House of Representatives. 26th Congress, 1st Session. Publication of Documentary History of United States by Matthew St. Clair Clarke and Peter Force. (H.doc.102). Washington: Government Printing Office, 1839-1840. (Serial Set 365).

CLARK, WILLIAM (1770-1838)

Schwartz and Ehrenberg (1980) note that William Clark worked with Thomas Jefferson and Meriwether Lewis to plan the Lewis and Clark expedition (p. 221). William sketched maps of part of the continent of North America while in the party of Lewis and Clark in 1804, 1805, and 1806 (pp. 222-223, 227-229).

White (1906) identifies Clark as a soldier, explorer and the second territorial governor of Missouri. He was born in Caroline County, Virginia, the son of John and Ann (Rogers), the grandson of Jonathan and Elizabeth (Wilson) Clark. At fourteen years of age his family moved to the falls of the Ohio in Kentucky, the present Louisville. There his brother, George Rogers Clark, erected a fort in 1777. Around 1804 Clark moved to St. Louis Missouri where he was commissioned a second lieutenant of artillery by President Jefferson. He was ordered to join Capt. Meriwether Lewis on an expedition from St. Louis to the mouth of the Columbia River. Following that successful expedition, Clark went to Washington to receive land in return for his services. He was later engaged in activities culminating in the admission of Missouri to the Union and became Superintendent of Indian Affairs at St. Louis, appointed by President Monroe (V.XII, pp. 301-302).

(1886). The Canadian north-west. London: Print by Cassell & Co. 12p. 18cm.

Gratuities to officers and men in expedition to Pacific Ocean under Lewis and Clark. American State Papers: Military Affairs, V.XVI (Doc.68).

(1813). An interesting account of the voyages and travels of Captains Lewis and Clarke, in the years 1804-5, & 6. Giving a faithful description of the river Missouri and its source-of the various tribes of Indians through which they passed-manners and customs-soil-climate-commerce-gold and silver mines-animal and vegetable productions... To which is added a complete dictionary of the Indian tongue, by William Fisher, esq. Baltimore: Printed and published by P. Mauro.

(Dec.1914). Journal of Lieut. Wm. Clark - Wayne's campaign. Mississippi Valley Historical Review, 1(3), 418.

(1822). ...Letter from the Secretary of War, transmitting, (in obedience to a resolution of the House of Representatives of the 18th July,) information in relation to the superintendency of Indian affairs, in the territory of Michigan during the year 1820, and part of the year 1821... Washington: Printed by Gales and Seaton.

(Nov. 8,1806). Letter to his brother, George Rogers Clark, dated at St. Louis, September 23, 1806, and containing a summary of the Lewis and Clark expedition. In the Evening fire-side, 2(45), 358-360.

Lewis and Clark expedition, pt. 1: Correspondence of Captain Meriwether Lewis. American State Papers: Indian Affairs, V.XVI (Doc.113).

Lewis and Clark expedition, pt. 2: Statistical view of Indian nations in Louisiana Territory and territories on its boundaries, and ethnological observations, p.707. American State Papers: Indian Affairs, V.VII (Doc. 113).

Lewis and Clark expedition, pt. 3: Historical sketch of Indian tribes in Louisiana and ethnological observations, by John Sibley, p. 721. American State Papers: Indian Affairs, V.VII (Doc.113).

Lewis and Clark expedition, pt. 4: Geographical description of Red River and adjacent country, by John Sibley, p. 725. American State Papers: Indian Affairs, V.VII (Doc.113).

Lewis and Clark expedition, pt. 5: Geographical description of Washita River in Louisiana and adjacent country, by William Dunbar and Dr. George Hunter, p. 731. American State Papers: Indian Affairs, V.VII (Doc.113). Washington: Government Printing Office. 1805-1806. (Serial Set ASP07).

Lewis and Clark expedition, pt. 6: Meteorological observations of William Dunbar and Dr. George Hunter, p. 742. American State Papers: Indian Affairs, V.VII (Doc.113).

(nd). Lewis and Clark's expedition. Montana Historical Society Contributions, 1, 99-112.

King, Nicholas. (1805-6). A map of part of the continent of North America, between the 35th and 51st degrees north latitude, and extending from the 89° degrees of west longitude to the Pacific Ocean. Compiled from the authorities of the best informed travellers, by M. Lewis. Sketched by William Clark. Washington, DC: National Archives. map. 80x117cm.

(1950). A map of part of the continent of North America ... shewing [sic] Lewis & Clark's rout over the Rocky Mountains in 1805 on their route to the Pacific from the United States. New Haven, Conn.: Yale University Library. 86x144cm. Scale 50 miles to the inch.

(c.1795). A new mode of legislation, lately invented by Matthew Clay, representative from the County of Pittsylvania. (np). [The petition of ... sundry inhabitants in the County of Pittsylvania, for clearing the Banister River, with the names of the petitioners.]

(1812). New travels among the Indians of North America; being a compilation, taken partly from the communications already published, of Captains Lewis and Clark, to the President of the United States; and partly from other authors who traveled among the various tribes of Indians ... With a dictionary of the Indian tongue. Compiled by William Fisher, esqr. Philadelphia: James Sharan, J. Maxwell, printer.

U.S. House of Representatives. 19th Congress, 1st Session. Report of General Clark on preservation and civilization of Indians. (H.doc.124). Washington: Government Printing Office. 1825-1826. (Serial Set 138).

(1809). The travels of Capts. Lewis & Clarke, by order of the government of the United States, performed in the years 1804, 1805, & 1806, being upwards of three thousand miles, from St. Louis, by way of the Missouri, and Columbia rivers, to the Pacifick [sic] Ocean: containing an account of the Indian tribes, who inhabit the western part of the continent unexplored, and unknown before. With copious delineations of the manners, customs, religion, &c. of the Indians. Comp. from various authentic sources, and documents, to which is subjoined, a summary of the statistical view of the Indian nations, from the official communication of Meriwether Lewis... Philadelphia: Hubbard Lester.

Other

Allen, Paul. (1915). Stirring adventures "up the Missouri" with Lewis and Clarke, pioneers of the great Northwest. Akron, O: The Superior Printing Co.

Chandler, Katherine. (1905). The Bird-Woman of the Lewis and Clark expedition; a supplementary reader for first and second grades. New York & Boston: Silver, Burdett and Co.

De Voto, Bernard. (Ed.). (1953). The journals of Lewis and Clark. Boston: Houghton Mifflin.

Gass, Patrick. (1814). Tagebuch einer entdeckungs-reise durch Nord-America, von der mundung des Missuri an bis zum einfluss der Columbia in den Stillen ocean, gemacht in den jahren 1804, 1805 und 1806, auf befehl der regierung der Vereinigten Staaten, von den beiden capitans Lewis und Clarke. Uebers. von Ph. Ch. Weyland. Mit einer charte. Weimar: H.S. Privil Landes-industrie-comptoirs.

Gregg, Kate L. (1937). Westward with dragoons; the journal of William Clark on his expedition to establish Fort Osage, August 25 to September 22, 1808; a description of the wilderness, an account of the building of the fort, treaty making with the Osages and Clark's return to St. Louis; edited for schools. Fulton, MO: Printed by Ovid Bell Press. 97p. 2 maps. 20.5cm.

Lewis, Meriwether. (1812). Die reisen der Capitaine Lewis und Clarke; unternommen auf befehl der regierung der Vereinigten Staaten in den jahren 1804, 1805, und 1806, uber eine landerstrecke von mehr als 3000 meilen, von St. Louis, auf dem Missouri und Columbia, nach dem Stillen meer. Enthaltend: eine beschereibung der indianischen volkerstamme, welche den westlichen theil von Nord-Amerika, der uns bisher unbekannt und unentdeckt war, bewohnen. Samt einer statistischen uebersicht der Indianer nationen, aus dem official bericht von Meirwether Lewis. Friedrichstadt(P.): Gedruckt bey M. Bartgis.

Lewis, Meriwether. (1947-1948). Lewis and Clark in North Dakota; the original manuscript journals and the text of the Biddle edition during the time the expedition remained in North Dakota. With explanatory notes by Russell Reid. Bismarck, ND: (np).

Lewis, Meriwether, & Clark, W. (1905). History of the expedition under the command of Captains Lewis & Clark to sources of the Missouri, thence across the Rocky Mountains and down the river Columbia to the Pacific Ocean, performed during the years 1804-5-6, by order of the government of the United States. A complete reprint of the Bibble ed. of 1814, to which all the member of the expedition contributed with an account of the Louisiana Purchase, by Prof. John Bach McMaster, and notes upon the route... London: D. Nutt.

Lewis, Meriwether, & Clark, W. (1816-1817). Reize naar de bronnen van den Missouri, en door het vaste land van America naar de Zuidzee. Gedaan op last van de regering der Vereenigde Staten van America, in de jaren 1804,

1805 en 1806. Door de Kapiteins Lewis en Clarke... Uit het Engelsch vertaald door N. G. van Kampen... Te dordrecht: A. Blasse & zoon. [Accents omitted]

Lewis, Meriwether, & Clark, W. (1815). Travels to the source of the Missouri river and across the American continent to the Pacific Ocean. Performed by order of the government of the United States, in the years 1804, 1805, and 1806. By Captains Lewis and Clarke. Published from the official report, and illustrated by a map of the route, and other maps. A new ed.... London: Longman, Hurst, Rees, Orme, and Brown.

Lewis, Meriwether, & Clark, W. (1923). The unbroken wild (1804) from Lewis and Clark's journals. In Joseph Lewis French, The pioneer west; narratives of the westward march of the empire. Boston: Littel, Brown, and Co. pp.1-14.

McClelland, John Morris. (nd). Lewis and Clark in the Fort Columbia area. (np).

Thwaites, Reuben Gold. (Ed.). (1904-1905). Original journals of the Lewis and Clark expedition, 1804-1806; printed from the original manuscripts in the library of the American Philosophical Society and by direction of its committee on historical documents, together with manuscript material of Lewis and Clark from other sources, including note-books, letters, maps, etc., and the journals of Charles Floyd and Joseph Whitehouse, now for the first time published in full and exactly as written. New York: Dodd, Mead & Co.

CLEVELAND, MOSES (1754-1806)

Cleveland, Ohio is named for the man who surveyed the site at the mouth of the Cuyahoga River (Schwartz and Ehrenberg 1980). As one of the directors of the Connecticut Land Company, Cleaveland also a surveyor, took a party of men to survey the companies land purchase in the Western Reserve of the Ohio Valley. White (1893, V.VI, p.257-158) confirms the previous information.

Brown (1897) notes that Cleveland was born in Canterbury, Connecticut to Col. Aaron and Thankful (Paine) Cleveland and died in Canterbury, Connecticut. He graduated from Yale college in 1777 and began the practice of law in Canterbury. Cleveland was a prominent member of the Connecticut legislature, a brigadier general of the militia, and the employee of the Ohio Company which directed him to survey the Western Reserve: 3,800,000 acres. At the mouth of the Conneaut Creek he established Port Independence which grew into Cleveland, Ohio (pp.57-58).

COLES, THOMAS (C.1806)

A Chart of the Coast of North Carolina between Cape Hatteras & Cape Fear was prepared in 1806 by Jonathan Price and Thomas Coles (Schwartz & Ehrenberg 1980). The survey was requested by President Thomas Jefferson and the U. S. Congress because of the hazards that the Outer Banks presented to U. S. shipping interests (p.221). Guthron (1984) reports that Price and Coles were authorized to utilize a revenue cutter, the Diligence, to carry out the surveys (p. 5).

(1806). A chart of the coast of North Carolina between Cape Hatteras & Cape Fear from a survey taken in the year 1806 by Thomas Coles and Jonathan Price pursuant to an act of Congress. (np). Scale not given. 62x88cm.

(1806). As above - North Carolina: (np).

COLLES, CHRISTOPHER (1737-1821)

Colles surveyed from Lake Ontario to the Hudson River via the Mohawk River for the New York legislature. Brown (1897) reports that Coles was born in Ireland. He was educated as an engineer and came to America, lecturing from 1772 through 1774, in Philadelphia and New York city about pneumatics, water supply for cities, and inland lock navigation. In 1777 he was employed by the Continental Army to provide instruction in gunnery. A telegraph was operated by him at Fort Clinton in 1812 and he claimed to have built the first steam engine built in America (V.2: p.122).

Colles is identified as the publisher of the first of a systematic series of American road maps (Schwartz and Ehrenberg 1980, p. 209). He is also given credit for the proposal of a system of canals connecting the Hudson River and Lake Erie. Colles is also the inventor of a perambulator, "a device that measures mileage by adding up the revolutions of a wheel trailed behind a carriage" (p. 210).

According to his biographer, "Colles is best known today for his contributions to cartography, an interest that held his attention for a relatively brief period. To historians, bibliophiles, and cartographers, in particular, Colles is remembered primarily for his Survey of the Roads of the United States of America, 1789" (Ristow 1961, p. 43). He was born of English ancestry in Dublin, Ireland on May 9, 1737. Near Kilkenny he attended school under Dr. Richard Pococke.

Pococke was a recognized cleric and oriental scholar. Under him, Colles studied mathematics, geography, engineering, and natural philosophy. After graduation he worked with an uncle, William, on a canal project linking Kilkenny and Inistioge. He was appointed, in 1761, clerk of the River Nore Navigation Board.

In 1764 he married Anne Keough, with whom Colles had several children. Between 1764 and 1770 he completed one cartographic activity, but decided in 1770 to move his family to America. He arrived in Philadelphia in 1771 (Ristow 1961, p. 6).

Immediately upon arriving in America Colles advertised his skills in the Pennsylvania Chronicle. Among those skills appears to have been that of lecturer in geography. He performed a series of those, on January 13, 1772. He, on February 17 of the same year, presented the "Geographic Lecture" in the rooms of the American Philosophical Society (White 1892, V.IX, p.271).

Near the end of 1773 or early in 1774 Colles and his family moved to New York City. There he continued his lectures on technical and scientific subjects: "His contributions in this field did, however, help promote and advance the cause of public education" (Ristow 1961, p.13). While in New York Colles noted the city's lack of water supply. He prepared a broadside that proposed a water system addressing it, "To the Worshipful The Mayor, Aldermen, and Commonality, Of the City of New-York, in Common Council convened." The Corporation of the City met and approved his plan.

The city proposed the advancement money for the water system to Colles on November 4, 1776, but it was defeated. When he moved to New York, Colles embraced the patriots cause and made friends with John Lamb. None the less, he and other American patriots left New York as the British Army approached.

Because of his sympathies, Colles was forced to move many times during the Revolution. He was not an active participant in the Revolution because of his age (38 years) and because of his physical frailness. He did however contribute as a lecturer on gunnery to artillery troops of the Continental Army. Additionally he worked on several projects fortifying the Hudson Highlands and constructing barriers to British ships.

Following the Revolution, Colles was again preoccupied with inland waterways and cartographic projects. The primary project appears to have been his road maps some of which he developed during travels in Pennsylvania and upper New York during the Revolution.

His activities continued to address matters relating to geography: Proposal relative to navigation on the Ohio; Proposal for removing obstructions on the Mohawk river; Improvement of the Inland Navigation between Albany and Oswego, and others. Though some of those projects were funded, Colles found the financial conditions of the new United States unable to support many of his ideas.

On October 6, 1821 the remains of Christopher Colles were laid to rest in the yard of St. Paul's Episcopal Church in New York. During his latter years he had been engaged with inventing and proposing new engineering projects, but none of those brought him fame or fortune. "Historians, however, credit the resourceful engineer with having the first publicly to introduce to the country a practical system of telecommunication," telegraph between New York City and Sandy Hook (Ristow 1961, p .89).

(1813). <u>An account of the astonishing beauties and operations of nature, in the minute of creation, displayed by the solar microscope.</u> New York: S. Wood.

(1815). As above- 44p. 14cm.

(1816). As above-

(1813). <u>Description of the numerical telegraph. For communicating unexpected intelligence by figures, letters, words, and sentences, with directions for writing the correspondence either public or private, and shewing [sic] the</u>

manner of working the machine with perfect accuracy and despatch. Brooklyn: Printed by Alden Spooner. 19p. 16.7cm.

(1794). The geographical ledger and systemized atlas; Being an united collection of topographical maps, projected by one universal principle, and laid down by one scale, proposed to be extended to different countries as materials can be procured. Containing I. Alphabetical references for pointing out the situation of lakes, islands... II. An alphabetical index, referring to the different parts of the map... III. An actual survey of a number of roads... exactly engraved upon copper. New York: Printed by John Buel. viii,[6]p. maps. 20.5cm.

(1808). Proposal of a design for the promotion of the interests of the United States of America... by means of inland navigable communications. New York: Printed for the author by L. Wood. 22p. map. 20cm.

(1774). Proposal for furnishing the city of New York with a constant supply of water. New York: H. Gaine. fol.

(1789). Proposals for publishing a survey of the roads of the United States of America. New York: (np).

1785). Proposals for the speedy settlement of the waste and unappropriated lands of the western frontiers of the state of New York, and for the improvement of the inland navigation between Albany and Oswego. New York: Printed by Samuel Loudon. 14p. 19.5cm.

(1789). A survey of the roads of the United States of America. References C. Tiebout, sculpt. New York: (np). 86 maps.

Other

Brown, Ralph H. (September 1938). Materials bearing upon the geography of the Atlantic Seaboard, 1790 to 1810. Association of American Geographers: Annals, 28, pp.201-231.

CONDY, JONATHAN WILLIAM (1770-1828)

Tucker (1985) names Condy as the author of a single geographic publication. No other information or publications have been located that might provide additional information about the man.

(1796). A description of the river Susquehanna, with observations on the present state of its trade and navigation, and their practicable and probable improvement. Illustrated with a map and an appendix. Philadelphia: Printed by Zachariah Poulson, Jr. 60p. map.

COOK, JAMES. (CARTOGRAPHER) (C.1778)

Cook is identified as having worked with Holland to survey British colonies. There is no evidence that he either became a United States citizen or returned to Britain. No other information is to be found about the man.

(1775). Map of North and South Carolina accurately copied from the old maps of James Cook, Pub in 1771, and of Henry Mouzon, in 1775.

(1773). A map of the province of South Carolina with all rivers, creeks, bays... and provincial lines...Jams Cook...July 7, 1773. Thos. Bowen, Sculpt, 1773...Plan of Buefort on Port Royal; Plan of George town; Plan of Camden: Draught of Port Royal Harbour: Plan of Charlestown and Chart of the bar and Harbour of Charlestown. (np). 81.5x79.5 cm

Other

Jefferys, Thomas. (1778). The American atlas; or, a geographical description of the whole continent of America; wherein are delineated at large, its several regions, countries, states, and islands; and chiefly the British colonies, composed from numerous surveys, several of which were made by order of government. By Major Holland, Lewis Evans, William Scull, Henry Mouzon, Lieut. Ross, J. Cook, Michael Lane, Joseph Gilbert, Gardner, Hillock, &c. &c. Engraved on forth-nine copper-plates, by the late Mr. Thomas Jefferys... London: Printed by R. Sayer and Bennett.

COTTINEAU DE KLOGVEN, DENNIS LOUIS (1745-1808?)

Cottineau de Klogven is one of those persons who published geography textbooks in the United States and geography related books elsewhere. The examination of numerous biographies did not reveal anyone by that name, however, a grave with the name Captain Denis N. Cottineau was found in Savannah, Georgia.

That person engaged in the 1779 fight between the Bon Homme Richard and the Serapis which ultimately brought John Paul Jones fame. This particular Cottineau was commissioned in the Continental Navy. He commanded the Pallas during the historic engagement succeeding in capturing the H.M.S. Countess of Scarborough to surrender. Following the engagement, he settled in the French West Indies. During the slave insurrection he fled to Pennsylvania and joined a colony there. Cottineau later came to Savannah where he stayed with Abbie Carles. He died in Savannah.

We suspect the two Cottineau were related, but that is not certain. It may also be possible that the two men are the same person.

(1806). Geographical compilation for the use of schools; being an accurate description of all the empires, kingdoms, republics and states, in the known world: with an account of their population, government, religion, manners, literature, universities, history, civil divisions, ecclesiastical hierarchy, principal cities (with an account of their importance, remarkable monuments, illustrious citizens, commerce and population) &c. &c. &c. the whole arranged in catechetical form. Compiled from the best American, English, and French authors, by D.L.C. teacher of geography. In two volumes.. Baltimore: Printed for the compiler by John West Butler. 2v. 16.5cm.

[Examined by the author in the University of South Carolina's South Caroliniana Library]

(1831). An historical sketch of Goa, the metropolis of the Portuguese settlements in India: with an account of the present state of that celebrated city, and of the surrounding territories under its immediate jurisdiction, collected from the most authentic sources. Madris: W. Twigg. 177p. folded map. 20cm.

(1922). History of Goa; a rare and important work, republished for a souvenir of the exposition at Goa, December 1922. Bombay: B. X. Furtado. 130p. folded maps (1 colored). 18cm.

(1810). The new geographical catechism, for the use of children... Charleston, SC: D. Longworth and W. P. Young. 79p. 16cm.

CRAMER, ZADOK (1773-1813)

Schwartz and Ehrenberg (1980) report that Cramer published navigational aids for the inland waterways of the United States. He published in 1801, the Ohio and Mississippi Navigator with directions and notes, but no charts. An 1806 edition added thirteen page-size charts (p. 221). Tucher (1895) gives Cramer's dates as 1773-1814.

Ristow (1985) reports that Cramer was born in New Jersey. His parents were Quakers. His early years were spent in Washington County, Pennsylvania. There he learned to be a bookbinder. That led to his publication of almanacs and the Ohio Navigator which helped establish his reputation. The Navigator was called, "one of the most useful guide-books ever published and the detailed descriptions, revised from time to time, make it especially useful to the historical student today" (pp.236-237).

(1801). The navigator; or, The traders' useful guide in navigating the Monongahela, Allegheny, Ohio and Mississippi rivers... Pittsburgh: author.

(1806). As above- ...containing an ample account of these much admired waters... a concise description of their towns, villages, harbours, settlements, &c. with particular directions how to navigate them, in all stages of the water... Illustrated with thirteen accurate maps of the Mississippi, and one of Pittsburgh. The 5th ed., much improved and enl. To which is added, an account of Louisiana... Pittsburgh: From the press of Zadok Cramer. 2[3]-94p. illus. maps. 17cm.

(1808). As above- ...and of the Missouri and Columbia rivers, as discovered by the voyage under Captains Lewis and Clark. 6th ed. improved and enl. Pittsburgh: Pub by Zadok Cramer and sold at his book store; from the press of Cramer & Spear. 3[5]-156p.

(1810 & 1811). As above- 7th ed....: Cramer, Spear & Eichbaum. 295p.

(1814). As above- 8th ed. 360p.

(1817). As above- 9th ed.

(1818). As above- 10th ed.

(1821). As above- 11th ed. iv,[5]-283p. 18.5cm.

(1824). As above- 12th ed.

(1802). ...The Ohio and Mississippi navigator, comprising an ample account of those beautiful rivers, from the head of the former, to the mouth of the latter. A particular description of the several towns, posts, caves, ports, harbours, &c. on their banks, and accurate directions how to navigate them, as well in times of high freshes [sic], as when the water is low. A description of its rocks, riffles, shoals, channels, and the distances from place to place. Together with a description of the Monongahela and Allegheny rivers. First taken from the journals of gentlemen of observation, and now minutely corrected by several persons who have navigated those rivers from fifteen and twenty years. 3d cor. ed. Pittsburgh: Printed by John Scull for Zadok Cramer. 40p. 17cm.

Other

Leahy, Ethel Carter. (1931). Who's who on the Ohio River and its tributaries, by Ethel C. Leahy... the Ohio River from the ice age to the future, history, biography, statistics. Cincinnati, OH: The E. C. Leahy Publishing Co.

CUMING, FORTESCUE (1762-1828)

Other than the single publication noted by Tucher, there appears to be no information about this person (1985). No entries were located in The National Union Catalog Pre-1956 Imprints (The American Library Association 1976).

(1810). Sketches of a tour to the Western country, through the states of Ohio and Kentucky; a voyage down the Ohio and Mississippi rivers, and a trip through the Mississippi Territory, and part of West Florida. Commenced at Philadelphia in the winter of 1807, and concluded in 1809. With notes and an appendix, containing some interesting facts, together with a notice of an expedition through Louisiana. Pittsburgh: Printed & published by Cramer, Spear & Eichbaum. 504p.

CUSTIS, PETER (C.1806)

Custis contributed to a learned society journal and co-authored a book. He appears to have been an explorer engaged by the United States government. No other publications nor any biographical information has been found about the person.

(1806). Observations relative to the geography, natural history, etc. of the country along the Red River in Louisiana... Philadelphia Medical and Physical Journal, Supplement no. 1, art. 7, p. 43-50.

Freeman, Thomas, & Custis, Peter. (1806). An account of the Red River, in Louisiana, drawn up from the returns of Messrs. Freeman & Custis, to the War Office of the United States, who explored the same in the year 1806. Washington: (np).

CUTLER, MANASSEH (1742-1823)

The first map to show the Seven Ranges of Townships comprising those proposed by the Federal government in the area west of Pennsylvania was produced by Cutler in 1787 (Schwartz and Ehrenberg 1980, p. 209). Encyclopedia Britannica (2002 Cutler, Manasseh) indicates that he conducted investigations in astronomy, meteorology, and botany. He was the leading author of the Northwest Ordinance of 1787. He was elected to Congress in 1801.

Cutler was born in Killingly, Connecticut, the son of Hezekiah and Savanna (Clark) Cutler, and died in Hamilton, Massachusetts. He graduated from Yale in 1765 and taught school for a year at Dedham, Massachusetts going on to become a minister, lawyer, politician, and entrepreneur. Cutler opened a neighborhood reading school and provided seamen with instruction in navigation and lunar observations. He was intimately engaged in the settling of Ohio and opened a school at Marietta, Ohio that became Marietta College. He was a member of the American Philosophical Society (Britannica.com; Brown 1897, V.2, p.303; White 1891, V.III, pp.70-71).

He was among the group that ascended Mt. Washington in 1784 and he estimated its height at 10,000 feet: 3707 feet in error. Pattison (1957) notes that Cutler was "the first man to incorporate stream lines from the public land surveys into a map" (p. 164). He issued the advertisement and drew up the verbal descriptions of public land surveyors on behalf of the Ohio Company.

In 1787 he issued a map to go with the advertisement. His map of the first four ranges of federal townships was quite accurate. Pattison asserts that the Ohio River and its right bank tributaries were in proper rectified form, although much of the rest of his map was inaccurate (Pattison 1957, p. 165).

(1784). An account of some of the vegetable productions naturally growing in this part of America, botanically arranged. Memoirs of The American Academy of Arts and Sciences, V.1, pt. 2, art.24, pp.396-493.

(1903). As above- Cincinnati, OH: J. U. & C. G. Lloyd. [3]-8,396-493p. 26cm.

(1888). Contemporary description of Ohio in 1788. In Ohio Archaeological and Historical Society. Publications. 2d ed. V.1, Columbus. pp.82-108.

Note. Translated from a French edition which appeared in Paris in 1789.

(1789). Description du sol, des productions, &c. &c. de cetter portion des Etats-Unis, situee entre la Pensylvanie, les rivieres de l'Ohio & du Scioto, & le Lac Erie. Traduite d'une brochure imprimee a Salem, en Amerique, en 1787. Paris: (np). 30p. 22cm. [Accents omitted]

(1787, 1893 & 1896). ... Description of Ohio, by Manasseh Cutler. An explanation of the map which delineates that part of the federal lands comprehended between Pennsylvania west line, the rivers Ohio and Scioto, and Lake Erie; confirmed to the United States by sundry tribes of Indians, in the treaties of 1784 and 1786, and now ready for settlement. Salem: Printed by Dabney & Cushing; Boston: Directors of the Old South Work. 12p. 20cm.

(1918). The first map and description of Ohio, 1787. By Manasseh Cutler. A bibliographical account, with reprint of the "Explanation," by P. Lee Phillips... Washington: W. H. Lowdermilk & Co.

(1787). A map of the Federal Territory from the western boundary of Pennsylvania to the Scioto River laid down from the latest informations... Boston: Printed by Adams and Nourse. 48x63cm.

(1872). New Jersey, Pennsylvania and Ohio, in 1787-8 passages from the journals of Menasseh Cutler communicated by Joseph F. Tuttle. (np). 24p. 23cm.

(1874). As above - Read before the society, May 16, 1872. New Jersey Historical Society - Procedures, Ser. 2, 3, 75-96.

(1888). Ohio in 1788. A description of the soil, productions, etc., of the portion of the United States situated between Pennsylvania, the rivers Ohio and Scioto and Lake Erie. Tr. from the French with notes and introduction by John Henry James. Columbus: A.H. Smythe. 104p. 17.5cm.

(nd). On the sea anemones of Swallow House or Cave in the Rocks at Nahant, near Lynn, MA. (np).

(1786). Selections from a note-book entitled: A description of the animals of North America taken from actual observations. Boston: (np). 8p.

(1785). Some of the vegetable productions naturally growing in this part of America. Boston: Adams and Nourse.

Other

Belknap, J., Peck, W. D., & Cutler, M. (1784). The history of New Hampshire. Philadelphia and Boston: (np). V.3; Description of White Mts., pp.39-54; forest trees and plants, pp.96-127; native animals, pp.146-187; stones, fossils, minerals, pp.188-196.

Cutler, William P., & Cutler, Julia P. (1888). Life, journals and correspondence of Rev. Manasseh Cutler, LL. D. 2 vols. Cincinnati: Robert Clarke & Co.

Humphrey, J. E. (Feb. 1898). Menasseh Cutler. Amer. Nat., 32, pp.75-80.

Lloyd Library. (1903). Menasseh Cutler. Bulletin no. 7. Pref. 8 p. port. Cincinnati, OH: author.

Peck, W. D., & Cutler, M. (1795). Methods of preserving animals and their skins collected from various authors. Massachusetts Historical Society Coll. 4, pp.8-14.

Phillips, Philip Lee. (1918). The first map and description of Ohio, 1787, by Manasseh Cutler: A bibliographical account. Washington: W. H. Lowdermilk & Co.

DARBY, WILLIAM (1775-1854)

By 1811 Darby undertook the making of a map and statistical account of Louisiana and its adjacent parts. Since he found existing maps unsatisfactory, he undertook a tour of the northern part of what became the state of Louisiana. His work on the Louisiana map was interrupted by the War of 1812. During the War of 1812 Darby served as topographer on the staff of General Andrew Jackson.

According to Ristow (1985) Darby was born in Lancaster County, Pennsylvania. His parents were Irish who settled in the region just before the American Revolution. They migrated to Ohio in 1786. Darby is reported as having a minimal formal education, but he was a reader who took a teaching position. When his father died in 1799, he moved to Natchez, Mississippi. There he became a planter until he was wiped out by a fire. Thereafter Darby worked as a Deputy surveyor for the U.S. General Land Office (pp.143-144). Although his work was not published until after 1812, Darby was an active contributor to American geography prior to 1812.

(1818). Emigrant's guide to the western and southwestern states and territories. New York & New Orleans: James Olmstead and Benjamin Long.

(1821). Memoir of Florida. Philadelphia: T. H. Palmer.

(1816). Map of the state of Louisiana with part of the Mississippi Territory. Philadelphia: John Melish.

DAVIDSON, ROBERT (1750-1812)

Brown (1900) notes that Davidson was an educator born in Elkton, Maryland and died in Carlisle, Pennsylvania. He graduated from the University of Pennsylvania in 1771 where he was appointed chair of history and belles-lettres and licensed to preach in 1774. While at the university he was professor of history, geography, chronology and rhetoric. He was elected to the American Philosophical Society in 1783 (V.2, p.355). His early education was at Newark Academy, Delaware.

During 1774 he was ordained by the Presbyterian church. In 1784 he became pastor of the Presbyterian church at Carlisle, Pennsylvania. In 1796 Davidson was chosen Moderator of the General Assembly of the Presbyterian Church.

In July, after the Battle of Bunker Hill, Davidson delivered a sermon before several companies of the Continental Army, "And many fell down, for the war was of God." When Philadelphia was occupied by the British in 1777, he retired to Delaware to avoid persecution for his patriotic views (Stephen & Lee 1917). Johnson and Malone (1928) note that his tenure as Professor at Dickinson College was as Professor of History, Geography, Chronology, Rhetoric and belles-lettres.

(1787). The elements of geography, short and plain. Designed as an easy introduction to the system of geography. In verse, by Robert Davidson, esq. Designed for the use of schools. With or without seven copper plates. London: T. Wilkins. 60p. 16.5cm.

(1786). As above- Added t. - p.: Geography epitomized; or, a tour round the world: being a short but comprehensive description of the terraqueous [sic] globe: attempted in verse, (for the sake of the memory;) and principally designed for the use of schools. By an American. Philadelphia: J. Crookshank and London: T. Wilkins.

(nd). An epitome of geography in verse. (np).

(1784). Geography epitomized; or, a tour round the world: being a short but comprehensive description of the terraqueous [sic] globe: attempted in verse, (for the sake of the memory;) and principally designed for the use of schools. By an American. Philadelphia: Printed and sold by Joseph Curkshank. 60 p. 16.5cm.

(1786). As above- London: (np).

(1791). As above- Burlington, VT: Neale & Lawrence. 64 p. 15.5cm.

(1803). As above- Morris-Town: Printed by H. P. Russell. 17cm.

(1805). As above- Stanford: D. Lawrence, H. Hull, and J. F. Hull. 72 p.

(c.1790). Geography epitomized. A short but comprehensive description of the terraqueous [sic] globe, in verse, to assist the memory. Leominster, Mass: Printed for C. Whitcomb. 60p. 20.5cm.

(c.1800). As above- [Leominster, Mass.] Print for Chapman Whitcomb...[by Adams & Wilder]. 3-60 w = 11 cm H = 18.7 cm

DAVIES, BENJAMIN (1774-1806)

Other than the following publications, nothing is known about Benjamin Davies:

(1805). A new system of modern geography, or a general description of the most remarkable countries throughout the known world; their respective situations, extents, divisions, cities, rivers, mountains, soils, and productions; their commerce, manners, customs, laws, and religion; together with their principal historical events, and political importance in the great commonwealth of nations: Compiled from the most modern systems of geography, and the latest voyages and travels, and containing many important additions to the geography of the Untied States that have never appeared in any other work of the kind. Illustrated with eight maps. 2d ed., Philadelphia: Jacob Johnson & Co. 628 p. 8 folded maps. 17.5cm.

(1813). A new system of modern geography: or, a general description of the most remarkable countries throughout the known world. Compiled from the most modern systems of geography, and the latest voyages and travels, and containing many important additions to the geography of the United States. 3d ed., carefully corr. and rev. Philadelphia: Johnson and Warner. 447p. 8 folded maps. 19cm.

(1815). As above- 4th ed.

(1810). A short and easy introduction to universal geography; by way of question and answer: containing a general description of the earth, with a brief account of the situation, natural and political state of all the principal empires, kingdoms and republics throughout the known world. To which is prefixed a definition of all the common geographical terms, and a number of useful problems performed on the terrestrial globe: Intended for the use of young pupils in the science of geography. Philadelphia: Bennett & Walton.

Note. The following reveals some information from a copy examined by the author: 216p. No illustrations. One table regarding U.S. states and types of legislatures. "Author of Modern Geography, etc."

2nd Edition - Corrected and improved.

W = 9cm. H = 14.5cm.

[The statement regarding copyright is interesting:]

Be it remembered, that on the twelfth day of July, in the thirty-fourth year of the Independence of the United States of America, A.D. 1809, Benjamin Johnson of the said district, hath deposited in this office the title of a book, the right whereof he claims as proprietor, in the words following, to wit. (p.2)

(1794). Some account of the city of Philadelphia, the capital of Pennsylvania, and seat of the Federal Congress; of its civil and religious institutions, population, trade, and government; interspersed with occasional observations. Philadelphia: Printed by Richard Folwell, for the author. 93p. 18cm.

Other

Babovich, W. M. (1992). The blind men and the elephant go south: a historiographical essay. New Orleans, LA: Wayne M. Babovich & Associates.

DAVIS, WILLIAM (1771-1807)

Virginia's first state map was prepared in 1807 by William Prentis and William Davis, in collaboration with Bishop Madison, from information provided by local surveyors (Schwartz & Ehrenberg 1980, p.224). Madison collaborated with William Prentis of Petersburh and with Davis. Between them they produced and presented their map to the Virginia General Assembly, 1805 (Ristow 1985, p.121).

(1802). A complete treatise on land surveying, by the chain, cross and offset staffs only. In three parts... The second edition, with very considerable additions and improvements. By William Davis... London: Printed by Knight and Compton for the author. xii,351,[1]p. illus. 7 folded plates. diagrs. tables. 22cm.

(1805). As above- xii,359p. 8 folded plates. 21.5cm.

Other

(Ed.). (1819). The mathematical principles of natural philosophy, by Sir Isaac Newton; translated into English by Andrew Motte: to which are added, Newton's system of the world; A short comment on, and defence of, the principia, by William Emerson; with The laws of the moon's motion according to gravity, by John Machin... A new ed. (with the life of the author; and a portrait, taken from the bust in the Royal Observatory at

Greenwich) carefully rev. and cor. by William Davis... London: Printed for Sherwood, Neely, and Jones.

DEAN, JAMES (1777-1849)

Searches have uncovered no information, other than the following publication.

(1808 & 1809). An alphabetical atlas, or gazetteer of Vermont; affording a summary description of the state, its several counties, towns, and rivers. Calculated to supply, in some measure, the place of a map; and designed for the use of offices, travellers [sic], men of business, &c. Montpelier: Printed by Samuel Goss - For the author. 43[1]p. (Tucher 1985)

DEWITT, MOSES (C.1793)

Ristow (1985) identified Moses as a cousin of Simeon DeWitt. Moses and John L. Hardenburgh prepared surveys of the townships in the Military Tract.

(1793). The Military Tract in north central New York State. (np).

DEWITT, SIMEON (1756-1834)

DeWitt was surveyor-general of New York and Geographer to the Army of the United States. He was born in Wawarsing, Ulster County, New York on December 25, 1756 to Dr. Andries and Jannetje (Vernooy) DeWitt. His schooling consisted of what a British colony could offer, thereafter he was placed under the guidance of Rev. Dr. Romeyn of Schenectady who prepared him for college, Queens (now Rutgers) College in New Jersey. DeWitt received his bachelors degree in 1776 and his masters in 1788. During the revolution he joined the Ulster County Battalion, fought Burgoyne's troops, and witnessed the surrender of Cornwallis. DeWitt was recommended to Gen. Washington, by his uncle - Gen. James Clinton, as Geographer to the Army. In 1778 he took the position and was appointed assistant to Col. Robert Erskine, then Geographer-in-Chief. When Erskine died in 1780, he became head of the department.

DeWitt was appointed to the position of surveyor-general for New York State on May 13, 1784. In that position he was engaged in delineating the boundary between New York and Pennsylvania. The map of New York produced from that effort published in 1802 the best example of what the state was at that time. He also completed one of the earliest state-subsidized maps (Schwartz & Ehrenberg 1980, p.219). He was also appointed surveyor general for the United States, but declined to continue in that position for the state of New York (Pattison 1957, p. 200). The University of New York was served by DeWitt from 1798 until his death. He was a member of the American Philosophical Society who contributed to its Transactions (1809) and he served as vice-president for the Lyceum of Natural History (Johnson & Malone 1928, V.5, pp.274-275).

(1803 &1875). An atlas of the state of New York. [see David H. Burr]

(nd). Establishment of a meridian line in the city of Albany. Luceum of Natural History, Transactions.

(1806). A map of Ithaca and the adjoining land belonging to Simeon De Witt. (np). map. 56x85cm. Scale: 5 chains to an inch.

(1792). A map of the state of New York. By Simeon De Witt, Surveyor-General. Albany: 12th New York District Copyright.

(1802). As above- Engraved by G. Fairman. Colored map on 6 sheets. 67cm.

(1804). As above- Contracted from his large map of the state. Colored map 57x69cm.

(1809). Observations on the eclipse of the Sun, June 16, 1806, at Albany. American Philosophical Society, Transactions.

(nd). On a plan of a meteorological chart. Lyceum of Natural History, Transactions.

(1801). Respecting a plan of a meteorological chart, for exhibiting a comparative view of the climates of North America, and the progress of vegetation. Transactions of the Society instituted in the state of New York for the promotion of agriculture, arts and manufactures, 1, 88-92.

(nd). A table of variations of the magnetic needle. Transactions of the Albany Institute.

(nd). Observations on the functions of the Moon, Deduced from the eclipse of 1806. Transactions of the Albany Institute.

(nd). A description of a new form of rain gauge. Transactions of the Albany Institute.

Other

Beck, T. Romeyn. (1852). Eulogium on life and services of Simeon DeWitt. Transactions of the Albany Institute. II, 313-315.

Evans, Thomas G. (1886). The DeWitt family of Ulster County, New York. (np).

DOOLITTLE, AMOS (1754-1832)

Ristow (1985) notes that Doolittle was "of New Haven" (p.151). The map of Virginia, North Carolina, South Carolina and Georgia, dated 1788 by Joseph Purcell, was engraved by Doolittle (Schwartz & Ehrenberg 1980). It was included in Jedidiah Morse's American Geography.

Doolittle was born in Cheshire, Connecticut and died in New Haven, Connecticut. He was an engraver who produced prints of battles at Lexington and Concord along with a variety of maps, book illustrations, and portraits (Hopkins, 1964, p.241).

(1796). A Chart of George's Bank including Cape Cod, Nantucket and the shoals lying on their coast, with directions for sailing over the same & c. Surveyed by Capt. Paul Pinkham, engraved by Amos Doolittle of New Haven. New Haven: Edmund M. Blunt.

(1816). The city of Sandusky. New Haven: (np). 32x43cm. Scale: 20 poles to an inch.
(1818). A correct map of Connecticut from actual survey. New Haven: (np). $12^{7/8}$x$15^{7/8}$ inches. Scale: 6.8 miles to 1 inch.
Note. Engraved for Trumbulls History of Connecticut.
(1788). A display of the United States of America, to the patrons of arts and sciences, in all parts of the world, this plate is most respectfully dedicated... New Haven: (np). [Broadside]
(1883). Four drawings of the engagement at Lexingon and Concord, April 19, 1775. Reproduced with an explanatory text by E. G. Porter. Boston: (np). 9p. 4 pl.
(1903). As above- Boston: Goodspeed. 4 plates (each $11^{5/8}$x$17^{5/8}$inches).
(1838). Map of Ohio. Cincinnati: (np). $51^{1/2}$x48in.
(nd). Map of northern Ohio, showing townships. (np). 16x13inches.
(1814). A new and correct chart of the east end of Long Island Sound; from Oyster-Pond Point to Montagu, [sic] and from Watchhill to Black Point; including also the rivers Thames and harbour of New London, with the islands, bays, reefs, soundings & anchorages, taken by actual surveys of the American Squadron in 1811 and the British Squadron in 1813 & 1814 and compiled from the original drafts by Ephraim Chesebrough. Engraved by A. Doolittle. New Haven: (np).
(1799). A new display of the United States... New Haven: Printed and sold by Amos Doolittle.
(c.1790). View of Nassau Hall, Princeton. New Haven: (np). 9.2x13.5cm.

DOWSE, EDWARD (C.1785)

Dowse was one of eight surveyors who reported for duty in 1785 to the newly appointed Geographer of the United States, Thomas Hutchins who was beginning work on the surveying of America. He was there to represent New Hampshire, but was unknown to anyone in that state. His appointment was due to the fact that he was available and willing to do survey work under the direction of Thomas Hutchins. His only claim was that he had recently traveled in the western country. That and the fact that no one could be found in New Hampshire to take the post, left it open for a person such as Dowse (Pattison, 1957).

No entries were located in The National Union Catalog Pre-1956 Imprints (The American Library Association 1976).

DOYLE, DAVID (C.1805)

The only information we have thus far found about Doyle is the following publication:
(1805) Pinkerton's geography, epitomized for the use of schools. By David Doyle . Wherein the arrangement of the original is faithfully preserved. In the description of the United States, each state is more systematically

described, than in any work of the kind heretofore published. Philadelphia: Printed for Samuel F. Bradford by T. S. Manning.

Other

Weaver, William. (1824-1878). Papers of Weaver-Brady families. 1,500 folio items and 79v. OCLC: 32671863.

DRAYTON, JOHN (1766-1822)

Drayton was governor of South Carolina, jurist, and author. He was born near Charleston, South Carolina on June 22, 1766 and died there on November 27, 1822. Drayton was the son of William Henry and Dorothy (Golightly) Drayton and the grandson of Thomas Drayton who came from Barbados in 1671. Thomas settled near the Cape Fear River, North Carolina moving to South Carolina in 1680.

John Drayton's education began at the Nassau Grammar School in Princeton, New Jersey and after some work at the College of New Jersey he was sent to England where he completed his education, probably at the Inner Temple. After his stay in England he returned to Charleston where he began to practice law.

There in 1794 Drayton married Hester Rose and they had six children. He entered public life early, successfully achieving numerous posts culminating with his tenure as Governor of South Carolina. Drayton was in charge of the executive department when the governor under whom he was serving, died. Although not officially recognized as governor, he performed those duties until he was elected in 1800 for a term of two years.

Among Drayton's accomplishments was that he recommended to the General Assembly of South Carolina in 1801, the establishment of the South Carolina College (now University of South Carolina) at Columbia. He served as President of the Board at organizational meetings held in his home. His final post was that of judge of the United States court for the District of South Carolina which he held until his death (Brown 1897, V.2, p.517; Johnson & Malone 1928, V.3, pp. 444-445; White 1906, V.XII, pp.162-163).

(1803). Destruction of Carolina pine trees, by an insect... in a letter to Barton... Proceedings of the American Philosophical Society, 22, appx. p.342.

(1808). John Drayton's... Beschreibung von Sud-Carolina. Aus dem englishen und mit anmerkungen und zusatzen des uebersetzers begleitet. Mit einer charte. Weimer: Im verlag des F. S. Privil. Landes-industrie comptoirs. 406p. 21.5cm. (Tucher 1985) [Accents omitted]

(1807). Carolina florist... [in manuscript] Charleston, SC: (np), 307p., 8°. In the Library of the University of South Carolina, Columbia. Also in Wilson Gee, S.C. Botanists. Univ. S.C., Bulletin, no. 72, pp.26-27, Sept. 1918).

(1794). Letters written during a tour through the northern and eastern states of America. Charleston, SC: Printed by Harrison and Bowen. 141p. 3 plates. 22cm.

(1821). Memoirs of the American revolution, from its commencement to the year 1776, inclusive; as relating to the state of South-Carolina: and occasionally referring to the states of North Carolina and Georgia. Charleston, SC: Printed by A. E. Miller. 2v. 3 folded maps. 22cm.

(1802). A view of South-Carolina, as respects her natural and civil concerns. Charleston, SC: Printed by W. P. Young. 255p. 31 plates. 3 folded maps. 2 folded tables. 22cm. (Tucher 1985)

(1802). Drayton's view of South Carolina. University of Georgia - Microform MFC F273.D75

Other

Meriwether, Margaret Babcock. (Ed.). (1943). The Carolinian florist of Governor John Drayton of South Carolina... With water-color illustrations from the author's original manuscript and an autobiographical introduction. Columbia, SC: The South Caroliniana library of the University of South Carolina. 133p. 16x20cm.

DUMAIN, LOUIS (C.1806)

According to Guthron (1984), Dumain was charged by the Senate in 1806 to survey the coast of Louisiana. He completed the survey and provided "maps and a long detailed letter describing the channels and local conditions," dated July 20, 1807 (p.5).

No entries were located in The National Union Catalog Pre-1956 Imprints (The American Library Association 1976).

DUNBAR, WILLIAM (1749-1810)

Dunbar was a native of Natchez, Mississippi. He was directed by President Jefferson (1806), in his Message to Congress, to explore the Washita River. That was done to the latitude of 34° 31' 4" north, longitude 92° 50' 45" from Greenwich. He was accompanied on the trip by Dr. George Hunter. The journey lasted from October 16, 1804 to January 31, 1805 (Meisel, 1926).

He was born in Elgin, Scotland and died near Natchez, Mississippi. In 1771 he moved from Scotland to the British province of West Florida where he established a plantation. Following the destruction of that plantation he moved to Natchez where he established another. His use of scientific improvement to enhance farming caused his new plantation to prosper. Dunbar was the first surveyor general of his area and a corresponding friend of Jefferson. He also became a member of the American Philosophical Society (Hopkins, 1964, p.252)

Claim to land adjoining Natchez, by William Dunbar under grant from Spain, American State Papers: Public Land, V.XXVIII (Doc. 93 & 100).

Claim to land in and adjoining Natchez, American State Papers: Public Land, V.XXVIII (Doc. 109 & 123).

(1804). Description of the river Mississippi and its delta, with that of the adjacent parts of Louisiana... Transactions of the American Philosophical Society, 6, pt.1, (30), pp.165-187; and pt.2, (32), p.191-201.

(1806). Discoveries made in exploring the Missouri, Red River and Washita, by Captains Lewis and Clark, Doctor Sibley, and William Dunbar, Esq. With a statistical account of the countries adjacent. With an appendix by Mr. Dunbar. Natchez: Printed by Andres Marschalk. 177p. 12.5x20.5cm.

(1904). Documents relating to the purchase and exploration of Louisiana. I. The limits and bounds of Louisiana. By Thomas Jefferson. II. The exploration of the Red, the Black, and the Washita rivers. By William Dunbar. Printed from the original manuscripts in the library of the American Philosophical Society and direction of the society's committee on historical documents. Boston & New York: Houghton Mifflin & Co.

(1804 & 1806). Journal of a voyage commencing at St. Catherines Landing, on the east bank of the Mississippi, proceeding downwards to the mouth of the Red River, and from thence ascending that river, the Black River and the Washita River as high as the Hot Springs in the proximity of the last mentioned river. Boston: (np). 189,76p. portrait. map.

(1805). The leopard in New Mexico: 2) On a new species of cabbage... Philadelphia Medical and Physical Journal, 2, pt. 1, Sect. 3, p.160, 170.

Lewis and Clark expedition, pt. 5: Geographical description of Washita River in Louisiana and adjacent country, by William Dunbar and Dr. George Hunter, p.731, American State Papers: Public Land, V.VII (Doc. 113).

Lewis and Clark expedition, pt. 6: Meteorological observations of William Dunbar and Dr. George Hunter, p. 742, American State Papers: Public Land, V.VII (Doc. 113).

(1809). On the language of signs among certain North American Indians. By William Dunbar... of the Mississippi Territory, communicated by Thomas Jefferson... Philadelphia: (np). p.1-8.

(1900). Report to the Spanish government at the conclusion of his services in locating and surveying the thirty-first degree of latitude. In Mississippi Historical Society. Publications, 3, 185-205.

Other

Jefferson, T. (1806). Message from the President of the United States communicating discoveries made in exploring the Missouri, Red River and Washita, by Captains Lewis and Clark, Doctor Sibley, and Mr. Dunbar, with a statistical account of the countries adjacent... Feb. 19, 1806... printed by order of the Senate. Washington, DC: A. & G. Way, Printers. 178p.

Jefferson, T. (1807). Travels in the interior parts of America; communicating discoveries made in exploring the Missouri, Red river, and Washita, by Captains Lewis and Clark, Doctor Sibley, and Mr. Dunbar; with a statistical account of the countries adjacent. As laid before the Senate, by the President of the United States. In February, 1806, and never before published in Great Britain. London: R. Phillips.

King, Nicholas. (c.1805). Map of the Washita River in Louisiana from the Hot Springs, to the confluence of the Red River with the Mississippi... Based on the journal and survey of William Dunbar. Engraved by William Kneass. Philadelphia: (np).

Rowland, Mrs. Dunbar. (1930). Life, letters and papers of William Dunbar of Elgin, Morayshire, Scotland, and Natchez, Mississippi, pioneer scientist of the southern United States; compiled and prepared from the original documents for the National Society of Colonial Dames in America. By Mrs. Dunbar Rowland (Eron Rowland)... Jackson, MI: Press of the Mississippi Historical Society. 1,410p. 2 plates. 2 port. 24cm.

DUNBIBAN, DANIEL (C.1791)

Guthron (1984) indicates that Dunbiban contributed to Osgood Carleton's Pilot with his chart of Cape Hatteras (p.8). Guthron goes on to note that Dunbiban was a land owner in New Hanover and a merchant in Wilmington. The North Carolina Coast Chart was published in editions of The American Pilot of John Norman between 1791 and 1794 (p.95).

(c.1791). Chart of the coast of America from Cape Hateras to Cape Roman from the actual surveys of D. Dunbiban. Engraved by John Norman. (np). $21^{3/4}$x14inches. Scale 1:1,000,000.

DUPORTAIL, LOUIS LEBEGUE (1736-1802)

Duportail was Commandant of Engineers in the Continental Army from 1777 to 1783. He carried three surveying compasses and three alidades because he viewed them as "indispensable instruments of his profession" (Harley, et.al. 1978).

White (1899) describes Duportail as a soldier who was educated at a military school in France entering service as an engineer. Franklin and Deane were instrumental in engaging Duportail and three other French officers to serve the colonial service. He was with Washington at the Battle of Monmouth and at the siege of Yorktown. Duportail returned to France and was awarded a field-marshall's position. He was a friend of Lafayette and suffered a similar fate when Lafayette fell into popular displeasure (White 1892, V.IX, p.417).

(1789). Plan d'un projet de Port Royal a La Hougue. (np). In A.N. Isnard, Observations sur le principe qui a produit les revolutions... Bound in Melanges sur l'Amerique, 6, pam. 5. [Accent omitted]

Other

Louvet de Louvrai, Jean Baptiste. (nd). Love and patriotism; or, the extraordinary adventures of M. Duportail, late Major-General in the armies of the United States. (np).

DWIGHT, NATHANIEL (1770-1831)

Born in Northampton, Massachusetts on January 31, 1770, Dwight became a physician, educator, and geographer. He was the son of Maj. Timothy and Mary (Edwards) Dwight. He died in Oswego, New York. According to Brown (1900) Dwight was the first person to publish a school geography in the United States (V.2, pp.569-570). Brown was obviously unaware of the publications of Jedidiah Morse and some others.

Nathaniel Dwight was like his father and grandfather, a civil engineer. His name can be found on old plans as the surveyor - he was also employed by Massachusetts and Connecticut in making surveys and maps. He probably laid out the plans for Fort William Henry.

His early education was under the direction of his mother, "the gifted daughter of Jonathan Edwards," who was a talented teacher who conducted a regular school for her children. Dwight also attended the school at Greenfield Hill that was taught by his brother, Timothy, who later became president of Yale. Although Dwight did not attend college, he taught school and studied medicine. He practiced medicine in Hartford and he entered the ministry in 1812 settling in Westchester, Connecticut. Although he was not financially successful, Dwight was recognized as a good medical practitioner who was kind and generous to a fault; and a minister who was earnest and faithful.

He was tall and well built. In 1798 he married Rebecca Robbins of Wethersfield, Connecticut. They had eight children, four of whom died in infancy. His 1795 geography book was cited as "better calculated to impress the facts which it contains on the minds of children than any heretofore published" (V.3, pp.568-569). Dwight's chief accomplishments are those of writer, physician, and constructive suggestions for improving the conditions of the insane (Johnson & Malone 1928).

(1800). An account of the American cantharis or Meloe Americae. Connecticut Academy of Arts and Sciences, Papers on Natural History, 1, pt.1, pp.99-102.

(1795). The great question answered: A short but comprehensive system of the geography of the world: by way of question and answer. (np).

(1795). A short but comprehensive system of the geography of the world: by way of question and answer. Principally designed for children, and common schools..."3d Connecticut edition." Hartford, Conn.: Printed and sold by Hudson & Goodwin. 221p. 16.5cm.

(1796). As above- 225p.

(1795 & 1808). As above- Philadelphia: Printed for Mathew Carey. 214p. 17cm.

(1800 & 1808). As above- Hartford: Printed by Hudson and Goodwin. 214p. 18cm..

(1805 & 1807). As above- 216p. 10x17cm.

(1795). As above- Hartford, CT: Printed by Elisha Babcock.

(1796 & 1797). As above- Boston: Printed by Manning & Loring, for David West. 215p. 17.5cm.

(1801). As above- Boston: Printed by Manning & Loring, for West & Greenleaf. 17cm.

(1801 & 1803). As above- Elizabeth-town, NJ: Printed by J. Woods for Evert Duyckinck. 228p. 18cm.

(1805). As above- New York: E. Duyckinck.
(1813). As above- ...Sixth edition. 216p. 18.5cm.
(1808 & 1810). As above- 216p. 17cm.
(1796). As above- Albany: Printed by Charles R. & George Webster. 192p. 10.5x16.5cm.
(1809, 1813, & 1815). As above- Albany: Webster and Skinner. 208+p. 17cm.
(1806). As above- 215p.
(1804). As above- 216p. 19cm.
(1798 & 1799). As above- 187p. 12mo.
(c.1790). As above- 189p.
(1812). As above- Albany: Farrand & Green. 214p. 17cm.
(1805). As above- Northampton, MA: S. & E. Butler. 209p. 17cm.
(1806, 1807 & 1811). As above- 216p.
(1812 & 1814). As above- Northampton, MA: Simeon Butler.
(1815). A system of universal geography, for common schools: in which Europe is divided according to the late act of Congress of Vienna... New York: E. Duyckinck. 212p. 12mo.
(1816 & 1817). As above- Northampton, MA: Simeon Butler.
(1817). As above- Albany: Printed by Websters & Skinners. 210p. 17.5cm.
(1830). As above- New York: Printed by J. & J. Harper.

Other

Dwight, B. W. (1874). The history of the descendants of John Dwight of Dedham, Massachusetts. (np).

Sheldon, George. (1903). The journal of Capt. Nathaniel Dwight and its leadings. Boston: Alfred Mudge & Son Inc.

Note. The journal was written by the captain of Belchertown, Mass. during the Crown Point Expedition - 1755. This was the time of the building of Ft. Wm. Henry.

DWIGHT, TIMOTHY (1752-1817)

The inclusion of Dwight in the "First Generation" is done because of his publications providing additions to the body of knowledge of the New England area. White (1898) notes that Dwight was president of Yale College from 1795 through 1817. He was born at Northampton, Massachusetts. His father had the same name. Timothy Sr. was a graduate of Yale and a merchant. His mother was one of the daughters of Jonathan Edwards.

He was an educator, theologian, and poet. His poem, The Conquest of Canaan (1785) is regarded as the "first American epic poem" (Biographical Dictionary 2001. Timothy Dwight; Encarta Encyclopedia 2001. Dwight, Timothy; Encyclopedia Britannica 2002.)

Dwight taught two years at New Haven and for six years at Yale. He farmed, preached, and conducted school at North Hampton until 1783. Thereafter he became pastor at Greenfield, Connecticut. White provides considerably more detail for those interested (V.1, p.168-169 & 173).

(1903). ...Boston at the beginning of the 19th century. By Timothy Dwight. From President Dwight's "Travels in New England." Boston: Directors of the Old South Works. 24p. 20cm.
(1811). A statistical account of the city of New Haven. New Haven: Printed and sold by Walter and Steele. xi,83,[1]p. 22cm.
(1874). As above- "Reprint from the New Haven City Year Book, 1874."
(1811). A statistical account of the towns and parishes in the state of Connecticut. New Haven: The Connecticut Academy of Arts and Sciences, V.1(1). 83p. (Tucher 1985)
(1821). Travels: In New-England and New York. In four volumes. New Haven: Timothy Dwight; S. Converse Printer. V.1, 524p.; V.2, 527p.; V.3, 534p.; V.4, 527p. (Tucher 1985)
(1823). As above- Printed by Charles Wood.

Other

Sprague, William Buell. (1817). Life of Timothy Dwight. (np).

EDDY, JOHN H. (C.1810)

Eddy identified himself as "Geographer" on maps he published between 1811 and 1818. No biographic information about Eddy appears available. His maps were engraved by Tanner, Vallance, Kearny & Company of Philadelphia and printed by Samuel Maverick in New York. Ristow (1985) further notes that little information is available about the man (p.103).
(c.1800). A map of Friends settlement, in the township of Le Ray, Jefferson Co., New York. (np). $8^{1/4}$x$9^{1/16}$in. Scale: 100 chains to 1 inch.
(1812). Map of the country thirty miles round the city of New York. (np).
(1828). As above- Drawn by J. H. Eddy.
(1836). As above- Designed & drawn by J. H. Eddy of New York. New ed. revised and corrected by J. Disturnell. New York: J. Disturnell. circular. $20^{3/4}$inches in diameter.
(1816). Map of part of Africa drawn from the latest authorities. New York: (np).
(1811). Map of the western part of the state of New York showing the route of proposed canal from Lake Erie to Hudson's River. Newark, NJ: P. Maverick, engraver. colored map. 29x36cm. Scale: 15 miles to 1 inch.
(1818). The state of New York, with part of the adjacent states; engr. by Tanner, Vallance, Kearny & Co., Philadelphia. New York: Printed by S. Maverick. colored map. 110x96cm. Scale: 1 in. to 10 miles.

ELIOT, J. B. (1754-1813)

France declared its alliance with American patriots in 1778 and that was followed by the publication of maps by some American geographers (Schwartz & Ehrenberg 1980). Eliot's map of the American Revolution was one of those published in 1778, Carte du Theatre..., is the earliest known map to use the name United States. The title on the map indicates that at the time he was an American engineer (p.198).

Ristow (1985) notes that Eliot was "aide de camp du General Washington". Since his maps were published in France, it is likely that Eliot served as a liaison officer from the French Army in America, but there is no proof of that he became an American citizen (p.61-62).

(1778). Carte du theatre de la guerre actuel entre les Anglais et les Treize Colonies Unies de l"Amerique Septentrionale. Paris: (np). [Accent omitted]

(1781). Carte du theatre de la guerre actuel entre les Anglais et les Treize Colonies Unies de l"Amerique Septentrionale ...ou se trouvent les principaux camps et les e'poques des combats qui se sont donne's dans cette partie de l"Amerique. Paris: (np). 68.5x50cm. Scale: 7 leagues to an inch. [Accent omitted]

(1783). Carte generale des Etats Unis de l"Amerique Septentrionale, avec les limites de chacun des dits Etats. Paris: Mondhare. [Accent omitted]

ELLICOTT, ANDREW (1754-1820)

Ellicott was a civil engineer, cartographer and mathematician whose parents were Dutch and English Quakers. He was the eldest son of Joseph and Judity (Bleaker) born in Solebury, Bucks County, Pennsylvania on January 24, 1754. He died while he was professor of mathematics at West Point on August 28, 1820. Ellicott briefly attended school in Solebury and Philadelphia. When he was 21 years old, he married Sarah Brown of Newton and moved to his family home in Maryland. The town of Ellicott City was named for the family. During the Revolutionary War Ellicott joined the Maryland Militia and rose to the rank of major.

Beginning sometime around 1782 he published a series of almanacs. Ellicott was appointed a member of the team assigned to continue the Mason and Dixon Line from the point at which it was dropped by Englishmen in 1767. In 1785 he moved to Baltimore and taught mathematics at the Academy of Baltimore. In 1786 he served a term in the Maryland Legislature. During that time he was a member of the Pennsylvania Commissions for running the western and northern boundaries for the state. Ellicott's name was on one of the plans for Washington, which were redrawn by him. Ellicott did much of the surveying and marking out of the plan on the ground after L'Enfant's dismissal from the project. In 1790 he was given the rank of surveyor-general (Johnson & Malone, V.3, pp.89-90; White 1906, V.XIII, pp.470-471).

Brown (1897) reports that he guided the building of Fort Erie at Presque Isle in 1795 and laid out plans for Erie, Warren, and Franklin, Pennsylvania. In 1796 he was engaged with determining the boundary between the Untied States and Spanish possessions. He was elected a member of the American Philosophical Society (V.2, p.631).

Barton (1813) notes at several points that Ellicott was an associate of David Rittenhouse as well as Ewing, Hutchins, and others among the first generation of American geographers (pp.300-301, 308). He also engaged with surveying the boundary between the United States and Spanish-controlled Florida and the boundary between South Carolina and Georgia (The Biographical Dictionary 2001, Andrew Ellicott).

(1799). Account of the methods employed in surveying the city of Washington. Transactions of the American Philosophical Society, 6.

(1802). Astronomical and thermometrical observations made at the confluence of the Mississippi and Ohio rivers. Philadelphia: (np), 202p. 4°.

(1818). Astronomical observations, &c. communicated by Andrew Ellicott, esq. Read Nov. 16th, 1810. American Philosophical Society. Transactions. Philadelphia, 1818. 1, 93-101.

U.S. House of Representatives. 20th Congress, 2nd Session. Copies of letters, Andrew Ellicott, boundary Georgia and Florida. (H.doc. 96). Washington: Government Printing Office, 1828-1829. (Serial Set 186).

(1787). Ellicott's Virginia, Maryland, and Pennsylvania almanack [sic], for the year of our Lord, 1788; being leap-year, and the twelfth year of American independence... Winchester, VA: Printed and sold by Bartgis & Willcocks. Also to be had at Fredricks-Town, MD: The English and German printing office.

(1803). The journal of Andrew Ellicott, late Commissioner on behalf of the United States during part of the year 1796, the years 1797, 1798, 1799, and part of the year 1800: for determining the boundary between the United States and possessions of His Catholic Majesty in America, containing occasional remarks on the situation, soil, rivers, natural productions, and diseases of the different countries on the Ohio, Mississippi, and Gulf of Mexico, with six maps comprehending the Ohio, the Mississippi from the mouth of the Ohio to the Gulf of Mexico, the whole of West Florida, and part of East Florida. To which is added an appendix, containing all the astronomical observations made use of for determining the boundary... likewise a great number of thermometrical observations... Philadelphia: Printed by Budd & Bartram, for Thomas Dobson. 151p. 6 folded plates. 8 folded maps. 25.5x21cm. (Tucher 1985)

(1814). As above- Printed by William Fry. vii,[1],299,151p. 6 folded maps. 8 folded astronomical charts.

(nd). Letter to Benjamin Rush describing the falls, and another to Washington relating his encounter of the British Commandant regarding the southwestern boundary of New York. In Buffalo Historical Society Publications, XV, 834, XXVI, 22. [Johnson & Malone, V.3, 89-90]

(1780). The Maryland, Delaware, Pennsylvania, Virginia, and North-Carolina Almanack. And ephemeris, for the year of our Lord, 1781... Baltimore: Printed and sold by M. K. Goddard.

(1792). As above- ... For the year of our Lord 1793. Baltimore: Printed and sold by Samuel and John Adams. 36p. 18cm.

(1795). Miscellaneous observations relative to the western parts of Pennsylvania particularly those in the neighborhood of Lake Erie.

Transactions of the American Philosophical Society, 4(25), pp.224-229.
(1806). Observations on the solar eclipse of June 16, 1806. NS communications to the American Philosophical Society, 2, Mathematics and Astronomy.
(1792). Plan of the city of Washington in the territory of Columbia. Ceded by the states of Virginia and Maryland to the U.S. of America, and by them established as the seat of their government after the year 1800. Boston: Engraved by Sam Hill.
(1792). As above- Philadelphia: Engraved by Thackara & Vallance. $20^{3/4}$x$28^{1/8}$in.
(1796). as above- New York: I. Reid. 40.5x56cm.
(1796). Several methods by which meridional lines may be found with ease and accuracy: recommended to the attention of the surveyors in the United States. Philadelphia: T. Dobson. 31p. 1 plan. 12°.
(1782). The United States almanack. (np). [Johnson & Malone, V.3, 89-90]

Other

Chambers, S. T. (1940). Observations on the river Potomack, etc., by Tobias Lear; unpublished correspondence of George Washington and Tobias Lear about the observations; published text compared with an autograph copy sent to George Washington. Baltimore, MD: S. T. Chambers.
Lear, Tobias. (1793). Observations on the river Potomack, the country adjacent, and the city of Washington. New York: Printed by Samuel Loudon and Son.
Mathews, Catharine van Cortlandt. (1908). Andrew Ellicott: his life letters. Portraits. Plates. Maps. Facsimiles. New York: The Grafton Press.
U.S. Dept. of State. (1798). Message from the President of the United States, accompanying a report to him from the secretary of state, and sundry documents relative to the affairs of the United States and Spain... Philadelphia: Printed by W. Ross.

ERSKINE, ROBERT, COL. (1735-1780)

Erskine was an army officer, geographer, and surveyor-general, among three men identified as the first three American men to hold the title "Geographer". He was born in Dunfermline, Scotland on September 7, 1735. He died on October 2, 1780 of an illness contracted while he was in the field. Erskine received his early education at the Dunfermline Grammar School and entered the University of Edinburgh in 1748. Finances, however, became a problem and forced Erskine to interrupt his college studies. His name appears again on the rolls of the university in 1752, but there appears to be no record of graduation.

Erskine moved to London where he became involved in business and continued to supplement his education. He entered the hydraulic engineering field. With those skills he invented a centrifugal hydraulic engine and other appliances which resulted in his election to the Royal Society on January 31,

1771. In 1770 he was asked to represent a group of investors at the American Iron Company in Passaic County, New Jersey. He and his wife Elizabeth arrived in New York on June 5, 1771.

By 1774 he was actively sympathizing with the colonists. Erskine organized a military company and was commissioned captain of the Bergen County Militia. On July 27, 1777 he was commissioned surveyor-general to the Continental Army. His work and maps are credited as important factors in the ultimate American victory (Marquis 1963, Historical V. p.241; Johnson & Malone 1928, V.3, pp.180-181).

Harley, et.al. (1978) recognizes him as, "The first geographer to the Continental Army" (p.33). As such he presented to Washington his concept of the job:

> In planning a country a great part of the ground must be walked over, particularly the banks of rivers and roads; as much of which may be traced and laid down in three hours as could be walked over in one; or in other words a surveyor who can walk 15 miles a day may plan 5 miles...six attendants to each surveyor will be proper; to wit, two chain-bearers, one to carry the instrument, and three to hold flag staffs...Young gentlemen of mathematical genius, who are acquainted with the principles of geometry, and who have a taste for drawing, would be the most proper assistants for a geographer. (p.33)

No entries were located in The National Union Catalog Pre-1956 Imprints (The American Library Association 1976).

Schmidt (1966) provides a list of maps that were originally provided by Erskine himself, "A List of the-Rough Draughts of Surveys by Robert Erskine F. R. S. Geog^{r} A. U. S. and Assistants. Begun A. D. 1778" (Appendix).

Num^{r} Principal Places

1 Road from Suffrans to Junes, Kings Ferry, Fort Montgomery &c, on the other side, Peeks Kill, Tellers Point, Tarrytown &c. Sheets A. B. & C. D-Nichols Hill N. Rr. Butter Hill &c. E-

2 Road from Ft. Montgomery to West Point &c.

3 -from D^{o}. to Forrest of Dean &c.

4 -from Junes up the Clove.

5 -Clove continued to Newborrough.[23] [Newburgh]

6 Newborrough to Fishkill.

7 -Fishkill to Peekskill & West Point.

8 Roads from Peekskill to White Plains, Mamaroneck &c. &c.

9[24] White Plains. [note. The number is repeated]

10 D^{o}. from D^{o}. to East Chester.

11 Tuckeho Road.

12 Road from Tarrytown to Crotton River.

13 Sawmill River Road, Valentines Hill, Post Road.

14 Road from Haman's to Gen'l. Gates, W. Plains.
15 -from Artillery Park W. Plains to Dobb's Ferry R^{d}. D^{o}.
16 Cross Road above Dobb's Ferry and Post R^{d}. to Saw Mill.
17 Road between Sawmill & Tuckeho, N. & S.
18 Young's Tav^{n}. to White Plains.
19 Cross Roads to East Chester & above.
20 Cross Roads between Mamaroneck, Rye &c.
21 D^{o}. from White Plains to Kings Street, Wrights Mills.
22 From Andrew Purdees towards Davanport's, & from Carpenter's towards SingSing.
23 From Crotton Bridge towards N. River.
24 From Saw Pits to Stamford Bedford, Crotton.
24[25] & from Stamwich to Kings Street.
[note. The number is repeated]
25 A contraction of Ditto joined to N°. 59
26 R. from 15 Mile Stone near Suffrans to Ft. Lee, Hackensack, Closter, Tappan, Clarkstown, Haverstraw, &c. by Capt. John W. Watkins, A. B.
27 Roads about Clarkstown &c.-by D^{d}. Pye, Esqr.
28 from Bedford to Ridgefield, Danburry &c.
29 from Fishkill to Fredericksburg, Crotton Bridge Quaker Hill &c.
30 from Pines Bridge or Crotton towards Fredericksburg & Cross to Peekskill.
31 A contraction from Bedford to Danburry.
32 A Ditto. Peekskill to Pines Bridge-Join'd to No. 59.
33 Road from Stamford to Greenwich.
34 A contraction of D^{o}.
35 A D^{o}. from Fishkill to Danburry &c.
36 A D^{o}. from Newborrough to Fort Lee.
37 General contraction 2 miles to an Inch by R E.=$_{3}$ pieces & 1 D^{o}.
38 Road from Judge Jay's near Fishkill to Robertsons Mills &c.
39 Road from New Millford to Hartford & back to Bulls Iron Works, 11 sheets marked 39-A, B, C, D, E, F, G, H, I, K & L.
40 Course of Farmington R. the Bearings taken by Mr. Porter. 1765.
41 Road from Ringwood to Slotts, Maces &c.
42 from Do...to Pompton Suffrans &c.
43 from Woodbury to Crumpton, 5 sheets A, B, C, D, & E.
44 Contraction from Danbury to Crumppond.
45 D^{o}. from Robertsons Mill to Peekskill.
46 Road from Pompton towards Morristown.
47 -to Morristown.
48 -from Rockaway Bridge to Horseneck to Totawa Falls and up to Pompton, 3 P's.
49 Contraction from Quaker Hill to Hartford 1 mile to an inch

50 Road through Peekskill Hollow A. & B.
51 -near New Windsor through Chester to Maces.
52 from Albany to Schoharie & contraction 4 pieces
53 Road from Ringwood to Longpond towards Warwick.
1779
54 General contraction 4 miles to an Inch by R^{t} Erskine.
55 Road from Quibbletown to Amboy & places by bearings fr. Capt. Scull.
56 Road from Pompton Paramus Great Falls &c.-A & B Sheets.
57 Road from Bethlehem to Broad Ax and Chestnut Hill Pennsylvania A, B, C, D, & E.
58 Contraction of D^{o}.
59 D^{o}. about White Plains.
60 D^{o}. D^{o}. D^{o}.
61 Plan Charlestown, So Carolina-by Watkins.
62 Draught of the Southern Coast of Lake Erie A, & B a contraction of Do.
63 Map of Niagra R. on the straits between Erie & Ontario.
64 Map of the ship channel on Lake St Claire.
65 Part of the Susquehanna & Delaware-Ononoughquago &c.
66 Small Contraction 8 miles to an inch fr S. Dewitt.
67 Road from Forks Passaic to Scotch Plains A. B C & D.
68 From Chester to Easton, a contraction of the foll^{g}
69 from near Chester to Bethleham 10 sheets marked 1st, 2^{d} & 3^{d} &c.
70 Road from Duyckinks Mill to Germantown.
B. from Cross Roads through Pluckemin towards Morristown;
C. Past Baskenridge and from Bebouts towr Quibbletown;
D. from Mt Bethel M. Hse. to near Quibbletown, and from Quibbletown to Brunswick;
E. from Brunswick to Bound Brook;
F. from Bound Brook to Duyckinks Mill.
71 Road Duyckink's Mill to the South Branch Garison's, Somerset Bound Brook, from South Branch to Reddington Brokaws & from near headquarters to S^{o}. Branch.
72 From Duyckinks Mill White House Potterstown, Germantown Lamatunck
73 From Skeepack Road cross Coriells Ferry to Morristown.
74 From Boundbrook, Quibbletown...[26] Elizabethtown, Wood Bridge, Amboy...[26] Piscataway, Brunswick, Somerset and Vanvoghton's Bridge-A, B, C, D, E, & F.
[[26] Parts of original list missing]
75 Morristown, Bottlehill, Chatham, Springfield, Elizabethtown, Raway, Westfield towards Short Hills A, B, & C.
76 Road from Headquarters to Middlebrook Pluckemin, &c.

77 Roads by Potterstown, Hickery Tavery, Muscanecunk Mountain, Reddingtown Johnstons Forge, Easton A, B, C, D, from letter B to the Road from Corriels to Morristown E from near Henry Camps on the road from Corriels to Morristown to Reddington.

78 From Short Hills Tav^{n}. to Spankton & Craigs Tav^{n}.

79 From Elizabeth Townpoint, Newark, Acquackanonck, Wesel, Totawa A B & C.

80 From near Garrisons Tavn. to past Gen Heards & from near Heards to Princetown, Somerset thru Kingston.

81 Rough contraction in the Jerseys. 8 miles to an inch.

82 Contraction from Elizabeth Townpoint to Totawa 2 miles an Inch.

83 Contraction to Easton & Corriels Ferry 4 miles 1 inch.

84 Road from Widow Van Ambrose to Forrest of Dean, & Cross roads from Smiths Clove to A & B.

85 From Smiths Clove to Chester.

86 From Junes to Archer & Warwick.-A. B. & B^{2}.

87 Roads from near Somerset Court house by Pennytown to the Different Ferrys to Trenton, to Bristol Philadelphia, Coopers Ferry, Mount Holly, Allenstown, Cranburry & Brunswick A, B, C, D, E, F, G; & H.

88 The county of Albany by Mr. Vromer & a contraction of D^{o}.

89 The River Schuylkill from Falls to Reading by D^{d}. Rittenhouse; with the depths, A, B; D^{o}. D a Contraction of D^{o}.

90 from near Warwick to Morristown, Booneton, & from Pompton by Charlotteburg Iron Works, Newfoundland to the Warwick Road A. B. C.

91 From Easton to Wioming, Northumberland to D^{o}. Contraction 2 miles 1 inch.

92 From Easton to Wioming to Tioga A, B, C, D, E, F, G, H, 2 inches a mile; J, K, L, M, 1 inch a mile.

93 From Newborrough to Albany and from Albany to Fishkill; 20 sheets A, B, C, D, E, F, G, H, J, L, M, N, O, P, Q, R, S, T, U, V.

94 Contraction of above from Albany to Newborrough & Fishkill 2 miles an inch.

95 From forks Susquehanna to Wioming A, B, C, D, E, F, G, H, J, K.

96 Rout from Fort Sullivan to Chemung, & to Newton A, B, B^{2d}, C, D, E, F, G, H, J, K.

97 From Seneca Lake to Chenessee A. B. C.

98 Contractions in the Jerseys 2 miles an inch A. B. C, large paper, by Capt. Scull.

99 Projections of lat & departure from Closing &...?...the meeting of Surveys 2 miles to a inch A. B. C. D. Large paper by Capt. Scull.

100 Spherical projections 2 miles to an inch by R. E. 4 sheets, large paper.

101 From Hartford, New haven, Norwalk, Bedford &c. A, B, C, D, & E.
102 Contraction of D° 4 miles an inch by R. E.
1780
103 D°. of the rout of the Western Army 2 miles an Inch A, A2d, B, C, & D., by Lt Lodge.
104 Roads about Cam Morristown A, B, C, D, & E. From near Chatham to Horseneck.
105 Survey of Morristown by the Chain only by Rt. Erskine.
106 Contractions in the Jerseys 1 mile an inch by S. DeWitt A. B.
107 Roads between Chatham, Scotch Plains, Turkey & Springfield.
108 -from Col. Deys to the Ponds & Wykoff to Bellgrove.
109 From Little falls to Acquacknunck.
110 from Dobbs Ferry to Paramus.
111 Contractions in the Jerseys 2 miles an inch Amboy, Brunswick, Pluckemin, Morristown, Newfoundland, Charlotteburg, Paramus By R. E.
112 Width of N. R.[27] at Closter A. B. at Dobbs Ferry. [[27] North River]
113 Roads between Suffrans, Tappan, Kakiate, Paramus, Dobbs Ferry, Clarkstown, &c.
114 Contraction of D°. & sundry places rectified.[28] [[28] Item no. 114, is the last entry in the handwriting of Erskine in this "indes" at the New-York Historical Society. Nos. 115-116 have been entered in an autograph strikingly similar to that of Washington himself. The remainder of the charts are, with the exception of one or two, listed in DeWitt's writing.]
115 Country between Hudson & Delaware.
116 Between Passaic & Paramus, Hackensack & Zabriskies.
117 From Mt Pleasant to New Germantown & from Pluckemin to Black River A. B. & C.
118 From Morristown towards Sussex C. H. by J. Armstrong A. B. & C.
119 From Sussex C. H. to Pittstown & Ringoes and back past the Union & from Hackettstown to N Germantown. A, B, C, D, E, F.
120 from the Sawmill River Road towards the Brunks to near Thos. Thompsons.
121 From where a road branches off to Middle Bush to Trenton. A. & B.
122 The environs of Kingsbridge.
123 The Slote 4 ch. to an inch.
124 The route to York in Virginia.-A B C D E F G H J K L M N L P Q R S T U.
125 The route from Virginia. A, B, C, D, E, F, G, H, J, K, and a 1/2 sheet.

126 Camp Verplancks Point.
127 The line between Pennsylvania & Maryland.
128 Roads from New Windsor to Goshen, Florada, Chester, &c.
129 Moores Land at W. Point.

Note. Mr. A. J. Wall, the librarian of the New-York Historical Society, records the following numbers as "missing" from the collection of Erskine maps there preserved: Nos. 1 B, part of No. C, Nos. 2, 4, 8, 10, 14, 17, 25, 27, 28, 29, 30, 33, 34, 37, three pieces 39, 49,-62, 63, 64, 65, 76, 80, 83, 85, 88, 89, 94, 101, 109, 111, 112, 114, 115, 126, 129.

(pp.248-256)

Other Erskine information and correspondence may be found in the New Jersey Archives.

(1780). A dissertation on rivers and tides. Intended to demonstrate in general the effect of bridges, cuttings, removing of shoals and embankments; and to investigate in particular the consequences of such works on the River Thames... Respectfully addressed...to the consideratn [sic] of the nobility, gentry and public at large; particularly the proprietors of lands on the banks of the Thames, between Cricklade and Gravesend, with a view to the amendment of the ruinous condition of the navigation...London: G. Wilkie. 32p.

(1779). Map of the Highlands prepared for General Washington. (np).

(1779). A plan of the river Schuykill wherein are laid down the several shoals with the depth of water by each in inches, also the depth of the channel in general in the beginning of August 1773, together with the means proposed for improving the navigation of said river and an estimate of the expense. Copied from the original plan of David Rittenhouse, esq. by Robert Erskine. F.R.S. (np).

(1780). Plan of the Slote. No. 123. Taken September 1780. Wm. McMurray, Capt. Scale of sixteen perches to an inch.

(1780). Sketch map of roads about Springfield, New Jersey. (np). 55x38cm.

(1780). Westchester County, NY, and vicinity. (np).

EVANS, LEWIS (C.1778)

Evans is identified as having worked with Holland to survey British colonies. There is no evidence that he either became a United States citizen or returned to Britain. No other information is to be found about the man.

Other

Jefferys, Thomas. (1778). The American atlas; or, a geographical description of the whole continent of America; wherein are delineated at large, its several regions, countries, states, and islands; and chiefly the British colonies, composed from numerous surveys, several of which were made by order of government. By Major Holland, Lewis Evans, William Scull, Henry Mouzon, Lieut. Ross, J. Cook, Michael Lane, Joseph Gilbert,

Gardner, Hillock, &c. &c. Engraved on forty nine copper plates, by the late Mr. Thomas Jefferys...London: Printed by r. Sayer and Bennett.

EWING, JOHN (1732-1802)

Barton (1813) notes, "On the first day of January, 1790, Dr. Rittenhouse was elected one of the Vice-presidents of the American Philosophical Society; his colleagues, in this office, being the rev. William Smith, D. D. and John Ewing, D. D. both of them able and distinguished astronomers" (pp.342-343). Barton reveals that Rittenhouse, Ewing, and John Nancarrow were appointed on October 10, 1791, by Governor Mifflin of Pennsylvania, to "view and mark out a road" from Schuylkill to Lancaster and from Philadelphia to that same borough, and to provide topographical observations (p.361-362).

Brown (1897) identifies Ewing as an educator, born in Nottingham, Maryland who died in Philadelphia. He was the son of Irish Protestants who settled in Pennsylvania. He attended and graduated from the College of New Jersey in 1754 and was licensed to preach, taught ethics in the College of Philadelphia, and was professor of natural philosophy in the University of the State of Pennsylvania. It is noted that Ewing assisted Rittenhouse in his surveys of state boundaries, and was an expert mathematician and scientist (V.3, p.24) (White 1898, V.I, p.341).

(1769). An account of the transit of Venus over the Sun's disc, observed by the Rev. John Ewing, A.M. at Philadelphia in the province of Philadelphia, June 3d, 1769. 1p.l.48p. 25cm.

(1784). Memorandum book of Dr. John Ewing with account of a journey to settle the boundary of Pennsylvania, May, 1784. In Pennsylvania Archives. Harrisburg, 1907. 21.5cm. 6th ser., V.14, 1-20.

Other

Barton, Thomas. (1764). The conduct of the Paxton-men, impartially represented the distresses of the frontiers, and the complaints and sufferings of the people fully stated... With some remarks upon the narrative, of the Indian-massacre, lately publish'd. Interspers'd with several interesting anecdotes, relating to the military genius, and warlike principles of the people call'd Quakers: together with proper reflection and advice upon the whole. In a letter from a gentleman in one of the back countries, to a friend in Philadelphia... Philadelphia: Printed by A. Steuart, and sold by John Creaig.

Patterson, Rev. R. (1809). A plain elementary and practical system of natural experimental philosophy; including astronomy and chronology. By the late Rev. John Ewing, D.D., Provost of the University of Pennsylvania. Being the course of lectures delivered by him in that seminary, (enlarged and improved from time to time) for upwards of twenty years. Illustrated by copperplate engravings. The whole carefully revised and prepared for the press, with sundry explanatory and supplementary notes; by Robert

Patterson... With a biographical sketch of the author. Philadelphia: Hopkins and Earle. xxv, 168p. 22.5cm.

FAIRMAN, DAVID (C.1804)

Ristow (1985) mentions Fairman, brother of Gideon Fairman, as one of the engravers of maps drawn by Samuel Lewis (p.265).

(1804). A new and elegant general atlas, comprising all the new discoveries, to the present time; containing sixty-three maps. Drawn by Arrowsmith and Lewis. Philadelphia: John Conrad & Co.; Baltimore: M. & J. Conrad & Co.; Washington, DC: Rapin, Conrad & Co.; Petersburg, VA: Somervell & Conrad; Norfolk, VA: Bonsal, Conrad & Co. Engraved by Benjamin Tanner, William Harrison, Jr., and David Fairman.

FAIRMAN, GIDEON (C.1774)

Fairman was born in Newtown, Fairfield County, Connecticut. There he served an apprenticeship as a mechanic. He later moved to Albany and learned silver-plate engraving. His own business was opened in 1796 and was in operation until 1810. He then moved to Philadelphia and specialized in bank note engraving (Ristow 1985, p.78).

(1802). A map of the state of New York. By Simeon De Witt, Surveyor-General. Albany: 12th New York District Copyright. Engraved by G. Fairman. Colored map on 6 sheets. 67cm.

Other

Fairman, Gideon, & Inman, Henry. (1824). Illustrations of [Cooper's] The Pioneers. Philadelphia: Hall. 5 plates. 27cm.

FILSON, JOHN (C.1747-1788)

Filson was an explorer and historian who was born in East Fallowfield, Pennsylvania around 1747. He was killed by an Indian in October 1788 while on a surveying trip on the Little Miami river. He was the son of Davison Filson. John Filson was taught by the Rev. Samuel Finley, later president of the College of New Jersey.

Details about his life are sketchy, but it is known that he produced the first and remarkably accurate map of Kentucky, and he wrote the first history of Kentucky, which was inadequate, misleading, and often inaccurate, but the descriptive portions of the book were vivid and attractive. Townsend (1907) provides a brief biography noting, "When we take into consideration the little history the new state had to be written in 1784, and allow for the superior deserts of his map of Kentucky and his life of Boone, we must candidly say that the merits of his history have not been surpassed by those of any since written" (p.68). The Biographical Dictionary (2001. John Filson) revealed some of the same information.

He spent a year in Lexington, Kentucky teaching at a private school. Filson may have had Revolutionary service, but no record of it has been found.

He did, however, arrive in Kentucky with Virginia military warrants for land, given to participants in the American Revolutionary War for their service. Those he could have acquired as a result of his service or in some other manner from other people. Filson traveled to Pennsylvania and then into the Illinois Country. He published a prospectus of a town that he and some associates hoped to lay out in Ohio, first known as Losantiville, now known as Cincinnati. (Johnson & Malone 1928, V.3, pp.382-383).

Brown (1897) reports that Filson left the following unpublished MSS [manuscripts]:

> A Diary of a journey from Philadelphia down the Ohio and up the Wabash River to Port St. Vincent in the spring and summer of 1785; An account of a trip from Vincennes to Louisville by land in August, 1785; a journal of two voyages from the Falls of the Ohio to Port St. Vincent on the Wabash River, etc., and an account of an attempted trip by water from Vincennes to Louisville, August, 1786, the attack upon him by the Indians on the Wabash and his subsequent trip to the Falls of the Ohio by land. Dr. Lyman C. Draper was the preserver of the MSS. and used them in preparing his Life of Gen. George Rogers Clark. (V.3, pp.87-88)

(1785). The adventures of Colonel Daniel Boon, one of the first settlers at Kentucke [sic]: containing the wars with the Indians on the Ohio, from 1769 to 1783, and the first establishment and progress of the settlement on that river. Written by the Colonel himself. To which are added, a narrative of the captivity, and extraordinary escape of Mrs. Francis [sic] Scott, an inhabitant of Washington County Virginia; who after the murder of her husband and children, by the Indians, was taken prisoner by them; on the 29th of June, 1785. Norwich, CN: Printed by John Trumbull. 24p. 18.5cm.

(1787). As above- ...Adventures of Colonel Daniel Boon, one of the original settlers at Kentucke: containing the wars with the Indians on the Ohio, from 1769 to 1784, and the first establishment and progress of the settlement on that river. Written by the Colonel. In The American museum, or Repository of ancient and modern fugitive pieces... Philadelphia: (np). V.2, pp.321-328.

> Written in the form of an autobiography of Boone, "from his own mouth," by John Filson and first published in the appendix to his Discovery, settlement, and present state of Kentucke. Wilming, De..: R. T. Durrett, John Filson, 1884, and the Preface to Filson's Discovery.

(1793). As above- In The discovery, settlement, and present state of Kentucky... London: (np). pp.34-49. 23.5cm.

(1794). As above- In Brers's Almanac and ephemeris... for the year of our Lord 1795. Hartford: (np). pp.19-26. 17.5cm.

(1797). As above- In Imlay, Gilbert. A topographical description of the western territory of North America... London: (np). pp.338-359. 22cm.

(1785). Bergkemmetten des Obersten Daniel Boon, elnes der ersten Kolonisten in Kentucke worinn eine Nachricht von den Kreigen mit den Indianern am Ohio von 1769 bis aut jetxige Zeiten, und von der ersten Einrichtung und dem Fortgang der Niederlassungen an selbigen Flusse enthalten ist. Historisches Portefecalle zur Kenntnis der gegenwartigen und vergangenen Zeit. Wien: Jahrg. pp.31-53. 12°. [Accents omitted]

(1784). The discovery, settlement, and present state of Kentucke [sic]: and an essay towards the topography and natural history of that important country: to which is added, an appendix containing, I. The adventures of Col. Daniel Boon, one of the first settlers, comprehending every important occurrence in the political history of that province. II. The minutes of the Piankashaw Council, held at Post St. Vincents, April 15, 1784. III. An account of the Indian nations inhabiting within the limits of the thirteen United States... IV. The stages and distances between Philadelphia and the Falls of the Ohio; from Pittsburgh to Pensacola and several other places. The whole illustrated by a new and accurate map of Kentucke [sic] and the country adjoining, drawn from actual surveys. Wilmington, Del.: Printed by James Adams. 113p. 21.5cm. (Tucher 1985)

The map to accompany this work, "This map of Kentucke drawn from actual observations... By John Filson." was published separately: Philadelphia: Engraved by Henry D. Pursell, & printed by T. Rook, for the author.

(1793). As above- New York: Printed and sold by Samuel Campbell. 204p.

(1793). As above- Illustrated with a large whole sheet map of Kentucky from actual surveys, and a plan with a description of the rapids of the river Ohio by Captain Thomas Hutchins... London: Printed for J. Stockdale. 68p. folded map. 21cm. (Tucher 1985)

(1795). As above- In Imlay, Gilbert. A topographical description of the western territory of North America... 2d ed. London: (np). pp.269-415. 22cm.

(1795). As above- In Imlay, Gilbert. A topographical description of the Western Territory of North America... 2d ed. New York: (np). pp.1-110. 17cm. (Tucher 1985)

(1797). As above- In Imlay, Gilbert. A topographical description of the western territory of North America... 3d ed. London: J. Debrett.

(1785). Histoire de Kentucke, nouvelle colonie a l'onest de la Virginie: contenant, 1o. La decouverte, l'acquisition, l'etablissement, la description topographique, l'historique dnaturelle, &c. du territoire: 2o. la relation historique du colonel Boon, un des premiers colons, sur les guerres contre les naturels: 3o. l'assemblee des Piankashaws au Poste Saint Vincent: 4o. un expose succinct des nations indiennes qui habitent dans les limites des treize Etats-Unis, de leurs moeurs & coutumes, & des reflexions sur leur origine; & autres pieces: avec une carte. Ouvrage pour servir de suite aux Lettres d'un cultivateur americain. (Tr.). le l'anglois, de m. Jon Filson; par m. Parraud... Paris: Chez Buisson. 234p. 19.5cm. (Tucher 1985) [Accents omitted]

(1823). Life and adventures of Colonel Daniel Boon, the first white settler of the state of Kentucky... written by himself. To which is added, a narration... of his life... until... his death... Annexed, is an eulogy on Col. Boon, and choice of life, by Lord Byron. Brooklyn: Printed for C. Wilder. 36p. 20.5cm.
(1824). As above- Providence: Printed by H. Trumbull. 18.5cm.
(1828). As above- Brooklyn: Printed for C. Wilder. 17cm.
(1916). As above- New York: Reprinted.
(1932). As above- Tarrytown, NY: Reprinted W. Abbatt.
(1784). Map of Kentucke... Philadelphia: T. Rook. 47x44cm.
(1956). As above- 50x45cm.
(1821). Narrative of the adventures of Col. Daniel Boon, from his first arrival in Kentucky in 1769 to the end of the year 1782. In Samuel Metcalf. A collection of some of the most interesting narratives of Indian warfare in the West. Lexington, KY: (np). [7]-35p. 21cm.
(1790). Reise nach Kentucke und nachrichten von dieser neu angebauenten landschaft in Nordamerika, von John Filson. Aus dem englischen ubersetzt. Leipzic: C. Weigel und Schneider. 124p. 20cm. [Accents omitted]
(nd). This map of Kentucky drawn from actual observation, is inscribed with the most perfect respect to the Honorable, the Congress of the United States of America and to His Excellency, George Washington, late Commander in Chief of their army. By their humble servant, John Filson. Boston: Heliotype Printing Co. 55.5x50cm.
(1784). As above- Philadelphia: Engraved by Henry D. Pursell. Printed by T. Rook for the author. 19.5x17.75in.
(1821). Voyage au Kentoukey, et sur les bords du Genesee, precede de conseils aux liberaux, et a tous ceux qui se proposent de passer aux Etats-Unis. Ouvrage accompagne d'une carte geographique, levee sur les lieux par l'auteur, en 1820. Paris: M. Sollier. [Accents omitted]

Other

Durrett, R. T. (1884). Life and writings of John Filson. (np).
Imlay, Gilbert. (1797). A topographical description of the Western Territory of North America: containing a succinct account of its soil, climate, natural history, population, agriculture, manners, and customs. With an ample description of the several divisions into which that country is partitioned... By Gilbert Imlay... Illustrated with correct maps of the Western Territory of North America; of the state of Kentucky, as divided into counties from actual surveys by Elihu Barker; a map of the Tenasee [sic] government; and a plan of the rapids of the Ohio. The 3d ed., with great additions. London: Printed for J. Debrett.

FISHER, GEORGE (1788-1861)

A text with geographic content was Fisher's contribution to American geography. He was born in Franklin, Massachusetts on March 17, 1788 and died in New York City on March 26, 1861. He attended Brown University and studied law. In 1816 he was admitted to the bar and began to practice law in Oswego County, New York. In 1818 he was appointed inspector of schools. Fisher was trustee for the village of Oswego in 1828 and 1833 becoming a member of the U. S. House of Representatives (a contested election) from New York in 1829 through February 5, 1830.

Wishing to provide education for his children, Fisher moved to France and spent five years there. He then returned to Oswego and became involved in real estate. Fisher became President of Northwestern Insurance Company before moving to New York City, where he died in 1861 (Marquis 1963, Historic Volume, p.250)

(1850, June 1). Letter to the Madisonian from George Fisher concerning the annexation of Texas to the United States. The Panama Echo. Saturday Morning. Note. George Fisher, 1795-1873.

(1831, February 10). Two letters from Fisher. Advertiser, no. 11. Note. The first addressed to the editors of the Advertiser of the Port of Matamoros, and the second to Godwin Brown Cotten, Editor of the Texas Gazette. Note. George Fisher, 1795-1873.

(1831, October 6). Letter: Estados Internos de Oriente, Mexico. His Excellency the Commandant General of the Eastern Internal States in official note... Letter-head: Maritime Customs-House of Galvezton. Signed: Jorge Fisher, Matamoros.

Note. George Fisher, 1795-1873.

The Preston Citizen. (1987). The Oneida Stake: 100 years of LDS history in Southeast Idaho. Preston, ID: author. 258p. 24cm. Note. George Fisher, 1795-1873.

(1787). The American instructor: or, young man's best companion. Containing spelling, reading, writing, and arithmetic... Instructions to write variety of hands, with copies both in prose and verse... Also merchants accounts, and a short and easy method of shop and book-keeping... Together with the method of measuring carpenters, joiners... and painters work... Likewise the practical gauger made easy...A compendium of the sciences of geography and astronomy. Also some useful interest-tables...Philadelphia: Print. Joseph Crukshank. 372p. 6 plates. 12°.

(1765). The instructor: or, American young mam's best companion. Containing, spelling, reading, writing, and arithmetic...to which is added this family's best companion...and a compendium of the sciences of geography and astronomy...also, several very useful tables. 18th ed. London: Printed for H. Woodfall. 384p. plates. diagrams. 17cm.

(1767). As above- 21st ed. Edinburgh: Printed by A. Donaldson. 382p. 18cm.

(1771). As above- 23rd ed. Edinburgh: Alston. 384p.

(1772). As above- 21st ed. London: Printed for J. Fuller.

(1773). As above- 22d ed. Edinburgh: A. Donaldson.

(1775). As above- 22d ed. London: Printed for J. Fuller. 17cm.

(1776). As above- New ed. Edinburgh: Gavin Also.
(1777). As above- 23d ed. London: Printed for H. Woodfall. 18cm.
(1779). As above- 23d ed. London: Printed for W. Strahan. 17cm.
(1780). As above- 25th ed. London: Bunney.
(1784). As above- 24th ed. London: W. Strahan.
(1785). As above- 30th ed. Worcester, MA: Isaiah Thomas; and Boston: J. Boyle, E. Battelle, W. Breen, B. Larkin, and J. Condy.

w = 10 cm H= 17 cm Several tables. One engraving opposite title page.

(1786). As above- Glasgow: J. Duncan. 375p. 17.5cm.
(1788). As above- London: Printed for W. Osborne & T. Griffin. 384p. diagrams. 18cm.
(1789). As above- 25th ed. London: Printed for J. Rivington and Son; J. Buckland; T. Longman, B. Law, J. Johnson, G. G. J. and J. Robinson, R. Baldwin, T. Vernor, W. Goldsmith, W. Woodfall, J. Bew, T. Lowndes, and C. Stalker.
(1791). As above- London: Printed for J. Martin. 394p. 16.5cm.
(1792). As above- A new edition. London: (np). 404p. 17.5cm.
(1794). As above- Walpole, NH: Isiah Thomas and David Carlisle; Worcester, Mass.: I. Thomas and Andrews; and Boston: D. West and E. Larkin, Jr.
(1797). As above- Wilmington: Printed and sold by Peter Brynberg. 360p. diagrams. 17cm.
(1797). As above- Glasgow: J. Duncan. 396p. tables. diagrams. 17.5cm.
(1799). As above- London: Printed for booksellers. 360p.
(1801). As above- Philadelphia: Printed by John Bioren for John Conrad; Baltimore: M. & J. Conrad & Co.; and Washington: Rapine, Conrad & Co. 348p. 17.5x10.5cm.
(1806). As above- 29th ed. London: Printed for J. Johnson. 324p. 18cm.
(1810). As above- 30th ed. London: Printed for F. and C. Rivington. 328p. 18cm.
(1810). As above- Philadelphia: Printed and sold by John Bioren. 300p. 18.5cm.
(1812). As above- 19cm.
(1822). As above- ...A new edition corrected and improved throughout by N. Downes. London: Printed for Dean & Munday. 288p. 18cm.
(1856). As above- ...A new edition, corrected by G. N. Wright. London: Dean.
(1833). The young man's best companion, improved; containing instructions for reading, writing (including the English grammar), arithmetic, merchants' and farmers' accounts; a variety of the most useful and approved forms of blanks now in use; mensuration, gauging according to the most modern and approved practice; and the art of duelling. Also a compendium of the sciences of geography and astronomy, and a table of specific gravities of bodies, with rules for ascertaining the name; to which is added several very useful tables. Philadelphia: Published by Thomas L. Bonsal. 370p. 18.5cm.

(1862). As above- London: Dean and Son. 296p.
(nd). As above- (np). 395p. 18x10cm.

Other

Fisher, George (Ed.). (nd). Cocker's arithmetick... (np).

FISHER, WILLIAM (C.1804)

Ristow (1985) notes that a William Fisher and others published The English Pilot, "the first great atlas of wholly English origin to deal exclusively with American waters..." (p.221). Numerous editions were published between 1671 and 1794. Therefore, at least one of the William Fishers would predate our first generation of American geographers. Other sources reveal the following which may have been products of a later William Fisher:

(1812). An interesting account of the voyages and travels of Captains Lewis and Clark, in the years 1804, 1805, and 1806. Giving a faithful description of the river Missouri... Interspersed with very entertaining anecdotes, and a variety of other useful and pleasing information remarkably calculated to delight and instruct the readers... Baltimore: Printed by Anthony Miltenberger; for the Purchasers. (Tucher 1985)

(1813). As above- 2d ed. Printed and published by P. Mauro. [1-4],7-266p. 2 fronts. 3 plates. (Tucher 1985)

(1812). New travels among the Indians of North America; being a compilation, taken partly from the communications already published, of Captains Lewis and Clark, to the President of the United States; and partly from other authors who travelled among the various tribes of Indians... Philadelphia: Published by James Sharan; J. Maxwell, printer. [1-8]. [7]-300p. (Tucher 1985)

FITCH, JOHN (1743-1798)

White (1893) reports that Fitch was a surveyor and inventor who produced, in 1787, the first serviceable United States steamboat (The Biographical Dictionary 2001, John Fitch; Encyclopedia Britannica 2001, Fitch, John). He was born in East (now South) Windsor, Connecticut, January 21st. His father was a farmer. When he was ten years old, Fitch was sent to a village school. There he developed a love of knowledge, devoting himself to arithmetic and geography. By the time he was seventeen he had learned surveying.

Fitch served with Washington at Valley Forge. Toward the end of the war he moved west with the intention of becoming a surveyor. He was captured by Indians while surveying in Kentucky.

In 1785, while he was living in Pennsylvania, Fitch came on the idea of developing steam power. He read extensively and developed a plan that evolved into a model. His model was a boat with wheels at the sides that was powered by a steam engine with three-inch cylinders and paddles. He pursued financing through the Pennsylvania Legislature and the Continental Congress, but met with no success. As a result he devoted his energies to the engraving of a map of the Northwest Territory.

His exploits with steam were, however not over. Fitch contracted with New York, Delaware, Pennsylvania, and Virginia for the right to use their waters for fourteen years for navigation by means of steam. He traveled to France in hopes of building a steamboat there, but he had to abandon the plan. There is speculation that his plans for the steam engine were passed from the hands of the American Consul at L'Orient in London to Robert Fulton. That became a sad part of Fitch's life. He returned to Pennsylvania where he became ill, dying in Bardstown from an overdose of opium supplied by his physician (White, 1893, V.VI. pp.63-64).

According to Ristow (1985), Fitch was the compiler of a map of the north western parts of the United States who used the proceeds from his map sales to finance experiments that resulted in the invention of the steamboat (p.68). His map drew extensively from McMurray's map of the region.

(1785). North West Parts of the United States. (np).

(1787). A map of the north west parts of the United States of America. (np). $29^{1/8}$x$21^{7/16}$in. Scale: 36.7 miles to 1 inch.

(1785). Prospectus and advertisement of his map of the N.W. parts of the United States. (np). 4°.

FLINT, ABEL (1765-1825)

Mapping the new United States of America depended on people who had some knowledge of surveying. Flint's books must have contributed to an understanding of that skill. His surveying books were in print for over sixty years.

(1804 & 1805). A system of geometry and trigonometry: Together with a treatise on surveying... Likewise, rectangular surveying...To the whole are added several mathematical tables... Hartford: Printed for Oliver D. Cooke, by Lincoln & Gleason. viii, [9]-168p. 22.5cm.

(1808 1812). As above- 2d ed.

(1813). As above- 3d ed.

(1813). As above- Boston: S. T. Armstrong.

(1818). As above- 4th ed. Hartford: Cooke & Hale.

(1825). As above- 5th ed. : O.D. Cooke and Co. 104,137p.

(1830). As above- 6th ed. iv, [7]-112,10,62,100p.

(1833). As above- ...A new edition enlarged with practical illustrations on surveying, including other branches connected therewith... By George Gillet... Also a concise treatise on logarithms, by F. P. Barnard...7th ed. (np).

(1835, 37, 38, 41, 44, 45 & 1847). As above- 8th ed. Hartford: Belknap & Hamersley; New York: Cooke & Co.

(1854 & 1858). As above ... New ed., rev., containing a new rule for correcting deviations of the compass, by L. W. Meech. Hartford: W. J. Hamersley. 160,112p. 21cm.

(1789). A treatise of practical surveying; which is demonstrated from its first principles - Fifth edition. Philadelphia: (np).

FOLIE, A. P. (C.1791)

Folie was among those French people who were forced to leave Saint-Domingue (the present Haiti) at the time of the 1791 forced exodus of whites from that island. He moved to Philadelphia. In the year 1794 he advertised for subscriptions of a map of Philadelphia (Ristow 1985, p.250). A draft of the plan was hung in the bookstore of Davies where subscriptions were taken. That event likely resulted in the map being referred to as "Davies's Map." Tooley (1985) identifies Folie as a "French Geographer" (p.208). Since we have no other information about the man, it is possible that he was described as a "French Geographer" because of his ancestry. As far as we know, he lived and worked in the United States from 1791 possibly continuing to his demise.

(1794). A grand plan of the city of Philadelphia and its environs; taken from actual survey by A. P. Folie, geographer from St. Domingo. (np).

(1792). Plan of the town of Baltimore and its environs dedicated to the citizens of Baltimore taken upon the spot by their most humble servant A. P. Folie, French Geographer. James Poupard sculpsit Philadelphia. (np): A. Folie.

FORSYTH, G.C. (C.1800)

No biographical or other bibliographic information has been located which might shed some light on who Forsyth was.

Geological, topographical and medical information concerning the eastern part of the state of Ohio... The Medical Repository, New York City, 12(4), pp.350-354.

FOSDICK, NICOLL (C.1805)

Guthron (1984) noted that Nicoll was from New London. He worked with John Cahoone to survey and produce a map of Long Island Sound in 1805.

(1805). Blunt's new and correct chart of Long Island Sound from Montauk Point to Frogs Point, including Fisher's Island Sound, and Watch Hill Reef. Surveyed in 1805 by John Cahoone, Newport and N. (icoll) Fosdick, New London. New Edition. Engraved by Peter Maverick. New York: E. M. Blunt. 64x20$^{3/4}$inches. Scale 1:130,000.

FRANKLIN, BENJAMIN (1706-1790)

Franklin was born in Boston to Joseph and Abiah (Folger) Franklin. He died in Philadelphia (Brown 1897, V.14, pp.174-178). His reputation as a patriot, diplomat, scholar and scientist are presently very available, so space will not be taken here to reveal much other than the fact that he was among many who added to the American public's knowledge of geography.

Portinaro and Knirsch (1987) report that Franklin and Timothy Folger prepared what is believed to be the "first chart of the Gulf Stream" (p.240). The chart was rendered around 1770.

The sheer volume of his publications prohibit their inclusion in these pages: The National Union Catalog Pre-1956 Imprints, (The American Library Association 1976, V.183) pages101-180 are devoted to Franklin. Therefore, only those publications that are clearly geographic in nature are included here.

(1788). Conjectures concerning the formation of the Earth, etc. in a letter...to the abbe' Soulavie, Sept. 22, 1782. Transactions of the American Philosophical Society, 3(1), pp.1-5.

(1859). Extracts from the works of Dr. Franklin, on population, commerce, &c. In J. R. McCulloch. (Ed.). A select collection of scarce and valuable economical tracts... London: (np). V.7, "Commercial pamphlets."

(1753). Letters relating to the transit of Mercury over the Sun, which is to happen May 6, 1753. Philadelphia: (np). 4p. 35.5cm.

(1790). Map of the Gulf Stream. Paris: Chez Le Rouge... $12^{3/8}$x$14^{3/8}$inches.

(1766). Physical and meteorological observations, conjectures, and suppositions. Read at the Royal Society, June 3, 1756. London: Society.

(1732, 1733, 1734 & 1758). Poor Richard, 1733. An almanack for the year of Christ 1733, being the first leap year.l Wherein is contained the lunations, eclipses, judgment of the weather, spring tides, planets motions & mutual aspects, Sun and Moon's rising and setting, length of days, time of high water, fairs, courts, and observable days. Fitted to the latitude of forty degrees, and a meridian of five hours west from London, but may without sensible error, serve all the adjacent places, even from Newfoundland to South Carolina. By Richard Saunders, philom. [pseud.]. Philadelphia: Printed and sold by B. Franklin. 24p.

(1790). Queries and conjectures relating to magnetism, and the theory of the Earth, in a letter to Mr. Bodoin. Transactions of the American Philosophical Society, 3,(4), pp.10-13.

(1790). Remarks upon the navigation from Newfoundland to New York... By Dr. Franklin... Directions for sailing between the Middle Ground and the Horse Shoe, at the entrance of Cape Henry... Boston: Printed by Benjamin Edes and Son.

FRASER, DONALD (1755-1820)

Fraser's inclusion in the first generation of American geographers is based on his publications. At this time it is difficult to absolutely ascertain whether or not he is the Donald Fraser of Scotland who became the teacher in New York. According to The National Union Catalog Pre-1956 Imprints Agnes Renton (Robson) Fraser was the wife of Donald (The American Library Association 1976, V.183, p.394). The following information was gleaned from an OCLC search under the subject heading Fraser, Donald:

(1807). A compendium of the history of all nations: exhibiting a concise view of the origin, progress, decline and fall of the most considerable empires, kingdoms, and states of the world, from the earliest times to the present period. Interspersed with a short account of the prevailing religions. Ornamented with a frontispiece... and three other plates. New York: Printed by Henry C. Southwick. 392p. 21cm.

(1801 & 1809). An essay on the origin, antiquity, &c. of the Scots and Irish nations, with an impartial sketch of the character of most of the nations of Europe. To which is added, an oration, lately delivered before the Caledonian Society, in this city. New York: Printed by Furman and Loudon. 32p. 20cm.

(1800). Essay on the origin of the Scots & Irish nations. New York: (np).

(1806). The history of man; or, A view of the virtues and vices of human nature in both sexes, with examples ancient & modern: together with some original essays on a variety of subjects, tending to promote the principles of virtue, prudence and politeness. The whole work being interspersed with useful and entertaining narratives. 1st American from 3d European ed. New York: Printed by Southwick and Hardcastle. 434p. 22cm.

(1790). History of man; or, the wonders of human nature in relation to the virtues, vices and defects of both sexes. 2d ed. Edinburgh: Bell. 2v. in 1.

(1814). An interesting companion for a leisure hour: or, an historical, geographical, and chronological compendium: containing a brief but comprehensive history of England, Ireland, Scotland, and Holland: together with a variety of curious articles, both miscellaneous and masonic, not generally known. New York: Printed by John Low 144p. 18.5cm.

(1791). The young gentleman and lady's assistant, partly original...in five parts viz. geography, natural-history, elocution, poetry - and miscellany. To which is annexed - A short system of practical arithmetic; wherein every example is wrought at large, and the whole, including the money of the United States, rendered easy to the meanest capacity. This work is divided into small sections for the convenience of schools. New York: Printed by Tho's Greenleaf. 273p. 17cm.

(1794). As above- 2d ed. Danbury: Printed for the author by N. Douglass. 216p.

Other

Sinclair, G. (1836). Answers to queries: Proposed April, 1834, by G. Sinclair, esq. (now Sir G. Sinclair) M. P. Convener to the Committee of the House of Commons on Patronage, to Donald Fraser...; to which is prefixed, their correspondence on that occasion. Edinburgh: M. Paterson. 38p. 22cm. Note. Published by request of several ministers of the United Secession Church. Donald Fraser, 1773-1841.

FREEMAN, JAMES D. D. (1759-1835)

Freeman was a respected American citizen. A Reverend Greenwood made the following comments at his funeral (1835, November 22). Mr.

Greenwood's sermon preaching in Kings Chapel - the Sunday after the funeral of James Freeman.

> Freeman was born in Charleston, Mass., April 22, 1759 - He attended Cambridge College 1773-1777. "He was then an excellent mathematician; was well acquainted with geography and history at large, and thoroughly so with geography and history of his own country;..." He was the first Unitarian minister in this county. Founder of and writer for the Massachusetts Historical Society.

Freeman is reported to be the first Unitarian minister in America, a patriot, and one of the founders of the Massachusetts Historical Society. He was born in Charlestown, Massachusetts on April 22, 1759 and died in Newton, Massachusetts on November 14, 1835. He attended public schools in Boston and entered Harvard College in 1773 and graduated in 1777. Freeman was the son of Constant and Lois (Cobb) Freeman. He married Martha Curtis on July 17, 1783. They appear to have had no children of their own, but they adopted one child.

Following his experiences in the Colonial Army, during which he was captured by privateers and detained for several months, he found opportunities to pursue his theological studies. He was invited in 1782 to be a lay-reader at King's Chapel in Boston, Episcopal congregation. He discarded the doctrine of the Trinity and many in the congregation followed suit. The Bishop was sent a copy of the new liturgy and a request for the ordination of Freeman as pastor. The Bishop declined, so the congregation chose to consecrate him with a "peculiar" service by his own wardens and people. As minister of King's Chapel, the 1st Unitarian Church in America (1787-1826) he was the first American minister to openly and explicitly avow and maintain purely Unitarian doctrines (The Biographical Dictionary 2001, James Freeman; Brown 1897, V.3, p.183; Johnson and Malone 1928, V.4, p.10-11; Marquis 1963, Historical Volume, p.191; White 1892, V.VII, p.445)

(nd). A description of Boston. (np).

(1802). A description of Chatham, in the county of Barnstable, September. In Massachusetts Historical Society Collection, 8, ser.1, p.142-154.

(1802). A description of Dennis, in the county of Barnstable, September. In Massachusetts Historical Society Collection, 8, ser.1, p.129-140.

(1815). A description of Duke's county, Aug. 13th, 1807. In Massachusetts Historical Society Collections, 3, Ser. 2, p.38-94.

(1802). A description of the eastern coast of the county of Barnstable, from Cape Cod, or Race-Point, in latitude 42" 5' [sic] to Cape Malebarre, or the Sandy Point of Chatham, in latitude 41" 33' [sic]. pointing out the spots, on which the trustees of the Humane Society have erected huts, and other places where shipwrecked seamen may look for shelter. October 1802. Boston: Printed by Hosea Sprague. 15p. 22cm.

(1815). Notes on Nantucket, August 1st, 1807. In Massachusetts Historical Society Collection, 3, ser.2, p.19-38.)

(1792 & 1793). Remarks on the American universal geography. By J.F... Boston: Apollo Press - Belknap and Hall. 61p. 21cm.

Other

Allen, William Winthrop. (nd). Review of Dr. Freeman's sermons. (np).

FREEMAN, THOMAS (C.1750-1821)

Freeman was a civil engineer, astronomer, and explorer who was born in Ireland and emigrated to America in 1784. He died at Huntsville, Alabama on November 8, 1821. He was appointed one of the surveyors for the new capital of the United States and surveyed the entire northern portion of the district, placing the boundary stones and doing the first topographic study of the city.

He then took a commission as surveyor to chart the boundary between the United States and Spain. On that assignment he encountered Andrew Ellicott who suspended Freeman from duty, but he was later cleared of charges leveled by Ellicott. He was then appointed to explore the Red and Arkansas rivers. His observations led to the accurate mapping of the lower Red River. Later he mapped the boundary between Tennessee and Alabama. In 1811 Freeman was commissioned Surveyor of Public Lands of the United States south of Tennessee. His biographers describe him as, "Courteous, a maker and keeper of friends, and of undoubted integrity and ability. He died poor, fighting the land speculators until the last" (Johnson & Malone 1928, V.4, pp.13-14).

(c.1806). An account of the Red River, in Louisiana, drawn up from the returns of Messrs. Freeman & Custis to the war office of the United States, who explored the same, in the year 1806. Washington, DC: (np). 63p. 22.5cm.

Peter Custis, joint author.
University of Georgia Special Collections
1. Red River of Louisiana
2. U. S. - Exploring expeditions

FURLONG, CAPTAIN LAWRENCE (C.1796)

Ristow (1985) reports that Furlong was given credit as the "compiler" of the American Coast Pilot, a product of Edmund March Blunt (p.227). Guthron (1984) noted that this was the first nautical publication by Edmund Blunt.

Ristow (1985) provides a description of the first edition written by Blunt and published in the Newburyport Herald:

> The American Coast Pilot containing the courses and distance from Boston to all the principal harbours, capes and headlands included between Passamaquoddy and the capes of Virginia with directions for sailing into and out of all the principal ports and harbours with the sounding on the coast; also a Tide Table shewing [sic] the time of the high water at full and change of the moon in all the above

places together with the courses and distances from Cape Cod to cape Ann to the Shoal of Georges and from said capes out to the south of East Channel and the setting of the current on the eastward and westward; also the latitude and longitude of the principal harbours and headlands, &c. by Capt Lawrence Furlong. Also courses, directions, distances etc. from the capes of Virginia to the River Mississippi from the latest survey, and observations (Approved by experienced Pilots and Coasters.) The first edition printed at Newburyport by Blunt and March. Sold by them and the principal book stalls in the U.S. 1796. (pp.227-228)

(1796). The American coast pilot. Newburyport, MA: Edmund Blunt.

GARDNER, (C.1778)

Gardner is identified as having worked with Holland to survey British colonies. There is no evidence that he either became a United States citizen or returned to Britain. No other information is to be found about the man.

Other

Jefferys, Thomas. (1778). The American atlas; or, a geographical description of the whole continent of America; wherein are delineated at large, its several regions, countries, states, and islands; and chiefly the British colonies, composed from numerous surveys, several of which were made by order of government. By Major Holland, Lewis Evans, William Scull, Henry Mouzon, Lieut. Ross, J. Cook, Michael Lane, Joseph Gilbert, Gardner, Hillock, &c. &c. Engraved on forty nine copper plates, by the late Mr. Thomas Jefferys...London: Printed by r. Sayer and Bennett.

GASS, PATRICK (1771-1870)

Gass was born in Falling Springs, Pennsylvania. He is identified as a soldier, and explorer who published the first account, as a member of the Lewis and Clark Expedition (Hopkins 1964, p.331).

(1938). Excerpt from journal of the voyages & travels of a corps of discovery, under the command of Capt. Lewis and Capt. Clarke [sic] of the army of the United States, from the mouth of the river Missouri through the interior parts of North America to the Pacific ocean, during the years 1804, 1805, and 1806. By Patrick Gass... With two engravings. San Francisco: Reprinted by Grabhorn Press for Carl Hoffman. 21cm.

(1904). Gass's journal of the Lewis and Clark expedition, by Sergeant Patrick Gass, one of the persons employed in the expedition. Reprinted from the edition of 1811, with facsimiles of the original title-page and the five original illustrations, a reproduction of a rare portrait of Gass, and a map of the Lewis and Clark route. With an analytical index and an introduction by James Kendall Hosmer... Chicago: A.C. McClurg & Co. iii,1,298p. 6pl. 22cm.

(1807). A journal of the voyages and travels of a corps of discovery, under the command of Capt. Lewis and Capt. Clarke of the Army of the United States, from the mouth of the river Missouri through the interior parts of North America to the Pacific Ocean, during the years 1804, 1805, & 1806. Containing the authentic relation of the most interesting transactions during the expedition. - a description of the country, and an account of its inhabitants, soil, climate, curiosities and vegetable and animal productions, by Patrick Gass, one of the persons employed on the expedition with geographical and explanatory note by the publisher. Pittsburgh: Printed by Zadok Cramer, for David M'Keehan. 8+9-292p. illus.

(1808). As above - iv,381p.

(1810, 1811 & 1812). As above - Philadelphia: Printed for Mathew Carey. 262p. 5 pls.

(1847). Lewis and Clarke's journal to the Rocky Mountains in the years 1804, 1805, and 1806; As related by Patrick Gass, One of the officers in the expedition. New editions with numerous engravings. Dayton: Published by Ells, Claflin, & Co. 238p.

(1814). Tagebuch einer Entdeckungs-Reise durch Nord-America, von der Mundung des Missouri an bis zum Einfluss der columbia in den stillen ocean, gemacht in den Jahren 1804, 1805, und 1806, auf Befehl der regierung der vereinigten staaten, von den beiden Capitans Lewis und Clarke. Uebersetzt von ph. ch. Weyland. mit einer Charte. Weimer: im Verlage des H.S. privil. x,362p. map. [Accents omitted]

(1810). Voyage des Capitaines Lewis et Clarke, depuis l'embouchure de Missouri, jusqu'a l'entree de la Colombia dans L'ocean Pacifique; Fait dans les annees 1804, 1805, et 1806, par ordre du gouvernement des Estats-Unis: Contenant le journal authentique des evenements les plus remarquables du voyage... Redige en Anglais par Patrice Gass, employe dans l'expedition; Et traduit en Francais par A.J.N. Lallemant... Paris: Chaz Arthus-Bertrand, Libraire. xviii,443p. map. [Accents omitted]

GAUGH, JOHN (C.1801)

With the exception of the following, according to The National Union Catalog Pre-1956 Imprints Gaugh was not published (The American Library Association 1976).

(1801). A collection of narrative pieces from ancient and modern history, with a short introduction to geography for the use of schools. Wilmington, DE: Printed & sold by Peter Brynberg.

Other

Gaugh, Chester B., Jr. (c.1992). The life and times of John G. Gaugh riverman extraordinary, 1826-1913: based on his personal account of experiences as steamboat engineer on the Ohio and Mississippi. Indianapolis, IN: author.

Note. Possibly related to the c.1801 John Gaugh.

GEDDES, JAMES (1763-1838)

Geddes made the preliminary surveys for the Erie Canal in 1808. Brown (1900) labels Geddes a "Representative" in New York. He was born in Carlisle, Pennsylvania and died in Onondaga County, New York. Prior to moving to New York, where he organized a company to manufacture salt, he attended a district school and worked on a farm.

His work record indicates that he was elected a magistrate in 1800, a member of the General Assembly in 1804 and 1821, was appointed an Associate County Justice in 1809, Judge of the Court of Common Pleas in 1812, and was elected a New York Representative for 1813-1815. In 1816 he was appointed Engineer of the Erie Canal, in 1818 Chief Engineer for the Champlain Canal, in 1822 Chief Engineer of the Ohio Canal, in 1827 he helped in locating the Chesapeake and Ohio Canal, and in 1828 he helped locate the Pennsylvania Canal (V.3, p.258).

Ristow (1985) reports that Geddes, an engineer with extensive experience in western New York, was authorized by Simeon DeWitt to make a survey of the Hudson River and the Great lakes. The state legislature of New York had authorized the DeWitt to undertake the survey to investigate the practicality of an inland waterway making the connection of the bodies of water (p.82). The ultimate result of that survey was the Erie Canal (Encyclopedia Britannica 2001, Geddes, James).

(1823). Canal report, made by James Geddes, esq., the engineer employed by the state of Ohio. By authority. Columbus, OH: Printed at the office of the Columbus Gazette by P. H. Olmsted. 14p. 22.5cm.

(1828). ...Chesapeake and Ohio Canal. Letter from the Secretary of War, transmitting estimates of the cost of making a canal from Cumberland to Georgetown. March 10, 1828. Read and laid on the table. Washington: Printed by Gales & Seaton. 100p. 22cm.

(1827). Interesting documents relative to the trade in coal, iron, lime, salt, gypsum, and other minerals on the proposed Chenango Canal. (np).

GIBSON, ROBERT (C.1790)

Pattison (1957) revealed one of four standard elements for doing land surveys and referred to Gibson when discussing "chains" (p.74-75). Prior to the time it had been normal to use ropes and wooden poles to measure land. Gibson provided a reliable standard for measuring land..

(1819). Mathematical tables: difference of latitude and departure; logarithms, from 1 to 10,000; and artificial sines, tangents, and secants. Baltimore: Fielding Lucas, Jun.; T. H. Palmer, printer. Philadelphia: 152p. Tables. 22cm.

(1811 & 1814). The theory and practice of surveying; containing all the instructions requisite for the skilful practice of this art... The whole corrected, newly arranged, and greatly enlarged, with useful selections, and a new set of accurate mathematical tables, by D.P. Adams. New York: E. Duyckinck. 324,184,xiiip. fold. pl. 22cm.

(1821). As above- By James Ryan. v,[2],360,184p. front., XIV fold. pl. 22cm.

(1828 & 1832). As above- New York: J. & J. Harper. 251,184p.

(1833). As above- vii,248,100p. XIV fold. plates. 24cm.

(1834 & 1835). As above- 248,62,100p. illus. 23cm.

(1838 & 1840). As above- Hartford: A. S. Barnes and Co. 2p. 1., [11]-248,73,91p. XIV fold. pl. 21cm.

(1768). A treatise of practical surveying; which is demonstrated from its first principles. 3d ed. Dublin: L. Flin and R. Fitzsimons. 319p. 21cm.

(1785 & 1789). A treatise of practical surveying; which is demonstrated from its first principles. Wherein every thing is useful and curious in that art, is fully considered and explained. Particularly three new and very concise methods for determining the areas of right lined figures arithmetically, or by calculation, as well as the geometrical ones heretofore treated of. The whole illustrated with copper plates. The fourth edition. With alterations and amendments, adapted to the use of American surveyors. Philadelphia: Printed and sold by J. Crukshank. viii,272,[1],90p. incl. tables. 12 pl. 20.5cm.

(1790). As above- Dublin: Cross. 326p.

(1792). As above- Philadelphia: Joseph Crukshank. viii,288p.

(1795). As above- 5th ed. Dublin: Printed by P. Wogan.

(1796). As above- 7th ed. Philadelphia: Crukshank.

(1798). As above- 8th ed. New York: W. A. Davis. 452p. 13 plates. tables. 21cm.

(1802). As above- Dublin: Wogan.

(1803). As above- 8th ed. Philadelphia: Crukshank.

(1803 & 1806). As above- 2d New York edition. New York: Printed for E. Duyckinck, by L. Nichols.

(1806). As above- 3d New York edition.

(1808 & 1810). As above- 9th ed. Philadelphia: Crukshank.

(1810). As above- Dublin: P. Wogan.

(1816). As above- by John D. Craig. Baltimore: F. Lucas, Jun. and Joseph Cushing.

(1818 & 1822). As above- Revised and corrected by John D. Craig.

Other

(1789). Tables of differences of latitude and departure. In Gibson, Robert. A treatise of practical surveying. Philadelphia: (np).

Trotter, Marcus. (1839). Gibson's treatise on land surveying: developing the most approved methods of surveying, viz. by the ciurumferentor [sic], theodolite and chain also trigonometrical surveying, together with a section on leveling exhibiting the sections and cross sections of a rail way: also tables of latitude and departure, natural sines, logarithmic tables, &c... New ed., by Marcus Trotter... Dublin: J. Cumming.

GILBERT, JOSEPH (C.1778)

Gilbert is identified as having worked with Holland to survey British colonies. There is no evidence that he either became a United States citizen or returned to Britain. No other information is to be found about the man.

Other

Jefferys, Thomas. (1778). The American atlas; or, a geographical description of the whole continent of America; wherein are delineated at large, its several regions, countries, states, and islands; and chiefly the British colonies, composed from numerous surveys, several of which were made by order of government. By Major Holland, Lewis Evans, William Scull, Henry Mouzon, Lieut. Ross, J. Cook, Michael Lane, Joseph Gilbert, Gardner, Hillock, &c. &c. Engraved on forty nine copper plates, by the late Mr. Thomas Jefferys...London: Printed by r. Sayer and Bennett.

GILLET, GEORGE (1771-1853)

Ristow (1985) reports that Gillet was born in Hebron, Connecticut. He was appointed surveyor general of Connecticut in 1813 following the publication of the map compiled with Warren. Gillet and Warren appear to have been among the first people to take notice of the growth of turnpike companies (p.97).

Gillet is one of those geographers who may more appropriately be assigned to the second generation of American Geographers, but is included because he was active prior to 1812 (Schwartz & Ehrenberg 1980). He and Moses Warren prepared a state map for Connecticut that was engraved by Abner Reed and published in Hartford by Hudson and Goodwin in 1812.

(1813). Connecticut, from actual survey, made in 1811; by and under the direction of Moses Warren and George Gillet: And by them compiled. published under the authority of the General Assembly, by Hudson & Goodwin. Engraved by Abner Reed, E. Windson. Hartford: (np).

(1833). A system of geometry and trigonometry, with a treatise on surveying...in which, also, the principles of rectangular surveying... are fully explained: with a complete series of mathematical tables, and the necessary explanations. By Abel Flint, A.M. A new ed., enl., with practical illustrations and surveying, including other branches connected therewith... By George Gillet Also, a concise treatise on logarithms, by F.A.P. Barnard... 7th ed. Hartford: Cooke and Co.

GLEASON, BENJAMIN (1777-1847)

Similar to many people included in this early group of geographers, Gleason is known only by one publication. That publication would have contributed to the knowledge United States citizens had about geography.

(1812 & 1814). Remembrancer. Geography, on a new and improved plan, topographically demonstrated, with maps, charts, and globes, by delineation, reference, and instruction...Lectures... Boston: Printed by Munroe & Francis. 148p. folded map. 19.5cm. [1814=2d ed.]

GODON, SILVAIN (C.1774-1840)

The following publications by Silvain Godon would place that person among those in the first generation of American geographers:

(1809). Mineralogical observations, made in the environs of Boston in the years 1807 and 1808, inclosed with a letter to John Davis. Memoirs of The American Academy of Arts and Sciences, 3, pt.1, art. 23, pp.127-154. 1 pl.

(1809). As above- Cambridge, MA: (np). 28p. 26cm.

(1807). Observations to serve for the mineralogical map of the state of Maryland. Transactions of the American Philosophical Society, 6, pt.2, (50), pp.319-323.

GOERCK, CASIMIR T. (C.1803)

Goerck, along with Joseph Mangin, was engaged in 1797 by New York City officials to survey and map the city. Their plan was engraved by Peter Maverick and published in 1803 (Ristow 1985).

(1803). A plan and regulation of the city of New York. New York: (np). 88x102cm. Scale: 400 ft. to 1 inch.

GOODENOW, STERLING (C.1811)

We only know that Goodenow authored the following text. No other publications or activity have been discovered regarding the man.

(1811). A brief topographical & statistical manual of the state of New York: Exhibiting the situation and boundries of the several counties - The cities, towns, mountains, lakes, rivers, creeks, &c... Albany: Published by J. Frary. 36p.

(1822). As above- 2d ed. New York: Published by E. Bliss and E. White: Printed by D. Fanshaw.

GRAHAM, G. (C.1801)

Schwartz and Ehrenberg (1980) identify a map that Graham drew. O. Carleton, J. Callender and S. Hill were the engravers of the map.

(1801). Map of Massachusetts proper compiled from actual surveys made by order of the General Court...By Osgood Carleton. Callendar and Hill, engravers. Boston: B. & J. Loring.

(1802). As above- Drawn by G. Graham; Engraved by J. Callender and S. Hill.

GRANT, J. (C.1784)

Grant is named by Samuel Holland as one of his deputies helping to survey New Hampshire. No other information is to be found about the man.

Other

Holland, Capt. Samuel. (1784). Topographical map of the state of New Hampshire, surveyed under the direction of S. Holland, Surveyor General of Lands for the Northern District of North America; by the following gentlemen, his deputies: T. Wright, G. Sproule, J. Grant, T. Wheeler, & C. Blaskowitz. London: Printed for W. Faden, geographer to the King. In New Hampshire. Geological Survey, 1868-1878. Geology of New Hampshire, Atlas. Scale: 8 miles to an inch.

GRAY, WILLIAM (C.1778)

Gray was a captain in the Fourth Pennsylvania Regiment of the Continental Army. He mapped Colonel William Butler's line of march against the Iroquois at Unadilla in 1778. He acted under orders from Robert Erskine (Harley, et.al. 1978). No additional information has been located about Gray.

U.S. House of Representatives. 22nd Congress, 1st Session. Patents to William Gray, William Davis, and George A. Scherpff (H.rp. 458). Washington: Government Printing Office, 1831-1832. (Serial Set 226).

GRIFFITH, DENNIS (C.1795)

Griffith published, in 1795, the best eighteenth-century map of Maryland (Schwartz and Ehrenberg 1980). It was also the first map of the state, engraved and published by John Vallance of Philadelphia on June 6, 1795. The map was the best in existence of the state until the 1840s publication by J. H. Alexander (p.118).

(1795). Map of the state of Maryland, laid down from an actual survey of all the principal waters, public roads, and divisions of the counties therein. Engraved by John Vallance. Philadelphia: John Vallance. 12°.

(1813). As above- Second edition. 2 sheets.

GRISCORA, JEAN (C.1803)

According to The National Union Catalog Pre-1956 Imprints Griscora was not published (The American Library Association 1976). The following publication is the only information we have about the person. Griscora is included in this text since books of the sort were popular during the 1784 - 1812 period and the information contained in their pages influenced the views and understandings of United States citizens.

(1803). Questions pertaining to geography and the introductory parts of astronomy. Burlington, N.J.: Printed by Stephen C. Vstick.

HALL, JOHN (1783-1847)

Barton (1813) notes Hall worked with Rittenhouse, Ewing, and Nancarrow to survey the turnpike between Philadelphia and Lancaster (pp.360-361). We have found no additional information about the man nor his work.

HARDENBERG, ABRAHAM (C.1800)

Barton (1813) reveals that Hardenberg worked with Ellicott, Rittenhouse, Porter, W. Morris and others during 1787, surveying the boundary line between Pennsylvania and Maryland (pp.241 & 246). The National Union Catalog Pre-1956 Imprints did not list any publications under the name (The American Library Association 1976).

HARDENBURGH, JOHN L.
(C.1791)

Moses DeWitt and Hardenburgh executed surveys of the townships in the Military Tract, north central New York state, which were completed and deposited with the commissioners of the General Land Office in 1791 (Ristow, 1985). The township maps appear not to have been published, but they were used by DeWitt when he compiled his map of the north central part of the state (p.76).

HARRIS, CALEB (C.1795)

Harris produced, in 1795, a classic map of the state of Rhode Island which appears to have been included in Mathew Carey's American Pocket Atlas of 1796 (Schwartz and Ehrenberg 1980, p.215).

(1795). A map of the state of Rhode Island. Providence: Carter & W. Sam'l Hill, sculpt. $21^{3/4}$x$15^{3/4}$in. Scale: $2^{1/2}$ miles to 1 inch.

HARRIS, HARDING (C.1795)

Harris, according to Ristow (1985), drew the Charles Blaskowitz map of Rhode Island (p.99). No additional information about Harris has been found.

Blaskowitz, Charles. (1795). A map of the state of Rhode Island taken mostly from surveys by Caleb Harris. Drawn by Harding Harris and engraved by Samuel Hill. Providence: author. 53X41cm. Scale 1:175,000.

HARRIS, THADDEUS MASON (1768-1842)

Tucher (1985) identifies Harris as a producer of topography. Harris was a clergyman born in Charlestown, Massachusetts and died in Dorchester Massachusetts. He graduated from Harvard in 1787, studied theology, served as librarian of Harvard, and was minister of the First Parish of Dorchester Unitarian Church (Brown 1897, V.3, p.550).

White (1898) goes on to relate that Harris was offered the position of private secretary to General Washington. He accepted the position, but fell ill with smallpox which terminated his secretarial work. In 1791 he was appointed librarian to Harvard. His experience there prompted his theological studies which terminated in 1793 (V.VIII, p.194).

(1806). Account of pawpaw or cowry shells, found in Dorchester. The America Academy of Arts and Sciences, 3, pt. 1, art. 25, pp.159-160.

(1804). Chronological and topographical account of Dorchester. (np).

(1805). The journal of a tour into the territory northwest of the Alleghany Mountains; Made in the spring of the year 1803. With a geographical and historical account of the State of Ohio... Boston: Manning & Loring. 271p. 5 pl. maps.

(1833). Letter to John Vaughn - Relative to a "Description of the ruins of an ancient city discovered near Palenque".

(1789-1796). Massachusetts magazine; or, monthly museum containing the literature, history, politics, arts, manners and amusements of the age. V. 1-8; Jan. 1789-Dec. 1796. Boston: Printed by I. Thomas and E. T. Andrews.

(1793). The natural history of the Bible; or, a description of all the quadrupeds, birds, fishes, reptiles, and insects, trees, plants, flowers, gums and precious stones mentioned in the Sacred Scriptures; collected from the best authorities... Boston: I. Thomas & E. T. Andrews. 272p.

(1820). As above- Boston: Wells & Lilly. 476p.

(1824). As above- London: Tegg. xxxii,430p.

(1825). As above- London: G. B. Whittaker. 238p.

(1835). Sephora; a Hebrew tale, descriptive of the country of Palestine, and of the manners and customs of the ancient Israelites. Abridged and corr. from the London edition. Worcester, MA: Clarendon Harris. 254p. 21cm.

(1810). Topographical and historical sketch of the town of Lunenburgh. (np).

Other

Strum, Christopher Christian. (1801). Beauties of nature delineated; or, philosophical and pious contemplations on the works of nature, and the seasons of the year. Selected from Strum's Reflections, by Rev. Thaddeus M. Harris... 2d ed. Charlestown, MA: Samuel Etheridge.

HARRIS, WILLIAM (1765-1829)

William Harris, S.T.D., was the author of geography books, the fourth President of Columbia College, and a Protestant Episcopal clergyman. He was born on April 29, 1765 in Springfield, Massachusetts and died in New York where he was President of Columbia College.

He descended on his father's side from Robert Harris of Roxbury, Massachusetts, and on his mother's from William Pynchon, the founder of Springfield, Massachusetts. He entered Harvard at the age of seventeen and graduated in 1786. Subsequently he was licensed to preach as a Congregationalist and became principal of an academy at Marblehead. In October 1791 he accepted the Episcopal faith and was ordained a deacon and one week later a priest in New York City. Harris, in 1802 became rector of St. Mark's church in the Bowery of New York City. Near there he established a classical school. In 1811 he was made President of Columbia College, largely because of his affiliation with the Episcopal Church. Johnson and Malone (1928) report: "A contemporary declared him 'a remarkable man, not so much for any one feature of his character as for a happy combination of the several qualities of mind and heart which go to make the effective guide, teacher, and friend of young men'" (V.4, pp.325-326).

Among his accomplishments during his tenure of office at Columbia was the planning and opening of the Columbia Grammar School (Brown 1897, V.3, pp.551-552). White (1893, V.VI. p.344) confirm the above, but declared that Harris had only three publications, only one of which we have located..

(1808). Outlines of geography, natural, civil and political: concisely stating the natural divisions of the terraqueous [sic] globe ... to which is added a brief view of the solar system... Carlisle, PA: Printed by Alexander & Phillips. 161p. folded plate. 21.5cm.

HARRISON, SAMUEL (C.1804)

Ristow (1985) notes that Harrison was the engraver of William Clark's map (p.445). We have found little additional information about the man.

(1814). Map of Lewis and Clark's track across the western portion of North America from the Mississippi to the Pacific Ocean by order of the executive of the United States in 1804, 5 & 6. Philadelphia: Bradford & Inskeep.

Other

Harrison, William (1978). William Harrison, Sr. and sons, engravers, a check list of their works. South Yarmouth, MA: author.

HARRISON, WILLIAM (C.1770-1800)

White (1893) discloses that William Harrison was the grandfather of Gabrial who was born in 1818. The senior Harrison was a bank-note engraver to the Bank of England, and map engraver for the East India Company. The state bank of Pennsylvania, soon after the end of the Revolutionary war, requested that a "first-class" engraver be sent to Philadelphia from the Bank of England. William was the engraver. He remained six months, returned to England and brought his family to Philadelphia where he remained the remainder of his life (V.V, p.218).

A North Carolina map compiled by John Strother and Jonathan Price was prepared in 1789 and engraved by Harrison (Schwartz & Ehrenberg 1980, p.224). The absence of one degree of longitude caused distortions that were ultimately corrected by Henry S. Tanner in his New American Atlas of 1823.

Ristow (1985) quotes a note from the National Intelligencer which reveals that Harrison engraved Lafon's New Orleans map:

> The author has described the counties of the territory according to the new plan of division, which he has executed himself, by order of the legislature, he lays down the latitude and longitude of the principal points. And to render it useful to mariners, he has laid down the soundings along the coast to the Gulph of Mexico.
> The map is four feet wide and three feet high, its latitude extends from the mouth of the Mississippi to the limits of the territory. [The map] will be of the most expert engraving, executed by the celebrated William Harrison, of Philadelphia, and will probably be ready for delivery in the month of May, next. (pp.141-142)

(1806). Carte generale du Territoire d'Orleans. Orleans: (np). [Accents omitted]

HASSLER, FERDINAND RUDOLPH (1770-1843)

Hassler was a Swiss mathematician with extensive surveying experience, born at Aarau, Switzerland (Guthron, 1984) . He (Brown 1897) emigrated to America in 1805, bringing letters of introduction to Albert Gallatin, who secured him an appointment as assistant professor of mathematics at the U.S. Military Academy (V.3, p.582). His skills and intellect gained him membership in the American Philosophical Society. He worked with Albert Gallatin, Secretary of the Treasury, and Robert Patterson, Director of the Mint. They composed a letter to people, mostly members of the Society, regarding suggestions for carrying out a survey of the American coast from Maine to Georgia (p.17). Hassler became the first Superintendent of the U. S. Coast Survey - 1807. The survey was authorized by Congress on February 10, 1807, but was not begun until 1816.

He developed the polyconic map projection, 1825. He also pioneered geodetic triangulation to determine control points for mapping (Encarta Encyclopedia 2001. Hassler, Ferdinand).

(1842). Coast survey and weights and measures. Letter from the Secretary of the Treasury, transmitting a report of Professor Hassler, Superintendent of the Coast Survey and work on the construction of standard weights and measures. December 22, 1842. Washington, DC: 27 Cong., 3d sess. House. Doc. no. 25.

(1842). Coast survey of the United States. Philadelphia: (np). 4p. 23.5cm.

(1839). Letter from the Secretary of the Treasury, communicating the annual report of the Superintendent of the Coast Survey, and of the fabrication of standard weights and measures. Washington, DC: Blair & Rives, printers. 8p. 25cm.

(1825). Papers on various subjects connected with the survey of the coast of the United States. Communicated 3d March, 1820. American Philosophical Society. Transactions, 2, pp.232-420.

(1824). Papers on various subjects connected with the survey of the coast of the U.S. Philadelphia: (np). 4°.

(1828). A popular exposition of the system of the universe, with plates and tables. New York: G. & C. Carvill. xp.230p. 21.5cm.

(1834 & 1836). Principal documents relating to the survey of the coast of the United States; and the construction of uniform standards and weights and measures for the custom houses and states... New York: Printed by J. Windt.

(1838). Report from the Secretary of the Treasury, transmitting a report from the superintendent of the coast survey, and of the fabrication of standard weights and measures, showing the progress of those works during the present year. December 6, 1838. Washing: Blair & Rives, printers. 7p. 23cm.

(1838). Report from the Secretary of the Treasury, transmitting a report of F. R. Hassler, Superintendent of the Coast Survey, and of the fabrication of standard weights and measures. December 12, 1837. Washing: Blair & Rives, printers. 16p. 23cm. (U.S. 25th Cong., 2d sess. Senate. Doc. 79.)

(1842). Survey of the coast of the United States. Further rectification of the facts alleged in the discussion of Congress, in December, 1842. Washington: (np). 7p. 25cm.

Survey of coast, expenditure, American State Papers: Commerce, V.XV (Doc. 214).

Survey of coast of United states, instruments procured by F.R. Hassler, American State Papers: Commerce, V.XV (Doc. 198).

U.S. House of Representatives. 24th Congress, 2nd Session. Hassler, F. R., and assistants, coast survey, for arrears of pay and increase compensation (H.doc. 13). Washington: Government Printing Office, 1836-1837. (Serial Set 301).

U.S. House of Representatives. 27th Congress, 3rd Session. Report concerning accuracy of coast survey, with testimony of F. R. Hassler (H.rp. 170). Washington: Government Printing Office, 1842-1843. (Serial Set 427).

Other

Aarau, Zschokke at. (1882). Translation from the German of the memoirs of Ferdinand Rudolph Hassler. Aarau, Switzerland: 1877. With supplementary documents pub. 1882. Nice: Printing office V. E. Gauthier and Co. 561p. 18.5cm.

Roberdeau, Isaac. (1827). ...Observations on the survey of the sea coast of the United States. Washington, DC: (np).

HEART, JONATHAN (1748-1791)

Heart was a Revolutionary War participant and casualty of the American westward expansion following that war. He was born in Farmington, Connecticut in 1748 and was killed by an Indian musket ball near the source of the Maumee River in Drake County, Ohio.

Heart received a classical education, graduating from Yale College in 1768. He proceeded to teach in several district schools until about 1773. At that time he had accumulated a bit of money, so he started a mercantile business that was quite successful. At the outbreak of the Revolutionary War, he turned over the business to his brother, enlisting in 1775 as a private in Putnam's Battalion of Connecticut Malitia.

Fortune shone on Heart during the war and he was in 1777 commissioned a lieutenant and he married Abigail Riley. By 1780 Heart was promoted to captain. During 1782 he was under Washington at Newburgh and Verplanck's Point, New York when he was appointed Brigade Quartermaster of the Connecticut Brigade. At the conclusion of the war in 1783 he returned to Connecticut to find that his business had failed and his Continental currency was near worthless.

During that period he learned the art of surveying and was commissioned by Congress as a captain in the First American Regiment. He was responsible for building a fort on French Creek, about 150 miles above Pittsburgh, in 1787 named Fort Franklin. In 1789 he worked with Major Doughty in the erection of Fort Washington, near the present Cincinnati.

Heart served under General Harmar in the 1790 campaign against the Indian towns at the junction of the Miami and St. Joseph rivers. At that time he drew a plan of Major Wylly's battle-ground, the old French post, and the Indian towns in the neighborhood.

Captain Heart became Major Heart in 1791. During an engagement with the Indians of the Northwest, against whom he was fighting, he was said to have "handled his men with consummate skill and bravery" (p.X). His skill and bravery may have led to his death as he led a hand full of men attempting to cover the retreat of the army from a battle that left 632 killed and 264 wounded of an original body of 1400 soldiers. Although some may argue that Heart has no place among the first generation of American geographers, he was engaged in expanding what was then known about the geography of America. Samuel F. Haven, Secretary of the American Antiquarian Society in 1856 at Worcester, noted that a plan forwarded by General Parsons to President Stiles of New Haven was drawn by Captain Heart. That plan and an elaborate description is in the Columbian Magazine of May, 1787. Butterfield (1885) offers additional notes:

> In the winter of 1786, President Stiles had written to Dr. Franklin, requesting his opinion of the fortifications at Muskingum, etc., described by General Parsons and others. It may be presumed that the drawing was sent at the same time, as it would be required, of course, to enable Franklin to form a judgment of the nature and object of the structures. Franklin might naturally transfer it to the editors of the magazine in Philadelphia for publication, as a matter of general interest and curiosity. It does not appear as a communication from Captain Heart, and is inserted without note or comment. (p.xi)

While heart was in the army he traveled up and down of the Ohio and other rivers. He made observations about the soil, climate, and value of the lands and communicated that information to his friend, William Judd of Farmington.

(1792). A letter to B. S. Barton... containing observations on the ancient works of art, the native inhabitants, etc., of the western country. Transactions of the American Philosophical Society, 3(25), pp.214-222.

(1885). Journal of Capt. Jonathan Heart on the march with his company from Connecticut to Fort Pitt, in Pittsburgh, Pennsylvania, from the seventh of September, to the twelfth of October, 1785, inclusive; to which is added the Dickinson-Harmar correspondence of 1784-5; the whole illustrated with notes and preceded by a biographical sketch of Captain Heart by Consul Wilshire Butterfiels... Albany, NY: J. Munsell's Sons. xv,94p. 22x18cm.

Other

Butterfield, Consul Willshire. (1885). Journal of Capt. Jonathan Heart on the march with his company from Connecticut to Fort Pitt, in Pittsburgh, Pennsylvania, from the seventh of September, to the twelfth of October, 1784-1785, inclusive; to which is added the Dickinson-Harmar correspondence of 1784-5; the whole illustrated with notes and preceded by a biographical sketch of Captain Heart by Consul Wilshire Butterfield... Albany, NY: J. Munsell's Sons. 94p. 22x18cm.

Imlay, Gilbert. (1797). A topographical description of the western territory of North America: Containing a succinct account of its soil, climate, natural history, population, agriculture, manners, and customs. With an ample description of the several divisions into which that country is partitioned... Illustrated with correct maps of the western territory of North America; of the state of Kentucky, as divided into counties, from actual surveys by Elihu Barker; a map of the Tenasee [sic] government; and a plan of the rapids of the Ohio. 3d ed., with great additions. London: Printed for J. Debrett.

HENRY, ROBERT (C.1792)

Brown (1901) identifies a man by that name who was an "educator" born in Charleston, South Carolina in 1792. He attended and graduated from the University of Edinburgh in 1814, returning to the United States in 1816. It is possible the book belonged to this Robert Henry, but its authorship is uncertain (V.IV, p.21).

The name, Robert Henry, is hand scribed at the top of the page, so it may be the name of the book's owner. Authorship noted is:

By A Friend to Youth. (1803). An easy introduction to geography, with a brief view of the solar system, &c. By way of question and answer. Adapted to the use of schools and private tuition. [Following the authorship]

> "Next to Religion, what study is so important to youth, as the study of the Earth we inhabit, and of that glorious System, of which our Earth forms a part?" The shining worlds above-In beauteous order stand, Or in swift courses move-By his supreme command. Psalm 148, Watts' ver. Charleston, SC: Printed for the editor by J. J. Evans & Co.

[Found in the South Caroliniana Library at the University of South Carolina, Columbia.]

There was a Robert Henry (1718-1790) who published a number of histories, primarily about Great Britain and published in that nation. Another Robert Henry (1765-1863) is credited with publishing the following:

(1891). Narrative of the battle of Cowan's Ford, February 2st, 1781, by Robert Henry, and narrative of the Battle of King's Mountain, by Captain David Vance. Manuscript preserved by Robert Henry, Esq... Greensboro: NC: Reece & Elam, printers. 50p. 22.5cm.

(1899). Robert Henry's narrative. Historical papers of the Trinity College Historical Society. Durham, NC: Ser. III,16-24.

The final candidate for this geography publication is Robert Henry (1792-1856), although he was a citizen of South Carolina, he would have been only 11 years old at the time of its publication. His publications are dated from 1820 to 1850 and none of those are related to geography. Therefore, it is most likely that the noted publication is by someone who may never be known to those of us who seek to know who actually produced the little volume.

HILL, SAMUEL (C.1792)

Hill, in 1792, engraved the earliest folio plan of Washington (Schwartz & Ehrenberg 1980). Later that year, "The original engraving of Ellicott's survey by James Thackara and John Vallance was issued as a separate folio in Philadelphia" (p.212). Osgood Carleton's 1795 map of Maine depicted towns, rivers, lakes, roads, mills, iron works, and public buildings along with the distances from Boston and each county seat and was engraved by Hill and J. Callender (Schwartz & Ehrenberg 1980, p.215). Carleton enlisted the support of the Massachusetts State Legislature which resulted in a resolution in 1794 that directed each town to prepare an accurate town map (p.219).

(1801). Map of Massachusetts proper compiled from actual surveys made by order of the General Court... By Osgood Carleton. Engraved by Callendar and Hill. Boston: B & J Loring.

(1794). Plan & elevation of the Tontine Crescent, now erecting in Boston. In Massachusetts Magazine, 6, part 2, opposite p. 67.

(1795). A S.W. view of the college in Providence, together with the President's house & gardens. Providence: (np).

(1794). A view of the Green in Lexington where the British troops first fired on the Americans in 1775. In Massachusetts Magazine, 6(1), January.

(1789). View of the triumphal arch and colonnade, erected in Boston, in honor of the President of the United States [Washington], Oc. 24. In Massachusetts Magazine, 2, facing p. 3. Jan., 1790.

Other

Ellicott, Andrew. (1792). Plan of Washington, D.C. Engraved by Samuel Hill. Boston: Samuel Hill.

HILLS, JOHN (C.1765)

Hills was in the British Army prior to his discharge in 1784 at which time he decided to remain in the new United States. First settling at Princeton, New Jersey, in 1785 he relocated to New York City and from there he moved to Philadelphia. Once in Philadelphia he published a map of the city that was engraved by John Cooke of Hendon, Middlesex, near London. His map of the city published in 1808 was engraved by William Kneass (Ristow 1985). On at least one of his maps he refers to himself as a surveyor and draughtsman.

(1889). A collection of plan's &c. &c. &c. in the province of New Jersey... (np). 151p. 20 colored maps.

Note. In manuscript maps showing Sir Henry Clinton's operations in the Jerseys; by various draftsmen including John Hills, A. Dunham, J. Rue, J. Fisher, James Grant, T. Milliadge, B. Morgan, A. Sutherland, J. Williams, A. Dennis.

(1784). Plan of the attack of the forts Clinton & Montgomery, upon Hudsons River, which were stormed by His Majestys forces under the command of Sir Henry Clinton, K. B. on the 6th of Octr. 1777. Drawn from the surveys of Verplank, Holland & Metcalfe by John Hills. London: Wm. Faden. 67x52cm.

(1857). Plan of the city of New York and its environs...surveyed in 1782 and drawn in 1785. New York: Engraved for Valentine's Manual.

(1797). A plan of Philadelphia and its environs surveyed by John Hills in the summers of 1801, 2, 3, 4, 5, 6, & 7. (np).

(1808). A plan of the city of Philadelphia and environs. (Shewing [sic] the improved parts.). Engraved by John Cooke, of Hendon, Middlesex, near, London, Phil. (np). $37^{1/2}$x$27^{7/8}$in.

(1798). A plan of the city of Philadelphia and environs, (Shewing [sic] the improved parts.) is dedicated to the Mayor, Aldermen, and citizens thereof, by their most obedient servant, John Hills, surveyor and draughtsman, May 30th, 1796. Republished. London: J. Boydell. 72x106cm. Scale: 600 ft. to an inch.

(1881). A plan of the city of Philadelphia and environs. (Shewing [sic] the improved parts.) May 30th, 1796. Republished. Philadelphia: Sam L. Smedley. 26x$36^{1/4}$in.

(1796). This plan of the city of Philadelphia and its environs, shewing [sic] the improved parts. Philadelphia: author.

(1778). A plan of part of the provinces of Pennsylvania and east & west New Jersey, shewing the operations of the Royal Army under the command of their excellencies Sir Willm, Howe & Sir Henry Clinton, Kts. B. from the landing at Elk River in 1777, to the embarkation at Navisink in 1778. London: Willm. Faden. 23x73cm.

(1784). A plan of the surprise of Stoney Point, by a detachment of the American Army commanded by Brigr. Genl. Wayne, on the 15th July, 1779, also of the works erected on Verplanks Point, for the defence of Kings Ferry, by the British forces on July, 1779. From surveys of Wm. Simpson and D. Campbell by John Hills. London: Wm. Faden. 50x70cm.

(1785). A plan of York Town and Gloucester in the province of Virginia, showing the works constructed for the defence of those posts by the British Army under the command of Lt. Genl. Earl Cornwallis; together with the attacks and operations of the American and French forces commanded by Genl. Washington and Count Rochambeau, to whom the said posts were surrendered on the 17th October 1781. From actual survey in the possession of Jno. Hills. London: Wm. Faden. 71x54cm.

(1780). Sketch of the position of the British forces at Elizabeth Town Point after their return from Connecticut Farm in the province of East Jersey under the command of his excellency. Lieut. Gen. Knyphausen on the 8th June 1780. London: Willm. Faden. 62x52cm.

(1784). Sketch of the skirmish at Petersburg, between the Royal Army under the command of Major Genl. Phillips, and the American army commanded by Major Genl. Stewben, in which the latter were defeated, April 25th 1781. London: Wm. Faden. 27x36cm.

(1777). Sketch of the surprise of German Town by the American forces commanded by General Washington, October 4th 1777. London: Wm. Faden. 46x53cm.

(1784). As above- 50.5x57.5cm.

HOLLAND, CAPT. SAMUEL (1728-1801)

Barton (1813) notes at several points that Holland was an associate of David Rittenhouse as well as Ellicott. He worked with Rittenhouse to establish the north-eastern corner of Pennsylvania (pp.238 & 241).

(nd). Carte du cour du fleuve de Saint Laurent en Canada. Boston, New York, St. Lawrence: (np). 38x148cm. Extension 12x14cm. [Accents omitted]

(1798). A chart of the coast of New England, from the South Shoal to Cape Sable, including Georges Bank. From Holland's actual surveys. Boston: W. Norman. 39.5x32.5inches. Scale: none.

(1935). Holland's description of Cape Breton Island and other documents. Compiled with an introduction by D. C. Harvey, archivist. Halifax, N.S.: Board of trustees of the public archives of Nova Scotia. 168p.

(nd). Journal of William Amherst in America, 1758-1760. (np).

(1794). A new and correct chart of the coast of New England and New York... London: (np).

(nd). Plan and project for fortifying the high commanding ground at the church of Dechambeaux, as also the Island of Richlieu. (Inset) Plan of the projected forts... Samuel Holland, capt of the 60 Rt. (np).

(1778). Plan de l'Ile de St. Jean au nord e l' Acadie et dans le Sud du Golfe de St. Laurent Saivant l'arpentage du Captain Anglois Holland... (np). 62x46.5cm. [Accents omitted]

(nd). Plan No. 1 to explain the report of the state of defense of the fortifications of Quebec. Shelburne Papers, 64, p. 196.

(nd). Plan No. 2 shewing [sic] the ground whereon the Citadel is proposed to built the ground lines of the present fortifications are colour'd red. . Shelburne Papers, 64, p.192.

(nd). Plan No. 3 of the citadel completely finished... Shelburne Papers, 64, p.189.

(nd). Plan No. 4 shewing [sic] the Bastion... and profiles of the whole by the different lines pointed out. Shelburne Papers, 64, p.187.

(nd). Plan (No. 5) shewing [sic] the half bastian with its foundations and the rampart towards the river side (as far as the Loophole-wall beginning at E. in Plan (No. 3) with its Casemates...and profile out by the yellow lines thru the work. . Shelburne Papers, 64, p.183.

(nd). Plan No. 6 shewing [sic] the profiles cut thru' the different works by the yellow lines seen on Plan No. 3. . Shelburne Papers, 74, p.177.

(1776). A plan of the city of New York. Drawn by Major Holland, Surveyor General. New York: (np). $12^{3/4}$x$14^{3/4}$inches. Scale: 469.2 feet to 1 inch.

(c.1777). A plan of the island of St. John with the divisions of the counties, parishes...likewise the soundings round the coast... London: W. Faden. 35.5x70.5cm.

(c.1768). The provinces of New York, and New Jersey; with part of Pennsylvania, and the governments of Trois Rivieres, and Montreal: Drawn by Capt. Holland. Engraved by Thomas Jefferys. London: R. Sayer & T. Jefferys. 67x50.5cm.

(1775 & 1776). As above- London: Sayer & Bennett. 51.5x135cm.

(1777). As above- ...Drawn by Major Holland... Corrected and improved from the original materials, by Governor. Pownall... Frankfort upon the Mayn: H. L. Broenner. 131x51cm.

(1776). The Seat of War in the Northern Colonies Newfoundland to Maryland. General map of the Northern Colonies: 1765-1774. London: (np).

(1784). A topographical map of the province of New Hampshire, surveyed...unto Samuel Holland Esq. by the following gentlemen his deputies... London: Printed for William Faden. 79x117.5cm.

(c.1790). A topographical map of the State of New Hampshire. (np). 20.5x14.5inches. Scale: none.

(1784). Topographical map of the state of New Hampshire, surveyed under the direction of S. Holland, Surveyor General of Lands for the Northern District of North America; by the following gentlemen, his deputies: T. Wright, G. Sproule, J. Grant, T. Wheeler, & C. Blaskowitz. London: Printed for W. Faden, geographer to the King. In New Hampshire. Geological Survey, 1868-1878. Geology of New Hampshire, Atlas. Scale: 8 miles to an inch.

Other

Jefferys, Thomas.(1778). The American atlas; or, A geographical description of the whole continent of America; wherein are delineated at large, its several regions, countries, states, and islands; and chiefly the British colonies, composed from numerous surveys, several of which were made by order of government. By Major Holland, Lewis Evans, William Scull, Henry Mouzon, Lieut. Ross, J. Cook, Michael Lane, Joseph Gilbert, Gardner, Hillock, &c. &c. Engraved on forth-nine copper-plates, by the late Mr. Thomas Jefferys... London: Printed by R. Sayer and Bennett.

HOOKER, WILLIAM JACKSON (1785-1865)

Schwartz and Ehrenberg (1980) report that Hooker was an engraver associated with the production of pocket guides to New York City. He also engraved many of the charts published by Edmund Blunt, from 1809 onward (p.220).

(1819). Analysis of the authorities upon which the dangers have been inserted in Blunt's new chart of the Atlantic, or Western Ocean. To which is added, an accurate table of the variations of the magnetic needle, as observed in different parts of that sea; with the dates of the respective observations... With an exhibition of facts, showing the progress of ice islands from Greenland to Newfoundland, &c. &c. Compiled by Edmund M. Blunt... New York: Printed for William Hooker, by J. Seymour. 76p. 22cm.

(1820). Charts and nautical books, published by Edmund M. Blunt, for William Hooker... New York: Printed by Clayton & Kingsland. 18p.

(c.1838). The coast of the United States of North America from New York to St. Augustine: Drawn and regulated according to the latest surveys and astronomical observations. By Edmund Blunt. Engraved and printed by W. Hooker. New York: E. & G. W. Blunt.

(1827). Hooker's map of the village of Brooklyn in the year 1827. For Bishop's Manual of the Corporation, for 1861... New York: Lith. of A. Brown & Co. 42x50cm.

(1824). Hooker's new pocket plan of the city of New York. Compiled & surveyed by William Hooker, A.C.S.A. Hydrographer & Engraver. Drawn, engraved, printed, published and sold by W. Hooker. New York: Hooker. 12.5x15inches.

(1825). Hooker's new pocket plan of the city of New York. Compiled & surveyed by William Hooker. New York: W. Hooker. 31x38cm.

(1836). New pocket plan of city of New York. New York: (np).

HOOPS, ADAM (C.1786)

Pattison (1957) notes that Hoops became Surveyor for Pennsylvania when Hutchins returned to Pittsburgh in June of 1786. Congress authorized resumption of the surveys of the east-west line which was begun, but not completed in 1785. Colonel Hoops was a professional surveyor, a friend of Hutchins, and a land speculator (p.134).

Other

(nd). Program of the unveiling of the memorial to Major Adam Hoops and the patriots of the War for Independence. New York: Daughters of the American Revolution, Olean Chapter.

(1821). Statement, &c. in regard to the purchase from the Holland Land Co. of 20,000 acres of land lying on the waters of the Allegany River, within the state of New York. (np). 40p. 21 cm.

HOWELL, READING (C.1790)

Howell used some of the information gathered by Nicholas Scull and his grandson, William Scull in addition to personal surveys to produce several maps of Pennsylvania (Ristow, 1985). Howell was a landowner and surveyor, "but little is known about his life or career" (p.108). He worked with several other surveyors to make drafts of the Delaware and Lehigh rivers. He was appointed, April 1790, along with two others to explore land near the headwaters of the Susquehanna, Lehigh, and Schuylkill rivers, for which he was payed a sum of three hundred pounds by the state assembly.

(1811 & 1816). Map of Pennsylvania. Philadelphia: (np).

(1834). ...A map of Pennsylvania, and the parts connected therewith, relating to the roads and inland navigation, especially as proposed to be improved by the late proceeding of Assembly. Washington, DC: Gales and Seaton. 46x66.5cm.

(1817). A map of the state of Pennsylvania. Philadelphia: Kimber & Sharpless. $20^{1/4}$x$32^{3/4}$inches. Scale: 10.3 miles to 1 inch.

(1792). A map of the state of Pennsylvania, with part of Lake Erie and Presque Isle, also, the Rivers Susquehanna, Ohio, Alleghany, Monongahela, and lesser streams, the mountains, roads, portages, etc. Philadelphia: author. 63x36cm.

(1792). Map of the state of Pennsylvania. London: Sold by James Phillips. 94x160cm.

(1815, 1816). Commissioners appointed to explore the head-waters of the rivers Delaware, Lehigh and Schuylkill, and the north-east branch of Susquehanna. ... The report of the Commissioners appointed to explore the head-waters of the rivers Delaware, Lehigh and Schuylkill, and the north-east branch of the Susquehanna. To His Excellency Thomas Mifflin, esquire, president of the Supreme Executive Council of the state of Pennsylvania. In Pennsylvania. General Assembly, 11815-16. House of Representatives. Journal... Harrisburg: General Assembly. 21.5cm. pp.19-27.

HUBBARD, JOHN (1759-1810)

Hubbard was multitalented, authoring material on a variety of subjects: reading, music, and geography. He is identified as an educator, lawyer, jurist, and author. He was born on August 8th, 1759, in Townsend, Massachusetts and died in 1810 in Hanover, New Hampshire. Hubbard was Professor of Mathematics and Natural Philosophy at Dartmouth University from 1804 until his death (Herringshaw 1898).

(1803). The rudiments of geography, being a concise description of the various kingdoms, states, empires, countries, and islands in the world... with an introduction explaining the astronomical table of the most important events which have happened from the creation of the world to the present day... Walpole, N.H. Printed for Thomas and Thomas by David Newhall. 232p. folded map. 17.5cm.

(1805,1807). As above- ... With an introduction explaining the astronomical part of geography. To which is added, a chronological table of the most important events which have happened from the creation of the world to the present day. Walpole, NH: Printed for Thomas & Thomas by George W. Nichols. 240p. 18cm..

(1808). As above- 4th ed. Troy, NY: Published by Wright, Goodenow & Stockwell.

(1811). As above- Walpole, NH: C. Felch.

(1814). As above- 6th ed. Barnard, VT: Published by Joseph Dix, I. H. Carpenter, printer. 240p. 16.5cm.

Other

Hubbard, William David, & Carter, Oliver Shelton. (1991). John Hubbard descendants from Body Camp Creek, Bedford County, Virginia. Roanoke, VA: Swift Print. 101 leaves. 29cm.

HUNTER AND BEAUMONT (C.1798)

Ristow (1985) lists Hunter and Beaumont as publishers of the Ohio Navigator. Zadok Cramer published his version in 1801 (p.236).

(1798). Ohio navigator. Frankfort, KY: Hunter & Beaumont.

HUTCHINS, THOMAS (1730-1789)

Hutchins is identified by Frederick Charles Hicks (1778) as "Geographer to the United States of America" appointed by Congress on November 12, 1782 (p.27) (The Biographical Dictionary 2001, Hutchins, Thomas). This appears to disagree with Brown's assertion that Morse was "the first American Geographer." Hutchins was born in Monmouth County, New Jersey in 1730 and died in 1789.

During that time he served with the British army in a variety of capacities, but primarily as a surveyor/map maker (Hicks 1778). Shortly thereafter hostilities broke out between the British American colonies and the mother country. Hicks notes that Hutchins declared himself a patriot during his stay in London:

> The outbreak of the war between Great Britain and the United Colonies while he was in London brought upon Hutchins the crisis of his life. He had risen to a post of honor in the British army

> solely through merit. He had hitherto exerted himself in a cause with which he sympathized. Now he was called upon either to relinquish the fruits of his life's endeavors, or to accept honors for fighting against the country of his birth. In this exigency he sacrificed himself unfaltteringly at the patriotic shrine. The persecutions to which he was subjected and the fortitude with which he bore them are related in the memorial which he presented to Benjamin Franklin, taking up the narrative in August, 1779: ...(p.23).

In 1789 Hutchins arrived in Pittsburg and in the company of Rev. John Heckewelder traveled to Marietta where he became ill. The illness described as "a gradual failing of the nerves, and an almost insensible waste of the constitution" was to be the death of Hutchins in Pittsburg on April 28, 1789. Among his many accomplishments, he supplied the maps and plates for the Account of Bouquet's Expedition, by Dr. William Smith, published in Philadelphia 1765 (White 1907, V.IX, p.267).

(1781). Description topographique de la Virginie, de la Pensylvanie, du Maryland et de la Caroline Septentrionale: Contenant les rivieres d'Ohio, Kenhawa, Siota, Cherokee, Wabash, des Illinois, du Mississippi, &c.; Le climat, le sol, les productions, tant animales que vegetales ou minerales; les montagnes, les rivieres, chemins... Accompagnee d'un plan des sauts de l'Ohio... Plus, un supplement, qui contient le journal de Patrice Kennedy's... Traduit de l'Anglois. Paris: Chez Le Rouge, geographe. 68,[4]p. 2 folded maps. [Accents omitted]

(January 28, 1786). Description of a remarkable rock and cascade, near the western side of the Youghiogeny River, a quarter of a mile from Crawford's Ferry, and about twelve miles from Uniontown, in Fayette County, in the state of Pennsylvania. Transactions of the American Philosophical Society, 2(4), pp.50-51.

(c.1787). A new map of the western parts of Virginia, Pennsylvania, Maryland and North Carolina; comprehending the River Ohio, and all the rivers which fall into it; part of the River Mississippi, the whole of the Illinois River, Lake Erie; part of the Lakes Huron, Michigan, etc. and all the country bordering on these lakes and rivers. Boston: By Thos Hutchins, Captain in the 60 Regiment of Foot, and Surveyor General to the United States. 33x40.5inches. Scale: (computed), 21.05 miles to 1 inch.

(1778). A topographical description of Virginia, Pennsylvania, New York, and North Carolina, Comprehending the rivers Ohio, Kenhawa, Sioto, Cherokee, Wabash, Illinois, and Mississippi, &c. The climate, soil and produce, whether animal, vegetable, or mineral... With a plan of the rapids of the Ohio... And an appendix containing Mr. Patrick Kennedy's journal... London: Printed for the author, and sold by J. Almon. [2]ii67[1]p. 2 folded maps.

(1787). As above- Boston: John Norman.

(1795). An historical narrative and topographical description of Louisiana and West Florida -Extracts. Augusta, GA: John Erdman Smith.

(1807). Hutchins reviewed. An almanac 1808. New York: Printed & sold by Smith & Forman.

(1781). Description topographique de la Virginie, de la Pennsylvanie, du maryland et de la Caroline Septentrionale: Contenant les rivieres d''Ohio, Kenhawa, Sioto, Cheroke'e, Wabash, des Illinois, du Mississippi &c...pub. par Thomas Hutchins...accompagne'e d'un plan des sauts de l'Ohio, d'un autre de tous les villages du canton des Illinois; une table des distances en milles entre le Fort-Pitt & l'embouchure de l''Ohio dans le Mississippi; plus, un supple'ment, qui contient le journal de Patrice Kennedy's, sur la rivie're des Illinois, & un e'tat ve'ritable des diffe'rentes nations & tribus des Indiens, avee le nombre des combattans...(Tr.). de L'anglois. Paris: Le Rouge. Also published in Boston: Printed and sold by John Norman. 68p. 3 maps. 19.5cm. Translated by G. I. LeRouge. [Accents omitted]

(1765). A general map of the country on the Ohio and Muskingham shewing [sic] the situation of the Indian-towns with respect to the army under the command of Colonel Henry Bouquet. $14^{1/8}$x$19^{1/8}$in. In Henry Bouquet. An Historical Account of the Expedition... Philadelphia: (np).

(1784). An historical narrative and topographical discussion of Louisiana and West Florida, comprehending the river Mississippi with its principal branches and settlements and the rivers Prel, Pascagoula...&c.... Philadelphia: Robert Aitken. 94p. 30cm.

(1795). As above- Augusta, GA: John Erdman Smith.

(1764). An historical account of the expedition against the Ohio Indians, in the year 1764. (Supposed author, see William Smith, 1727-1803).

(1778). A new map of the western parts of Virginia, Pennsylvania, Maryland and North Carolina, comprehending the River Ohio...Part of the River Mississippi, the whole of the Illinois River... London: author. map. 4 sheets. 87.5x107.5cm.

(c.1787). A new map of the western parts of Virginia, Pennsylvania, Maryland and North Carolina; comprehending the River Ohio, and all the rivers which fall into it; part of the River Mississippi, the whole of the Illinois River, Lake Erie; part of the Lakes Huron, Michigan, etc. and all the country bordering on these lakes and rivers. Boston: By Thos Hutchins, Captain in the 60 Regiment of Foot, and Surveyor General to the United States. 33x40.5inches. Scale: computed, 21.05 miles to 1 inch.

(c.1787). Plat of the seven ranges of townships being part of the territory of the United States N.W. of the river Ohio which by a late act of Congress are directed to be sold...W. Barker sculp... Surveyed in conformity to an ordinance of Congress of May 20th, 1785. Under the direction of Thomas Hutchins late Geographer to the United States. Philadelphia: Matthew Carey. 24.2x13.6in. (over all 26.9x18.7in.).

(1809). As above- "Scale of four miles to an inch." Listed as map no. 23, in Carey's American Atlas. Philadelphia: M. Carey.

(1778). Proposals for publishing by subscription a map of the interior parts of North America. (np). 8°.

(1781). Partie occidentale de la Virginie, Pennsylvanie, Maryland et Caroline Septie., la rivie're d'Ohio, et toutes celles qui s'y jettent, partie de la rivie're des Illinois... Paris: Le Rouge. map. 47x58.5cm. Scale: 40 miles to 1 inch. [Accents omitted]

(1762). A tour from Fort Cumberland northwestward round part of the lakes Erie, Huron and Michigan...to Fort Pitt. (np). $21^{1/2}$x$16^{3/4}$in.

(1778). A topographical description of Virginia, Pennsylvania, Maryland, and North Carolina, comprehending the rivers Ohio, Kenhawa, Sioto, Cherokee, Wabash, Illinois, Mississippi, &c. the climate, soil and produce, whether animal, vegetable, or mineral; the mountains, creeks, roads, distances, latitudes, &c. and of every part laid down in the annexed map. Published by Thomas Hutchins, captain in the 60th Regiment of Foot. With a plan of the rapids of the Ohio, a plan of the severl [sic] villages in the Illinois country, a table of the distances between Fort Pitt and the mouth of the Ohio, all engraved upon copper. And an appendix, containing Mr. Patrick Kennedy's journal up the Illinois River, and a correct list of the different nations and tribes of Indians, with the number of fighting men, &c. London: Printed by author, sold by J. Almon. 143p. 2 folded plans. 21cm.

(1787 & 1797). As above- Boston: Printed and sold by John Norman.

(1904). As above- Cleveland, OH: The Arthur H. Clark Company.

(1765). A topographical plan of that part of the Indian-Country through which the army under the command of Colonel Bouquet...See Hutchins - A general map of the country on the Ohio. Philadelphia: (np).

(1784). (co-author, Thomas Mendenhall). Traverse tables of difference at latitude and departure, constructed to every quarter of a degree of the ...Particularly suited to the purposes of land surveying. Philadelphia: Printed & Sold by Joseph Curkshank.

(1878). Western Pennsylvania in 1760. Pennsylvania Magazine of History, 2, p.149.

Other

Bond, Beverly W., Jr. (Ed.). (1942). The courses of the Ohio River taken by Lt. T. Hutchins, anno 1766, and two accompanying maps. Cincinatti: Historical and Philosophical Society of Ohio.

Gordon, Capt. Harry. (1788). Map of the Ohio River. np - Table of distances from Fort Pitt to the mouth of the Ohio.

Hicks, Frederick Charles (Ed.). (1904). A topographical description of Virginia, Pennsylvania, Maryland, and North Carolina; reprinted from the original ed. of 1778. Cleveland: The Burrows Brothers Company; and Cleveland: Arthur H. Clark Co. 143p. foulded map. 2 plans. 25cm.

Imlay, Gilbert. (1797). A topographical description of the western territory of North America: containing a succinct account of its soil, climate, natural history, population, agriculture, manners, and customs. With an ample description of the several divisions into which the country is partitioned...Illustrated with correct maps of the western territory of North America; of the state of Kentucky, as divided into counties, from actual surveys by Elihu Barker; a map of the Tenasee[sic] Government; and a plan of the rapids of the Ohio. 3d ed. London: Printed for J. Debrett.

Historical narrative of Louisiana and West Florida by Thomas Hutchins.

Quattrocchi, Anna Margaret. Thomas Hutchins, 1730-1789. Unpublished Ph.D. dissertation. Dept. of History, University of Pittsburgh.

Tucker, Sara Jones. (1942). Indian villages of the Illinois Country. Springfield, IL: State of Illinois.

A reproduction of Hutchins' map, tinted is Plate XXIX.

HUTTON, ISAAC (C.1794)

Hutton was the engraver for a map of Albany, New York that was done by Simeon De Witt under the direction of the Mayor, Aldermen, and Commonality (Ristow 1985, pp.78 & 250).

De Witt, Simeon. (1794). A plan of the city of Albany. New York: author.

IDDINGS, JAMES (C.1804)

The National Union Catalog Pre-1956 Imprints listed only the following publication for Iddings (The American Library Association 1976).

(1804). The monitor's instructor, or a system of practical geography, of the United States of America, in particular ... the whole rendered familiar for the use of schools. Wilmington, DE: Printed by William Black. 323p. 15cm.

Part of the text is in verse.

IMLAY, GILBERT (1755-1796)

Imlay was born in Monmouth County, New Jersey in 1754 and buried on the Island of Jersey in 1828. His life appears to have been quite an adventure.

He was a lieutenant in the American army. It appears from the title by which he was known, captain, that he attained that rank during the war. At the end of the war he moved to Kentucky and purchased a tract of land. While in Kentucky he became Deputy surveyor and engaged in land speculation. Financial problems, probably related to his speculations on land, forced him to leave the state and perhaps North America.

The next record of Imlay is the publication in London of his A Topographical Description of the Western Territory of North America. By March of 1793 he was engaged in French political affairs. Imlay allied himself with Brissot and others who were engaged in scheming to seize Louisiana from Spain. During that time he addressed at least two communications regarding the project to the United States Committee of Public Safety: Observations du Captain Imlay (translated in Annual Report of the American Historical Association for 1896, I, pp.953-954).

Once again Imlay found himself in financial difficulties that influenced his liaison with Mary Wollstonecraft. She used her resources to straighten out his affairs in Scandinavian countries. He was ungenerous to Wollstonecraft and she attempted suicide at least twice (Johnson & Malone 1928, V.3, pp.461-462).

Ingpen (1908), in his publication of Wollstonecraft's love letters, provides further insight into the character of Imlay. A letter to her sister describes Wollstonecraft's initial impression of Imlay, "You know that I am safe, through the protection of an American, a most worthy man who joins to uncommon tenderness of heart and quickness of feeling, a soundness of understanding, and reasonableness of temper rarely to be met with. Having been brought up in the interior parts of America, he is a most natural, unaffected creature" (p.xviii). Wollstonecraft appears, however, not to have been formally wed to Imlay. They, none the less, had a daughter, Frances (Fanny), born on May 14, 1794. Unfortunately, Imlay provided no evidence that he was concerned with her prosperity. Wollstonecraft attempted to persuade Imlay to return to her and to provide some support for her and their child, but support did not materialize.

(1793). A description of the Western Territory of North America; containing a succinct account of its climae [sic], natural history, population, agriculture, manners and customs, with an ample description of the several divisions into which that country is partitioned, and an accurate statement of the various tribes of Indians that inhabit the frontier country. To which is annexed, a delineation of the laws and government of the state of Kentucky. Tending to show the probable rise and grandeure of the American empire. In a series of letters to a friend in England. Dublin: Printed for W. Jones. 249p. 17cm.

First edition, London, 1792, published under the title: A topographical description of the western territory of North America.

(1793). ...Nachrichten von dem westlichen lande der Nordamerikanischen freistaaten, von dem klima, den naturaprodukten, der volksmange, den sitten und gebrauchen desselben, nebst einer angabe der indianischen volkerstamme, die an den grazen wohnen, und einer schilderung von den gesetzen und der regierung des staates Kentucky. In briefen an einen freund in England. Aus dem englischen ubers., mit vielen anmerkungen und bestimmungen der naturlichen produkte, von E. A. W. Zimmerman... Berlin: In der Vossischen Buchhandlung. 168p. 21cm. [Accents omitted]

(1792). A topographical description of the Western Territory of North America: containing succinct account of its soil, climate, natural history, population, agriculture, manners, and customs. With an ample description of the several divisions into which that country is partitioned, and an accurate statement of the various tribes of Indians that inhabit the frontier country. To which is annexed, a delineation of the laws and government of the state of Kentucky. Tending to show the probable rise and grandeur of the American empire. In a series of letters to a friend in England. London: F. Debrett. 247p. 20.5cm.

(1793). As above- ... And an essay towards the topography, and natural history of that important country. By John Filson... By George [i.e. Gilbert] Imlay. 2d ed., London: Printed for J. Debrett. 433p. 2 folded maps. 22cm.

(1797). As above- ... By Gilbert Imlay, A captain in the American Army during the war, and Commissioner for laying out lands in the Back Settlements. 3d ed., with additions. London: F. Debrett.

Note: [opposite title page - foldout map] "A map of the Western Part of the Territories belonging to the United States of America..."

Other

Wollstonecraft, Mary. (1908). The love letters of Mary Wollstonecraft to Gilbert Imlay, with a prefatory memoir, by Roger Ingpen. Illustrated with portraits. Philadelphia: J. B. Lippincott Co., etc.

Wollstonecraft, Mary. (1879). Mary Wollstonecraft; letters to Imlay, with prefatory memoir by C. Kegan Paul. London: C. K. Paul & Co.

IRVING, WASHINGTON (1783-1859)

Irving was a contemporary of those people who can be identified as among the first generation of American geographers. He however does not appear to have published anything of a geographic nature during the period. Also of note is the fact that titles with topography, atlas, geography, do not appear to be among the titles of Irving's books, he did produce numerous books regarding exploration and travel. Only a sample of those is noted here. It is apparent that he had only a superficial knowledge of geography, history, French and Latin: Although the same may be said of many of those who are included in these pages. White (1891) asserts that Irving is, "The Father of American Literature" (V.III, pp.17-19).

It is of note that Irving published in 1809, A History of New York, that was a satire about Swedes, Yankees, colonial historians, Dutch settlers in New Amsterdam, red-breeched Jefferson and his democrats, English, French, and Spanish literature. That volume may be identified as having elements of human geography though it is referred to as "the first great book of comic literature written by an American" (Johnson & Malone 1928, V.3, pp.505-511).

Note. The following are merely single listings of publications that went through numerous editions. The list is only of those publications that may have elements of geography within them.

(1837). The adventures of Captain Bonneville. Portland, OR: Binfords & Mort. 357p. illus. by Robert A. Reynolds. 24cm.

1. The West - Description & travel.
2. Northwestern states - Description & travel.

(1809). A history of New York, from the beginning of the world to the end of the Dutch dynasty. Containing many surprising and curious matters, the unutterable ponderings of Walter the Doubter, the disastrous projects of William the Testy, and the chivalric achievments of Peter the Headstrong, the three Dutch governors of new Amsterdam; being the only authentic history of the time that ever hath been, or ever will be published. By Diedrich Knickerbocker... New York: Inskeep & Bradford; Philadelphia: Bradford & Inskeep. 2v. 18.5cm.

(1837). The Rocky Mountains: or, scenes, incidents, and adventures in the far west; digested from the journal of Capt. B. L. E. Bonneville... and illustrated from various other sources. Philadelphia: Carey, Lea, & Blanchard. 2v. folded maps. 20cm.

1. The West - Description & travel.
2. Northwestern states - Description & travel.

(1828). The history of the life and voyages of Christopher Columbus. New York: F. DeFau. 2v. fronts. plates.

2. America - Description & Exploration - Spanish.

JACKSON, JOHN GEORGE (1777-1825)

John George Jackson was a surveyor of public land in the territory north of the Ohio River, a member of the Virginia House of Burgesses, and a Democratic representative from Virginia in the 8th through the 14th congress. Brown reports (1901) that he was born in Clarksburg, Harrison County, Virginia and died there. He was the son of Col. George Jackson and grandson of John and Elizabeth (Cummins) Jackson, who emigrated from Maryland in 1750. President Monroe appointed him a judge of the Western District of Virginia. He married Polly Payne, the sister of President Madison's wife, Dolly. Following the death of Polly he married the daughter of Governor Meigs of Ohio (V.IV, p.327).

Jackson was one of the deputy surveyors appointed under the Land Act of 1800 to work under the direction of Rufus Putnam (Pattison 1957, p.207). The National Union Catalog Pre-1956 Imprints listed no publications by Jackson (The American Library Association 1976).

JEFFERSON, THOMAS (1743-1826)

Brown (1959) provides a sketch attributed to Jefferson, "A sketch of the new states proposed by Jefferson and others, 1783-1784" (Plate 52; Pattison 1957, p.23-25). His map suggested that states be given the following names: Sylvania, Michigania, Chersonesus, Assenisipia, Metropotamia, Illinoisia, Saratoga, Polypotamia, Pelisipia, and Washington. Congress approved the Jeffersonian states: "Resolve of Congress of April 23d, 1784, 'the land, all of it lying northwest of the Ohio'" (p.127). Although officially approved, the

Jeffersonian states did not come into being because of conflicts resulting from exploiting and vested interests of those already holding lands in those proposed states.

Ristow (1985) identifies Jefferson's engagement with matters geographic. He is mentioned on pages 28, 30-32, 49, and 151. A biography of Jefferson is to be found in White (1891, V.III, pp.1-5).

Jefferson's Notes on the State of Virginia, first published in Paris and London in 1787, describe the geography, climate, population, communities, laws, colleges, manners, and commerce of Virginia (Schwartz & Ehrenberg 1980, p.209).

Census, petition of American Philosophical Society, American State Papers: Miscellanious, V.XXXVII (Doc. 121).

(1797). A memoir on the discovery of certain bones of a quadruped of the clawed kind in the western parts of Virginia. Transactions of the American Philosophical Society, 4(30), pp.246-260.

(1806). Message from the President of the United States communicating discoveries made in exploring the Missouri, Red River and Washita, by Captains Lewis and Clark, Doctor Sibley, and Mr. Dunbar, with a statistical account of the countries adjacent... Feb. 19, 1806... printed by order of the Senate. City of Washington: A. & G. Way, printers. 178p.

(1853). Notes on the state of Virginia. Illustrated with a map, including the states of Virginia, Maryland, Delaware and Pennsylvania. A new edition, prepared by the author, containing notes and plates never before published. Richmond: J. W. Randolph [Chas. H. Wynne, printer]. iv[4]-275[1]p. folded map. 4 maps.

(1785-1786). Notes on the state of Virginia; written in the year 1781, somewhat corrected and enlarged in the winter of 1782, for the use of a foreigner of distinction, in answer to certain queries proposed by him respecting I. Its boundaries... 2. Rivers... 3. Sea ports... 4. Mountains... 5. Cascades and caverns... 6. Productions mineral, vegetable and animal. 23 Histories... Paris: (np). [2]-391p. folded table. illus. map.

With manuscript note from Jefferson to F. Hopkinson similar to that described in Tompkins 96, and without cancels or added material.

Tucker (1990) notes the following while comparing a book by Belknap to the above title, "discussed subjects topically: geography, climate, Indian monuments and relics, commerce, transportation, education, religion, industry, population, literature..." (p.36).

(1787). As above- London: Printed for John Stockdale.

(1788). As above- Philadelphia: Printed and sold by Prichard and Hall.

(1794). As above- Philadelphia: Printed for Mathew Carey. [4]-336p.

(1800). Jefferson's notes, on the state of Virginia; With the appendixes..Complete. Baltimore: Printed by W. Pechin. 194,53p. folded table. illus. maps.

(1801). As above- New York: Printed M.L. & W.A. Davis; For Furman and Loudon.
(1801). As above- Philadelphia: R.T. Rawle, Publisher: John Thompson, printer. [1-4], [1]-374, 373-74, [375]-436, [1], [1]-56p. folded front map. folded plate. 2 maps. folded table.
(1802). As above- Boston: Printed by H. Sprague. 368p.
(1803). As above- Trenton: Printed by Wilson & Blackwell. 363p.
(1825). As above- Philadelphia: H. C. Carey & I. Lea; New York: E. Bliss & E. White; Boston: Wells & Lilly; London: John Miller. [1-4],[7]-344p.
(1829). As above- Boston: Wells and Lilly. [4]-280p.

Other

Lipscomb, Andrew a., and Bergh, Albert E. (Eds.). (1903). The writings of Thomas Jefferson. 20 vols. Washington: The Thomas Jefferson Memorial Association.
Luther, F. B. (April 1885). Thomas Jefferson as a naturalist. Mag. Amer. Hist., 13, pp.379-390.
Mitchill, S. L. (1826). A discourse on the character and services of Thomas Jefferson, more especially as a promoter of natural and physical science, pronounced by request before the N. Y. Lyceum of Natural History, on the 11th day of October, 1826. New York: Carvill. 67p.
Verner, Coolie. (January 1951). The maps and plates appearing with the several editions of Mr. Jefferson's "Notes on the state of Virginia." Virginia Magazine of History and Biography, LIX, 21-33.

JOHNSON, JACOB, REV. (1713-1797)

Johnson (1911), lists over fifty titles in which Jacob Johnson is mentioned. He was born at Wallingbord, Connecticut on April 7, 1713. Little is noted of his early life except that his father wanted him to be educated for the ministry of the Congregational Church. That resulted in him attending Yale College using a Berkeley Scholarship, graduating in 1740.

At the death of his father, in 1749, the younger Jacob received an estate of 1,351 pounds and two slaves. The father had been a successful farmer and like his father a member of the General Assembly. None the less, Jacob had difficulty in 1768 clothing his family.

Johnson participated at the signing of the treaty at Fort Stanwix when the dividing line between the English colonies and the Indian territories were established. For the territory, nearly 1,000 miles long, the Indians were given $10,000 in goods and money (p.16). The treaty and other related agreements angered Johnson who viewed the rights of the Indians as violated.

Johnson married Mary Giddings with whom he had nine children, four of whom grew to adulthood. Lydia (1756-1781) was married to Colonel Zebulon Butler. Jacob Williamson (1765-1807) was married to a lady named Bailey. Jehoiada Pitt (1767-1840+) in 1840 married Hannah Frazer. Christiana Olive (1769-1831) married William Russell, Jr.

The early spring of 1797 saw the approach of the end of Johnson. His grandson noted that Johnson had a vision of a heavenly messenger who informed him of his impending demise. The next morning Johnson picked up a mattock and spade then climbed a steep ridge. There he dug his own grave. Upon returning to his bed, Johnson (1911) proceeded, "He informed his son Jehoiada of what he had done, gave some directions for the funeral in a cheerful and unconcerned manner, and retired to rest; but ere the morning sun shone into his window the Angel of Death had passed by that peaceful cottage and breathed in the face of the good old man as he slept, and there was mourning in the little hamlet" (p.95).

(1806). Geographical and statistical and political amusement; by which may be obtained a general and particular knowledge of the United States. In a series of interesting games and maps designed for the purposes. Philadelphia: author.

Other

Johnson, Frederick C. (1911). Rev. Jacob Johnson, M.A. Pioneer preacher of Wyoming Valley (Wildes-Barre, PA) 1772-1790 first settled pastor, First Presbyterian Church, Wildes-Barre, PA. Wilkes-Barre, PA: (np).

JOHNSTON, ROBERT (C.1785)

Johnston was one of eight surveyors who reported for duty in 1785 to the newly appointed Geographer of the United States, Thomas Hutchins who was beginning work on the surveying of America. Appointed by the state of Georgia, which had plenty of western territory, Johnston was probably only interested in acquiring land for himself and may not have had the requisite credentials of a surveyor (Pattison 1957, p.128).

The National Union Catalog Pre-1956 Imprints provided no entries for our man (The American Library Association 1976).

JOHNSTON, W. (C.1795)

Guthron (1984) lists Johnston as an engraver of charts executed by Jonathan Price, and John Strother (p.8).

(1795). A description of Occacock [sic] Inlet, and of its coasts, islands, shoals and anchorage, with the courses and distances to and from the most remarkable places and directions to sail over the bar and through the channels, adorned with a map taken by actual survey. Engraved by W. Johnston. New Bern, NC: F. X. Martin.

(1796). To navigators this chart being an actual survey of the sea coast and inland navigation from Cape Henry to Cape Roman is most respectfully inscribed by Price & Strother. Engraved by W. Johnston. New Bern, NC: Price & Strother. 38x13$^{3/4}$inches. Scale 1:490,00.

JONES, T. W. (C.1802)

According to The National Union Catalog Pre-1956 Imprints this Jones and his co-author published only the following text (The American Library Association 1976). No biographical information about the person has been located.

(1802). The traveller's directory or a pocket companion, shewing [sic] the course of the main road from Philadelphia to New York and from Philadelphia to Washington. Philadelphia: Mathew Carey. 23p.

S. S. Moore, co-author.

KEARNY, FRANCIS (C.1780)

Ristow (1985) lists Kearney as one of the early American engravers. He was apprenticed to the engraver Peter Maverick in New York City. In 1810 he moved to Philadelphia. According to Stauffer, "Kearney achieved his greatest success in line, stippled, and aquatint engraving for magazines, annuals, and books" (p.191). John H. Eddy published maps of New York State in 1811 and 1818 engraved by Kearny and others (pp. 256& 295).

(1818). The state of New York, with part of the adjacent states; engr. by Tanner, Vallance, Kearny & Co., Philadelphia. New York: Printed by S. Maverick. colored map. 110x96cm. Scale: 1 in. to 10 miles.

KIMBER, EMMOR (C.1805)

Ristow (1985) identifies Kimber as publisher of Reading Howell's maps (p.110). He also published two of Samuel Lewis' maps (pp.266 & 446).

(1805). Kimber's almanac, for the year of Our Lord 1805. (np).

(1809). Elements of geography, principally compiled with a view to teach children, at an early age, the geography of the United States. Philadelphia: author.

Other

Lewis, Samuel. (1815). A new and accurate map of the United States of North America, exhibiting the countries, towns, roads &c. in each state. Carefully compiled from surveys and the most authentic documents. by Samuel Lewis. Philadelphia: Published by Emmor Kimber.

Lewis, Samuel. (1816). A new and correct map of the United States of North America. by Samuel Lewis. Philadelphia: Published by Emmor Kimber.

Howell, Reading. (1792). Map of the state of Pennsylvania. London: Sold by James Phillips.

(1806, 1816 & 1817). As above- Philadelphia: Emmor Kimber, publisher.

KING, NICHOLAS (1771-1812)

Klemp (1976) offers a map of the Mississippi drawn by King and engraved by Francis Shallus of Philadelphia. The map is correct as far as Little Falls, but distorted thereafter. In 1803, King prepared a manuscript map of the western portion of the North America as it was known prior to the explorations and mapping of Lewis and Clark. It was the, "First official map of the trans-Mississippi West compiled from data furnished King by William Clark of the Lewis and Clark expedition" (Schwartz & Ehrenberg 1980, pp. 218, 223-224).

(1888). The King plats of the city of Washington in the District of Columbia, 1803. Washington: N. Peters, Photo-Lithographer. 21 maps. 98cm.

(1805-6). A map of part of the continent of North America, between the 35th and 51st degrees of north latitude, and extending from the 89º degrees of west longitude to the Pacific Ocean. Compiled from the authorities of the best informed travellers, by M. Lewis. Washington, DC: National Archives.

(c.1804). Map of the Washita River in Louisiana from the Hot Springs to the confluence of the Red River with the Mississippi. Laid down from the journal & survey of Wm. Dunbar in the year 1804. (np). 23x88cm.

(1809). Peter manuscript atlas of Mount Pleasant and Mexico, D.C., 1788-1809. Washington: (np). [7] 145p. maps. 50cm.

Other

Pike, Zebulon Montgomery. (1807). An account of a voyage up the Mississippi River, from St. Louis to its source; made under the orders of the War Department, by Lieut. Pike, of the United States Army, in the years 1805 and 1806. Compiled from Mr. Pike's journal. Washington: (np).

KNEASS, WILLIAM (C.1781)

Kneass was born in Lancaster, Pennsylvania on September 25. He worked for about thirty years as an engraver. Between the years 1818 and 1820 he was in partnership with James H. Young. John Hills' map of the Philadelphia, published in 1808, was engraved by Kneass (Ristow 1985).

(1808). A plan of the city of Philadelphia and environs surveyed by John Hills in the summers of 1801, 2, 3, 4, 5, 6, & 7. (np).

LAFON, BARTHOLMEW (C.1765)

Lafon was born in France and immigrated to New Orleans in the early 1790s. It appears that he was trained as an architect and engineer because in 1796 he was entrusted to design public baths for New Orleans. Though the baths were never constructed, he was also engaged with architectural activities and planning urban districts. In 1804 and 1805 he carried out extensive surveys of the Louisiana Purchase for the United States. Those surveys produced two maps, one consisting of seventeen sheets (Ristow 1985, p.140-141).

(c.1808). Annuaire Louisianais; commencant a l'equinoxe de mars 1808, et se terminant a celui de mars 1809. New Orleans: author. [Accents omitted]

(1806). Calendrier de commerce de la Nouvelle-Orleans, pour l'annee 1807. New Orleans: Jean Renard. 118cxxp. 15cm. [Accents omitted]

(1806). Carte generale du territoire d'Orleans comprenant aussi la Floride occidentale et une portion du territoire de Mississipi. Dressee d'apres les observations les plus recentes. Orleans: Nelle. 82.5x123cm. [Accents omitted]

LAMBERT, SAMUEL (1769-1832)

Guthorn (1984) reveals that Lambert was a member of the Salem East India Marine Society before becoming the proprietor of a navigation school in Salem, Massachusetts. He sold instruments, glasses, charts and other goods prior to publishing his first manuscript in 1812, A New Chart of Massachusetts Bay Drawn from the Latest Authorities. Lambert established a school in Salem, Massachusetts in which students could study the following as advertised on a broadside:

> Informing his friends and the public that he continues to teach navigation and the lunar observations, to find the latitude by double altitudes of the Sun, by one altitude of the Sun taken near noon, by the meridian altitude of the Moon, star or planet, by an altitude of the Pole Star, taken at any hour of the night, to find the latitude on shore, to regulate a watch or time-keeper on shore, the construction use of Mercator's Chart, &c.&c. The projection and calculating of eclipses of the Sun and the Moon, and occulations of fixed stars by the Moon, and the method of determining the longitude by observing them, with a number of other problems in nautical astronomy, useful to a navigator. Charts and plans of harbours copied with neatness and dispatch. Book-keeping by single and double entry, with a manner of making out all kinds of accounts. (p.9)

(1820). Information useful for navigators. Compiled by Samuel Lambert, teacher of navigation, etc. Salem, MA: Sold by Cushing & Appleton, S. Lambert and J. W. Archer, T. C. Cushing, Printer.

(1822). A new chart of the coast of North America from Cape Hatteras to Egg River including Chesapeake Bay drawn from the latest authorities by Samuel Lambert, Hydrographer. Engraved by A. Bowen. Salem, MA: Cushing & Appleton & S. Lambert. 37x26$^{1/2}$inches. Scale 1:73000.

(1821). A new chart of the coast of North & South Carolina, Georgia & E. Florida. Drawn from the latest authorities by Samuel Lambert. Salem, MA: Cushing & Appleton, & S. Lambert. 35$^{3/4}$x24inches. Scale 1:1,030,000.

(1812). A new chart of Massachusetts Bay drawn from the latest authorities by S. Lambert. Salem: Cushing & Appleton & S. Lambert. 23x36.5inches. Scale: 3.7 geographical miles to 1 inch.

(1813). A new chart of Nantucket Shoals & George's Bank, with the adjacent coast, drawn from the latest authorities. Salem: (np). 23$^{3/4}$X36$^{1/4}$inches. Scale: Mercator's.

LANGHAM, ELIAS (C.1800)

Langham was one of the deputy surveyors appointed under the Land Act of 1800 to work under the direction of Rufus Putnam (Pattison 1957, p.

207). No other information about Langham has been located. The National Union Catalog Pre-1956 Imprints provided no entries for Langham (The American Library Association 1976).

LATHROP, JOHN, JR. (1772-1820)

Lathrop's contributions to the knowledge of geography was through his publications. He was born in Boston, Massachusetts on January 13, 1772. He attended and was graduated from Harvard in 1789 and entered the law office of Christopher Gore. There he began to practice law in Dedham, Massachusetts in 1797. Lathrop settled in Boston and wrote for the Boston Gazette.

The next ten years were spent in Calcutta, India, where he established a school and wrote for local newspapers. In 1809 Lathrop returned to the United States where he taught school in Boston. During the period he wrote, lectured, and delivered public addresses. Lathrop later moved to Georgetown, District of Columbia where he continued to teach and write. He was able to obtain a position with the post-office which he occupied until poor health caused him to retire. Lathrop died in Georgetown, DC on January 30, 1820 (Johnson & Malone 1928, V.9, p.135; White 1892, V.VII, p.135).

(1797). An account of the deleterious effects of mephitic air, or marsh miasma, experienced by three men, July 27, 1797, in a well, on the Boston pier... Memoirs of The American Academy of Arts and Sciences, 2, pt.2, pp.81-84.

(1800). An account of the springs and wells on the peninsula of Boston, with an attempt to explain the manner in which they are supplied... Memoirs of The American Academy of Arts and Sciences, 3, pt. 1, art. 8, pp.57-68.

(1812). A compendious treatise on the use of globes, and of maps; compiled from the works of Keith, Ferguson, Adams, Hutton, Bryan, Goldsmith and other eminent authors; being a plain and comprehensive introduction to the practical knowledge of geography and astronomy... Boston: J. W. Burditt & Co. 183p. folded plates. folded diagram. 18cm.

(1821). As above- Boston: wells and Lilly and J. W. Burditt.

(1811). Synopsis of a course of lectures, on the following branches of natural philosophy, viz. matter, and its properties - mechanics - electricity - hydrostatics - pneumatics, and astronomy. - The introductory lecture to be delivered at the Exchange Coffee-House, on Monday the 4th Feb. 1811. Boston: Russell and Cutler, printers. 16p. 16cm.

LAY, AMOS (C.1780)

Ristow (1985) relates that Lay was probably born in New York. He was a New York map maker and publisher who published editions, some modified with as additional territory was added, in 1812, 1817, 1819, 1822, 1823, 1824, and 1826. He additionally published a U.S. map in 1819 and 1836 (p.99).

(1801). Map of the northern part of the state of New York. Jointly executed by Amos Lay and Arthur J. Stansbury. New York: (np).

(1812). Map of the northern part of the state of New York; compiled from actual survey. Newark, NJ: P. Maverik. 70 x 128cm.

(1820). Map of the state of New York with part of the states of Pennsylvania, New Jersey &c. Compiled corrected & published from the most recent authorities and accurate surveys...1817. 2d edition revised. New York: (np). $49^{3/4}$x$49^{1/4}$inches. Scale: (computed) 7 miles to 1 inch.

(1826). As above- 125x126cm.

(1814). A new correct map of the Seat of War in Lower Canada protracted from Hollands large map compiled from actual survey made by order of the provincial government. Laid down with many late additions and improvements by Amos Lay... Philadelphia: A. Lay & J. Webster. 55x85cm. Scale: 7 miles to the inch.

(1819). U. S. map. Designed & engraved by Thos. Starling. London: Amos Lay.

LEAR, TOBIAS (1762-1816)

Little is known about Lear other than the fact that he was secretary to General Washington.

(c.1799). Beschryving der rivier Potomack en de stad Washington in Noord-America. Uit het Engelsch vertaald. Amsterdam: H. de Vries. 1799?. 32p. 17cm (Also attributed to Andrew Ellicott).

Based on a Report of the Committee of the Merchants of Georgetown and Alexandria.

(c.1840). Last words of General Washington. Philadelphia: (np). 7p. 25.5cm.

(1794). Observations on the river Potowmac [sic], the adjacent country, and the city of Washington. (In The New York Magazine; or Literary repository. New York, 1790-1797, 21.5cm, v.5 p.210-215, 268-274)

As above - ...New York: Printed by Samuel Loudon and son, 1793, 29p. fold. map 18cm.

(Erroneously ascribed to Andrew Ellicott. The authorship of Lear is affirmed by a letter written by him to Washington Nov. 3, 1793, and Washington's answer Nov. 8, 1793. The letters are in the Library of Congress.)

As above - ...unpublished correspondence of George Washington and Tobias Lear about the observations; published text compared with an autograph copy sent to George Washington, edited by Samuel T. Chambers... Baltimore, MD: S. T. Chambers, 1940. 26p. incl. mounted front. (port.) 24cm.

LEWIS, MERIWETHER (1774-1809)

Lewis was half of the famous Lewis and Clark team, both captains in the American Army. Klemp (1976) provides a map of the tract that was prompted by Thomas Jefferson who gave orders to explore the Rocky

Mountains and find a way to the Pacific coast (plate 49). The expedition took place between 1804 and 1806 (The Biographical Dictionary 2001. Lewis, Meriwether; Encarta Encyclopedia 2001. Lewis, Meriwether).

White (1893) reveals that Lewis was an explorer born near Charlottesville, Virginia on August 18, 1774. He was the son of Capt. William and Lucy (Meriwether) Lewis. His father's brothers, Charles and Nicholas were prominent in the Revolutionary and Cherokee wars.

Lewis' mother was responsible for his education. He was sent to a Latin grammar school and learned botany on his family farm. His militia experience brought him a captaincy in the Norwestern Campaign during the Whiskey Rebellion. Thomas Jefferson, took on Lewis as his private secretary and ultimately became one of the principal explorers of the Missouri route to the Pacific Ocean (White 1893, V.V. p.122).

(1805). A box of plants, earths and mineral from... per Jefferson, who wishes the seeds sent to Mr. William Hamilton. Proceedings of the American Philosophical Society, Philadelphia, 22, appx. p.379.

Gratuities to officers and men in expedition to Pacific Ocean under Lewis and Clark, American State Papers: Military Affairs, V.XVI (Doc. 68).

(1814 & 1817). History of the expedition under the command of Captains Lewis and Clark, to the sources of the Missouri, thence across the Rocky Mountains and down the river Columbia to the Pacific Ocean. Performed during the years 1804-5-6. By order of the government of the United States. Prepared for the press by Paul Allen, Esquire. In two volumes. Philadelphia: Published by Bradford and Inskeep; and New York: Abm. H. Inskeep; J. Maxwell, printer. V.I: xxviii470p. 3 maps. 1 folded. V.II: ix[1]-522p. 3 maps.

(1844). As above- New York: Harper & Bros. V.1, [i]-vi, [i]-v[7]-371p. V.II: [i]-x[9]-395p.

Lewis and Clark expedition, pt. 1: Correspondence of Captain Meriwether Lewis, American State Papers: Indian Affairs, V.VII (Doc. 113).

Lewis and Clark expedition, pt. 2: Statistical view of Indian nations of Louisiana territory and territories on its boundaries, and ethnological observations, p.707, American State Papers: Indian Affairs, V.VII (Doc. 113).

Lewis and Clark expedition, pt. 3: Historical sketch of Indian tribes in Louisiana and ethnological observations, by John Sibley, p.721, American State Papers: Indian Affairs, V.VII (Doc. 113).

Lewis and Clark expedition, pt. 4: Geographical description of Red River and adjacent country, by John Sibley, American State Papers: Indian Affairs, V.VII (Doc. 113).

Lewis and Clark expedition, pt. 5: Geographical description of Washita River in Louisiana and adjacent country, by William Dunbar and Dr. George Hunter, p.731, American State Papers: Indian Affairs, V.VII (Doc. 113).

Lewis and Clark expedition, pt. 6: Meteorological observations of William Dunbar and Dr. George Hunter, p.742, American State Papers: Indian Affairs, V.VII (Doc. 113).
(1804). Observations on some fossil bones, in a letter to Jefferson. Proceedings of the American Philosophical Society, 22, appx. p.350-351.
(1814). Travels to the source of the Missouri River and across the American continent to the Pacific Ocean performed by order of the government of the United States in the years 1804, 1805, and 1806, by Captains Lewis and Clarke, published from the official report, and illustrated by a map of the route, and other maps. London: Longman, Hurst, Rees, Orme & Brown. 24+663p. maps.

Other

Greely, A. W. (1893). Meriwether Lewis. In his Explorers and travellers. New York: (np). pp.105-162.
Thwaites, Reuben Gold. (1904). A brief history of Rocky Mountain exploration with especial reference to the expedition of Lewis and Clark. New York: Appleton & Co. 289p.
Thwaites, Reuben Gold. (Ed.). (1904-1907). Early western travels, 1748-1846; a series of annotated reprints of some of the best and rarest contemporary volumes of travel, descriptive of the aborigines and social and economic conditions in the middle and far west, during the period of early American settlement; edited, with notes, introduction, index, etc. Cleveland: The A. H. Clard Co. 32v. illus. plans. maps.

LEWIS, SAMUEL (C.1754)

Lewis' first published map of Virginia was included in Guthrie's 1795 Geography Improved (Schwartz & Ehrenberg 1980, p. 229). It is also included in Carey's 1795 American Atlas. Gutherie's 1809 Geography contained a United States map by Lewis.

Ristow (1985) notes that Lewis produced over half the maps in the 1795 Carey text. He formed a partnership with Aaron Arrowsmith, an Englishman, publishing additional maps. Lewis drew all the American maps that Arrowsmith used. They were engraved by several people: Benjamin Tanner, William Harrison, Jr., and David Fairman (p.265).

(1795). American atlas. (np). Mathew Carey.
(1812). The British possessions in North America, from the best authorities. (np). map 38x44cm. Scale: 120 miles to a degree.
(1813). A correct map of the United States, with the West Indies from the best authorities. Philadelphia: (np). Scale of miles: 69.5 to a degree. 36.5x38inches.
(1813). General atlas. Philadelphia: (np). 58 maps. 44x27cm.
(nd). Georgia. (np). map 21x25cm. Scale: 1 inch equals 40 miles.
(1805 & 1812). Louisiana, drawn by S. Lewis. (np). map 25x21cm.

(1814). As above- Philadelphia: Carey. map 40x44cm. Scale: 1 inch to 10 miles.

(1796). A map of part of the N: W: Territory of the United States, compiled from actual surveys and the best information. (np). map 49x65cm. Scale: about 18 miles to the inch.

(1807). A map of the country from Raritan River in East Jersey to Elk Head in Maryland shewing the...operations...in 1776 & 1777. Drawn by S. Lewis...Engraved by I.H. Seymour. Philadelphia (1806-1807). In Marshall's The Life of George Washington. Philadelphia: (np). Plate V, $9^{7/8}$x$15^{1/2}$inches.

(1807). A map of country...the scene...of the Northern Army; including the wilderness through which General Arnold marched to attack Quebec. Engraved by F. Shallus. (Philadelphia, 1806 & 1807). In Marshall's The Life of George Washington. Philadelphia: (np). Plate Vi, $10^{3/8}$x$8^{5/8}$inches.

(1807). A map of part of Rhode Island shewing the positions of the American and British armies at the siege of Newport... 1778. Drawn by S. Lewis... Engrav'd by Benj'n Jones. Philadelphia. In Marshall's The Life of George Washington. Philadelphia: (np). Plate Vii, 16.5x10inches.

(1794). Map of the state of New Hampshire. (np).

(1807). A map of those parts of Virginia, North Carolina, & Georgia, which were the scenes of the most important operations of the Southern armies. Compiled by s. Lewis. Engrav'd by Francis Shallus, Philadelphia, 1806. In Marshall's The Life of George Washington. Philadelphia: (np). Plate VIII, $14^{1/2}$x$10^{3/4}$inches.

(1795). Map of the United States. Philadelphia: (np).

(1819). A new and correct map of the United States of North America; exhibiting the counties, towns roads &c. in each state. Philadelphia: E. Kimber. 8°.

(1804). A new and elegant general atlas, comprising all the new discoveries, to the present time; containing sixty-three maps. drawn by Arrowsmith and Lewis. Philadelphia: John Conrad & Company.

(1804). As above- Baltimore: M & J. Conrad & Company.

(1804). As above- Washington, DC: Rapin, Conrad & Company.

(1804). As above- Petersburg, VA: Somervell & Conrad.

(1804). As above- Norfolk, VA: Bonsal, Conrad & Company.

(1774). A new map of the West Florida, Georgia, & South Carolina, with part of Louisiana. Reduced and copied by Samuel Lewis, February 4th, 1774. (np).

(c.1812). North Carolina. Boston: (np). $7^{7/8}$x$9^{7/8}$inches. Scale of miles: 50 to a degree.

(1814). North Carolina, S. Lewis, del., Wm. Charles, sc. Philadelphia: M. Carey. $10^{7/8}$x$18^{1/4}$inches. Scale of miles: 22 to an inch.

(1818). North Carolina from the latest surveys. Philadelphia: W. Carey & Son. 11x$18^{1/2}$inches. Scale of miles: 60 to a degree.

(1807). A plan of New York Island, part of Long Island &c. shewing [sic] the position of the American and British armies, before, at and after the engagement on the Heights August 27th 1776. Drawn by S. Lewis. J. H. Seymour sculp. (Philadelphia 1806-7). In Marshall's The Life of George Washington. Philadelphia: (np). Plate II, $16x10^{1/4}$inches.

(1807). A plan of the country from Frog's Point to Croton River shewing [sic] the positions...on the White Plains on the 28th. (Philadelphia 1806-7). In Marshall's The Life of George Washington. Philadelphia: (np). Plate III, $16^{1/2}x9$inches.

(1807). Plan of the investment and attack of York in Virginia, Engraved by Francis Shallus. (Philadelphia, 1806-7). In Marshall's The Life of George Washington. Philadelphia: np. Plate IX, 9 1/8 x 8 3/8 inches.

(1807). A plan of the northern part of New Jersey shewing [sic] the positions of the ...armies after crossing the North River in 1776. (Philadelphia, 1806-7). In Marshall's The life of George Washington. Philadelphia: (np). Plate IV, 16x10inches.

(1807). Plan of the siege of Charleston in S. Carolina. Drawn by S. Lewis. Tanner, sc. (Philadelphia, 1806-7). In Marshall's The Life of George Washington. Philadelphia: (np). Plate X, $9^{1/2}x15$inches.

(1794). The province of Maine, from the best authorities. Philadelphia: (np). 36x24cm. Scale 1:1,750,000.

(1795). The state of New York compiled from the best authorities. (np). 41x51cm. Scale: 1 inch equals 20 miles.

(1795). The state of North Carolina from the best authorities &c. by Samuel Lewis, engraved by Vallance. Philadelphia: M. Carey. $11x18^{1/2}$inches. Scale of miles: about 50 to a degree.

(1818). The state of South Carolina from the best authorities &c. by Samuel Lewis, W. Barker sculp. Philadelphia: M. Carey & Son. $15^{1/4}x17$inches. Scale of miles: about 69.5 to a degree.

(1804). Tennessee. Boston: Arrowsmith and Lewis. $7^{3/4}x9^{5/8}$inches. Scale of miles: about 40 to a degree.

(1819). Traveller's guide...map of the U.S. incl. great portions of Mo. Territory. Philadelphia: (np).

Other

Arrowsmith, Aaron, & Lewis, S. (1804). A new and elegant general atlas, comprising all the new discoveries, to the present time; containing sixty-three maps, drawn by Arrowsmith and Lewis. Philadelphia: J. Conrad.

Morse, Jedidiah. (1819). The American universal geography; or, a view of the present state of all the kingdoms, states, and colonies in the known world. 7th ed. Boston: Lincoln & Edmands.

LINCOLN, BENJAMIN (1733-1810)

Lincoln was the author of several books, including some regarding geographic information (The American Library Association 1978). He was a

United States general during the American Revolution. He was also the first Secretary of War, 1781-1783. Lincoln suppressed Shays' Rebellion of 1787 (The Biographical Dictionary 2001).

His surrender, in Charleston, South Carolina 1780, brought him criticism. His success with Shays' Rebellion apparently brought some measure of forgiveness since he became Lieutenant Governor of Massachusetts and was the Collector for the port of Boston (Encyclopedia Britannica 2001. Lincoln, Benjamin)

(1793). A description of the situation, climate, soil and productions of certain tracts of land in the District of Maine, and Commonwealth of Massachusetts. Philadelphia: (np). 44p.

(1790). Observations on the climate, soil and value of the eastern counties in the district of Maine written in the year 1789 [also] copy of a letter on the religious state, 1790. Boston: Massachusetts Historical Society Collections, 4 pp.142-156.

(1898). Original papers relating to the Siege of Charleston, 1780, mostly selected from the papers of General Benjamin Lincoln, in the Thomas Addis Emmet Collection, Lenox Library, New York, and now first published <From the city Year Book for 1897.> Charleston, SC: Press of Walker, Evans & Cogswell Co.

LODGE, BENJAMIN (C.1779)

Lodge, in 1779, produced a series of maps to record the 'Route of the Western Army under Gen'l Sullivan.' He was acting under orders from Robert Erskine, Geographer of the United States (Harley, et.al. 1978, pp.24, 39). His skills as a "mapmaker" placed him at army headquarters where he worked under Nathanael Greene and John Sullivan (p.70). No other information about Lodge has been located. The National Union Catalog Pre-1956 Imprints provided no entries for Benjamin Lodge (The American Library Association 1976).

LUCAS, FIELDING, JR. (1781-1854)

Lucas was as a publisher and sometimes author of fine books, textbooks, and maps (Foster 1956). He was born on September 3, 1781 in Fredericksburg, Virginia, the boyhood home of George Washington. Little is known about the Lucas family.

Fielding Lucas Jr. in 1798, lived in Philadelphia working as an employee or apprentice at a stationery and book store. The only clue to his schooling is an album containing pen and pencil sketches dated 1801 and 1803, relating to some of the literary classics of the day. He, in 1810, married Elizabeth Mary Carrell, daughter of a merchant and prominent Catholic layman. Baltimore became the home of Lucas in 1804 when he took up the business of book selling. There he became partners with M. and J. Conrad and Company who were residents of Philadelphia. It is likely that Lucas was their Baltimore manager. Foster (1956) provides the following insight into the books stocked by booksellers of the day: "Humboldt's Political Essays on the Kingdom of New Spain, British Theatre, Bingley's History of Quadrupeds, Miss Porter's Scottish Chiefs, Morse's Geography, the American Gazetteer, and the Children of Abbey (p.172).

Lucas was also involved with Mathew Carey in the publishing business as well as through providing maps for one of Carey's atlases. Foster notes that Lucas may have been involved in some plagiarism, "The New and Elegant General Atlas was almost a copy of one issued in London in 1806 by Laurie & Whittle" (p.187).

(1820). Arkansas territory. (np). map. 21x28cm.
(1823). Canary Islands. Baltimore: F. Lucas. colored map. 30x38cm.
(1859). Chart of the Chesapeake and Delaware Bays. Corrected, 1859. Baltimore: (np). $39^{3/4}$x27inches. Scale: 5 7/8 miles to 1 inch.
(1812). A correct map of the Seat of War. Baltimore: John Conrad and Fielding Lucas, Jr. $15^{1/2}$x$22^{3/4}$inches.
(1823). A general atlas containing distinct maps of all the known countries in the world. Constructed from the latest authority. Baltimore: Fielding Lucas. 100 colored maps. 37.8cm.
(1826). As above- Baltimore: (np). 104 maps. 40cm.
(c.1830). As above- Baltimore: Fielding Lucas. 67 colored maps. 37cm.
(1822). Geographical, statistical and historical map of Florida. Drawn by F. Lucas, Jr., Engraved by Young & Delleker. Philadelphia: H. C. Carey & Lea. 11x$9^{1/4}$inches. Scale: 69.5 miles to a degree.
(1823). Louisiana. Baltimore: (np). colored map 29x44cm. Scale: $^{5/16}$in. to 10 mi.
(1827). Lucas' progressive drawing book... consisting chiefly of original views of American scenery, and embracing the latest and best improvements in the mode of instruction... Baltimore: F. Lucas, Jr. 3v in 1. illustrations. plates. diagrams. 26x36.5cm.
(1822). Maryland. Drawn by F. Lucas, Jr. In Carey & Lea atlas. (np).
(1819). Mathematical tables: difference of latitude and departure; logarithms, from 1 to 10,000; and artificial sines, tangents, and secants. Baltimore: F. Lucas, Jr. 152p. 22cm.
(1814). A new elegant general atlas, containing maps of the United States. Baltimore: F. Lucas, Jr.; and Philadelphia: P. H. Nicklin. 31 colored maps.
(1815) As above- 54 maps. 37x32cm.
(1817) As above- 57 colored maps.
(1824). As above- 21 colored maps (2 double). 37.5cm.
(c.1918). Palestine. Baltimore: F. Lucas, Jr.
(c.1822). Plan of the city of Baltimore: Compiled from actual survey. Baltimore: (np).
(1823). United States atlas. Baltimore: (np). 31 maps, various sizes and scales.

Other

Foster, James W. (1956). Fielding Lucas, Jr., early 19th century publisher of fine books and maps. Worcester, Mass.: American Antiquarian Society.

Latrobe, John Hazlehurst Boneval. (1832). Picture of Baltimore, containing a description of all objects of interest in the city; and embellished with views of the principal public buildings. Baltimore: F. Lucas, Jr.

LUDLOW, ISRAEL (C.1786)

Ludlow became surveyor for South Carolina in 1786 and was a part of Hutchins survey team following his return to Pittsburgh in June of 1786. Congress authorized resumption of the surveys of the east-west line which were begun, but not completed in 1785.

He was a young man from New Jersey. In his later years Ludlow was to become interested in the Symmes Purchase, perhaps an indication that his initial interest in surveying was speculation. Pattison (1957) reports that he surveyed more land in the Ohio Country than any other man prior to his death (p.134). He was from Morris County, New Jersey. Ludlow was one of the deputy surveyors appointed by the Land Act of 1800 to work under the direction of Rufus Putnam (Pattison 1957, p.207).

Other

(nd). Sketch of the life and times of Israel Ludlow... (np).

Teetor, Henry B. (July 1885). Israel Ludlow and the naming of Cincinnati. Magazine of Western History, II, 251-257.

LUKENS, JOHN (JESSE) (C.1800)

Barton (1813) notes that David Rittenhouse was first known because of the efforts of Dr. William Smith and Mr. John Lukens.

> Perhaps it might be said, with greater strictness, that the 'discovery' here spoken of, belonged solely to Mr. Barton; by whom it was communicated, very early, to his learned and reverend friend, Dr. Smith, - and through him, to the ingenious astronomical observer, Mr. Lukens, (afterwards Surveyor-General,) as well as some other distinguished characters of that time. (pp.119, 275, 300, etc.)

No other information about Lukens has been located. The National Union Catalog Pre-1956 Imprints provided no entries for John Lukens (The American Library Association 1976).

MACKAY, JAMES (C.1805).

The only James Mackay identified in Brown (1897, V.5, p.257) was a representative from Bladen County, North Carolina who lived from 1793-1853. Since the first mention of James MacKay was in 1805, the North Carolina gentleman is most likely not our person.

A French map dated 1805, prepared by Francois Perrin du Lac utilized information from the explorations of a James Mackay (Schwartz & Ehrenberg

1980). It depicted the lower portion of the Missouri River and the headwaters of the Missouri and Mississippi rivers. Other information involving Mackay's contributions to geography are derived from United States government publications. The National Union Catalog Pre-1956 Imprints provided no entries for Mackay (The American Library Association 1976).

Claim to land in Missouri, petition of James Mackay, American State Papers: Public Lands, V.XXX (Doc. 296, 297, 306, & 1022).

U.S. House of Representatives. 24th Congress, 1st Session. Delassus, Charles Dehult, Augustus Chouteau, James MacKay, and others, opinion of Supreme Court in Missouri land cases, (H.doc.148). Washington: Government Printing Office, 1835-1836. (Serial Set 289).

U.S. House of Representatives. 16th Congress, 1st Session. Mackay, James, claim to land, (H.rp.41). Washington: Government Printing Office, 1819-1820. (Serial Set 40).

Opinions of Supreme Court of United States upon certain claims of land in Missouri: John Smith vs. U.S.; Isabella Mackay, widow and John Zenon Mackay, and others, heirs of James Mackay vs. U.S.; Julie Coulard, widow, and James G. Soulard, and others, heirs of Antoine Soulard vs. U.S., American State Papers: Public Lands, V.XXXV (Doc. 1458).

U.S. House of Representatives. 15th Congress, 2nd Session. Petition of James Mackay for land in Missouri territory, (Hdoc.95). Washington: Government Printing Office, 1818-1819. (Serial Set 22).

MCCOMB, JAMES, JR. (C.1789).

A native of New York City, McComb was one of the earliest American engravers of maps (Ristow 1985). His first published work appears to have been plates for Colles's Survey of the Roads of the United States. His A New & Accurate Plan of the City of New York in the State of New York in North America was compiled by B. Taylor and engraved by John Roberts is described as:

> One of the most accurate and beautiful engraved plans of the city, and particularly interesting on account of its tiny bird's-eye views of some of the most important buildings, as well as for the clear idea which it gives of the country lying along the East River in the vicinity of Crown Point. (p.247-249)

(1797). A new and accurate plan of the city of New York in the state of New York in North America. Engraved by John Roberts. New York: (np).

(1789). Plan of the City of New York. In The New York directory. Engraved by Cornelius Tiebout. New York: (np).

MACLAIN, ALEXANDER (C.1785).

Barton (1813) reports that the survey of the western boundary line of Pennsylvania, begun in May 1785 by Rittenhouse, Ellicott, Porter, and Nevill with the help of B. S. Barton, was completed with the Lake Erie margin in 1786 by Alexander Maclain, esq. (pp.307-308). Beatrice Holloway used the name as a pseudonym. Here is the possibility of another woman who was engaged in geographic matters. None the less, no other information about Maclain has been found. The National Union Catalog Pre-1956 Imprints provided no entries for Maclain (The American Library Association 1976).

MCLEAN, ARCHIBALD (1733-1812)

Andrew Porter worked with McLean and other men with whom he appears to have completed surveying the boundary of Pennsylvania (Pattison 1957). Barton (1813) notes several people engaged in establishing the boundary line between Virginia and Pennsylvania: "This boundary-line was begun in May, 1785, by Messrs. Rittenhouse, Ellicott, Porter, and Nevill; assisted by the present Dr. Benjamin Smith Barton, then a youth about nineteen years of age, whose medical and other scientific acquirements rendered him, even at that early period of life, an useful associate of the commissioners" McLean was not mentioned (pp.306-307). McLean's primary occupation, not unlike many of his contemporary contributors to our knowledge of geography, was that of a clergyman and author of religious materials (The American Library Association 1978).

Other

Jones, William. (1823). The works of Mr. Archibald M'Lean, late pastor of the Baptist Church, Edinburgh; with a memoir of his life ministry, and writings by William Jones. London: W. Jones. 6v. in 7. 22cm.

MACLURE, WILLIAM (1763-1840)

Brown (1897) indicates that McLure was a geologist born in Ayr, Scotland the son of David and Ann. In 1796 he came to the United States on business, returning to Europe in 1803. Later returning to the United States he began a geological survey of the United States for the purpose of crossing the Alleghany Mountains. His geological findings about the Alleghany Mountains were presented to the American Philosophical Society.

McLure traveled extensively, established in Spain an agricultural school for the poor. In 1824 he returned to the United States and purchased land at New Harmony, Indiana for the purpose of establishing a similar agricultural system in this country. His expectations, however, were not fulfilled, so he moved to Mexico where he died in 1840 (V.5, p.292).

From his early experience in the United States came a geological map (1809) prepared by Maclure (Schwartz & Ehrenberg 1980). The map and his "Observations on the Geology of the U.S." appeared in the Transactions of the American Philosophical Society. He is credited with producing "the first true geological map of any part of North America and one of the earliest of such maps compiled" (Encyclopedia Britannica 2001. Maclure, William).

White (1906) reveals that Maclure traversed the Alleghany Mountains "no less than fifty times" in his research for the publication of Observations on the Geology of the United States, Explanatory of a Geological Map.

White also reports numerous other publications of Maclure (V.XIII, pp.368-369).

(1818). Essay on the formation of rocks; or, An inquiry into the probable origin of their present form and structure. Philadelphia: (np). 60p. 22.5cm.

(1832). As above- New Harmony, IN: author.

(1838). As above- Philadelphia: J. Wilbanks. 32p. 21 cm.

(1809). Observations on the geology of the United States, explanatory of a geological map. Transactions of the American Philosophical Society, 6, pt.2, (62), pp.411-428.

(1817). Observations on the geology of the United States of North America; with remarks on the probable effects that may be produced by the decomposition of the different classes of rocks on the nature and fertility of soils; applied to the different states of the Union agreeably to the accompanying map. Transactions of the American Philosophical Society, 1(1), pp.1-91, 1 col. pl. 1 col. map.

(1818). Observations on the geology of the United States of North America; with remarks on the probable effects that may be produced by the decomposition of the different classes of rocks on the nature and fertility of soils; applied to the different states of the Union agreeably to the accompanying map. Philadelphia: A. Small.

(c.1955). As above- Stanford, CA: Academic Reprints.

(1832). Observations on the geology of the West India Islands, from Barbadoes to Santa Cruz, inclusive. New Harmony, IN: author. 17p. 21 cm.

(1811). Suite des observations sur la geologie des Etats-Unis servant a l'expt cation de la carte ci-jointe. Paris: Courcier. 32p. 4°. map.[Accents omitted]

Other

Merrill, G. P. (1906). William Maclure. In U. S. Natural Museum, Annual Report, 1904, pp.217-219, port.

Morton, S. G. (1841). A memoir of William Maclure, late president of the Academy of Natural Sciences, Philadelphia. Philadelphia: Academy.

Owen, David Dale. (1840). Catalogue of mineralogical and geological specimens, at New-Harmony, Indiana. Collected in various parts of Europe and America, by William Maclure... Arranged for distribution at the request of Miss Maclure and Alexander Maclure, his executors. New Harmony, IA: (np). 15[1]p. 18cm.

MCMURRAY, WILLIAM (C.1784)

Ristow (1985) notes that McMurray proposed, in the Pennsylvania Packet, to publish a map of the United States by subscription. Unfortunately that publication was delayed and Abel Buell was able to publish the first U.S. map compiled, engraved and finished by an American. McMurray served as Assistant Geographer to the United States under Erskine and DeWitt. His principal wartime cartographic contributions were made in September 1780 (p.66-67).

(1784). The United States according to the definitive Treaty of Peace signed at Paris, Septr. 3d 1783. (np).

MADISON, BISHOP JAMES (1749-1812)

Madison was the first cousin of President James Madison. He was educated for the ministry and became President of William and Mary College in 1777. During that time, one of the responsibilities of the college was to train and license surveyors. Madison became aware of the lack of a good map of Virginia, but lacking the necessary skills to do the job himself, he collaborated with William Prentis of Petersburgh and with William Davis. Between them they produced and presented their map to the Virginia General Assembly, 1805 (Ristow 1985, p.121).

Brown (1897) reveals that Madison was the "first Bishop of Virginia and 4th in succession in the American Episcopate." He was born near Port Republic, Rockingham (then Augusta) County, Virginia and died in Williamsburg, Virginia. He married in 1779 Sarah Tate, the granddaughter of Secretary William Cocke. They had two children, James Gatesby and Susan Randolph. Susan married Robert G. Scott (V.5, pp.318-319).

White (1893) confirms the above going on to report that Madison was born in Augusta County, Virginia on August 27. He was a descendant of John Madison, a common ancestor of James Madison, the President of the United States. There is some disagreement regarding the name of the son, who White identifies as John Catesby Madison. His remains are in the chapel of the College of William and Mary (V.III, p.234 & V.VII, p.216).

(1780) A letter on the supposed fortifications of the western country, from Bishop Madison of Virginia to Dr. Barton. Philadelphia: (np). With William Dunbar.

(1807). A map of Virginia formed from actual surveys, and the latest as well as most accurate observations. Drawn by William Davis; Engraved by Frederick Bossler. Richmond: To the General Assembly of Virginia... by their Fellow Citizens, James Madison, William Prentis, William Davis, Proprietors. 114x175cm.

Other

Tucker, St. George. (nd). A letter to the Rev. Jedidiah Morse... (np).

MALHOUM, JOHN (1747-1821)

Malhoum's contribution to geographic knowledge in America appears to have been authorship of the following text. No other information about Malhoum has been located. The National Union Catalog Pre-1956 Imprints provided only one entry for Malhoum (The American Library Association 1976).

(1797). The naval gazatteer; or, seaman's complete guide...illustrated with a correct set of charts from the latest and best surveys...in two volumes. Boston: Printed & sold by W. Spotswood and J. Nancrede [1747-1821]

MANGIN, JOSEPH F. (C.1803)

Mangin, along with Casimir T. Goerck, was engaged in 1797 by New York City officials to survey and map the city. Their plan was engraved by Peter Maverick and published in 1803 (Ristow 1985). Additional information about Mangin has not been uncovered. The National Union Catalog Pre-1956 Imprints provided no entries for Mangin (The American Library Association 1976).

(1803). A plan and regulation of the city of New York. New York: (np).

MANSFIELD, JARED (1759-1830)

Brown (1897) reveals that Mansfield was a mathematician. He was born and died in New Haven, Connecticut the son of Stephen and Hannah (Beach) Mansfield; grandson of Jonathan and Sarah (Alling) Mansfield. He attended and graduated from Yale College in 1777. He traveled extensively in Ireland, England, and Wales. Upon returning to from his travels, he moved to Philadelphia, Pennsylvania and took charge of a friends' Latin grammar school. Mansfield returned to New Haven and taught students of both sexes (1795-1802). In 1802 he was appointed captain of the Corps of Engineers and stationed at West Point. Between 1803 and 1812 he became U.S. surveyor-general and was promoted to major and then lieutenant-colonel (V.5, pp.348-349).

White (1891, V.III, p.214) further notes that Mansfield married Elizabeth, the daughter of David and Mary (English) Phipps. His interest in astronomy prompted him to import special astronomical instruments from London, and to establish the first observatory in the United States in his own house near Cincinnati, Ohio.

Mansfield, Surveyor General for the United States in 1803, was successor to Rufus Putnam (Pattison 1957, p.201). He resigned his post as mathematics professor at the United States Military Academy at West Point. In 1802 he published a book, Essays, Mathematical and Physical... New Haven, CN: (np). Demonstrating his skill as an observer and mathematician, he calculated the orbit of a comet that appeared in 1807.

He, according to Pattison (1957), established a practical solution to the conflict between rectangular and convergency surveying. The problem was:

> North-south lines, which are not parallel by definition, were called for by law as the boundaries of all townships, yet at the same time the townships were required to be of consistent width, so as to retain their rectangularity and uniformity of size. A compromise was necessary, and Mansfield's intersecting master lines proved a basis for it. By beginning at intervals of six miles, measured from the central—or "principal"—meridian, independent meridians could be laid down as township boundaries. To prevent these boundaries, when prolonged, from affecting the widths of the townships excessively, new, supplementary base lines could be run out from the principal meridian. Along these supplementary base lines, new

north-south lines could be initiated, at the proper six-mile intervals. (pp.210-212)

(1783). Astronomical tables & mathematical questions. Primus incepit, J.M.A.S. 1781, renovavit & revisit, March 1st, 1781... (np). 11,19p. diagrs. 19.5cm.

(1810). ...A calculation on the orbit of the comet, which lately appeared; together with some general observations on comets... In Connecticut academy of arts and sciences. Memoirs, 1, pp.103-110. New Haven.

(1806). Map of the state of Ohio taken from the returns in the office of the surveyor general by Jared F. Mansfield. Washington, DC: author.

(1810). ... Of the figure of the earth... In Connecticut Academy of Arts and Sciences. Memoirs., 1, pp.111-118. New Haven. 22.5cm.

MANSFIELD, JOHN F. (C.1806)

Other than the one map, no information about Mansfield has been uncovered. The National Union Catalog Pre-1956 Imprints provided no entries for Mansfield (The American Library Association 1976).

(1806). Map of the state of Ohio taken from the returns in the office of surveyor general. (np). Scale: 1 in. to about 10 miles.

MARSHALL, JOHN (1755-1835)

Tooley (1985) notes that Marshall relied "heavily" on the British publication of Edmund Burke and included a map in his Life of Washington (p.272). Harley, et al. (1978) also asserts that Marshall depended on other publications for his four-volume Life of George Washington (p.112). He was the fourth Chief Justice, of the U.S. Supreme Court.

Marshall was born in Germantown, Virginia. His early education was provided by his parents, under pioneer conditions. He participated in the American Revolution as a lieutenant and then a captain. His formal education took place at the College of William and Mary where he studied law and was admitted to the bar in 1780. That led Marshall into politics and membership in the Virginia Assembly from 1782 to 1791 (Encarta Encyclopedia 2001. Marshall, John)

(nd). Atlas to Marshall's life of Washington. See his The Life of George Washington. (np).

(1804). A compendious view of the colonies planted by the English on the continent of North America. In Life of Washington, 1. London and Philadelphia.

(1824). A history of the colonies planted by the English on the continent of North America, from their settlement, to the commencement of that was which terminated in their independence. Philadelphia: A. Small. xv486p. 23cm.

(1805 & 1809). Het leven van George Washington, uit deszelf oorspronglijke papieren, onder toezigt van Bushrod Washington zamengesteld door John

Marshall...Uit het Engelsch door J. Wernick... Haarlem: A. Loosjes, pz. 10v. in 6. 10 maps. 21cm. [Accents omitted]

(1804). The life of George Washington: compiled under the inspection of the Honourable Bushrod Washington, from original papers bequeathed to him by his deceased relative, and now in possession of the author; to which is prefixed, an introduction containing a compendious view of the colonies planted by the English on the continent of North America, from their settlement to the commencement of that was which terminated in their independence. Philadelphia: C. P. Wayne. 5v. 8°.

(1804 & 1807). The life of George Washington, Commander in Chief of the American forces during the war which established the independence of his country, and first President of the United States. Compiled under the inspection of the Hon. Bushrod Washington from original papers...to which is prefixed an introduction, containing a compendious view of the colonies planted by the English on the continent of North America. London: Printed by R. Phillips. 5v. fold maps. 20.5cm.

(1804, 1805 & 1807). As above- Philadelphia: C. P. Wayne.

(1831, 32, 33, 34, 35, 36, 38, 39 & 1840). As above- Philadelphia: J. Crissy.

(1843). As above- Philadelphia: J. Crissy, and Thomas, Cowperthwait.

(1845). As above- Philadelphia: J. Crissy.

(1846). As above- Philadelphia: J. Crissy, and Markley, and Thomas Cowperthwait & co..

(1848). As above- Philadelphia: J. Crissy, and Markley, and Thomas Cowperthwait & co..

(1848). As above- Philadelphia: E. C. Markley, & son.

(1850). As above- 2d ed. Philadelphia: J. Crissy, and Markley, and Thomas Cowperthwait & co..

(1854). As above- Philadelphia: J. Crissy, and Markley.

(1859). As above- New York: Derby & Jackson.

(1860). As above- New York: Derby & Jackson.

(1925). As above- New York: W. H. Wise & Co. 5v. 23 cm.

(1926). As above- Fredericksburg, VA: The Citizen's guild of Washington's boyhood home.

(1930). As above- New York: Walton Book co. 2v. maps.

(1839). The life of George Washington, written for the use of schools. 3d ed. Philadelphia: J. Crissy. x379p. 19cm.

(1839). As above- 10th ed. Philadelphia: J. Crissy.

(1840). As above- London: Printed for T. Allman. 440p. 13cm.

(1842). As above- 10th ed. Philadelphia: J. Crissy.

(1844). As above- Philadelphia: J. Crissy.

(1847). As above- 12th ed. Philadelphia: J. Crissy, and Markley.

(1849). As above- 20th ed. Philadelphia: J. Crissy, and Markley.

(1852). As above- 20th ed. Philadelphia: J. Crissy, and Markley.

(1853). As above- 20th ed. Philadelphia: J. Crissy, and Markley.

(1857). As above- New York: Derby & Jackson, Cincinnati: H. W. Derby..
(1858). As above- New York: Derby & Jackson, Cincinnati: H. W. Derby..

Other

Adams, John Stokes (Ed.). (1937). An autobiographical sketch by John Marshall; written at the request of Joseph Story and now printed for the first time from the original manuscript preserved at the William L. Clements library, together with a letter from Chief Justice Marshall to Justice Story relating thereto. Ann Arbor: The University of Michigan Press. xxiii48p. 26.5cm.

Brockenbrough, John W. (Ed.). (1837). Reports of cases decided by the Honourable John Marshall, late Chief Justice of the United States, in the Circuit court of the United States, for the district of Virginia and North Carolina, from 1802 to 1833 [i.e. 1836] inclusive. Philadelphia: James Kay Jun. & Brother; and Pittsburgh: John I. Kay & Co.

Cotton, Joseph P. Jr. (1905). The constitutional decisions of John Marshall: ed., with an introductory essay by Joseph Cotton. New York: G. P. Putnam's Sons. 2v. 23.5cm.

Oster, John Edward. (1914). The political and economic doctrines of John Marshall, who for thirty-four years was Chief Justice of the United States. And also his letters, speeches, and hitherto unpublished and uncollected writings. New York: The Neal Publishing Co. 4p. 369p. 23cm.

Warren, Charles. (1942). The Story-Marshall correspondence (1819-1831). New York: New York University School of Law.

MARTIN, ABSALOM (C.1785)

Martin was one of eight surveyors who reported for duty in 1785 to the newly appointed Geographer of the United States, Thomas Hutchins beginning work on the surveying of America. He was appointed for New Jersey after soliciting the job.

Pattison (1957) Made the following observation:

> Martin undoubtedly went west as an agent for prospective investors, but the success with which he performed his mission on their behalf is unclear. We can recognize in Martin's appointment, however an early expression of an interest in western lands on the part of New Jersey men, which culminated, a few years later, in the purchase of a large tact in the southwest corner of present-day Ohio by Judge John Cleves Symmes. (p.126)

Martin was one of the deputy surveyors appointed under the Land Act of 1800 to work under the direction of Rufus Putnam (Pattison1957, p.207). He served under Hutchins on the Seven Ranges survey. His duties for the Act were to survey the U. S. Military Reserve. The National Union Catalog Pre-1956 Imprints provided no entries for Martin (The American Library Association 1976).

MARTIN, DAVID (C.1796).

Ristow (1985) reports that Martin engraved five maps in W. Winterbotham's Atlas. No other information regarding Martin has been located. The National Union Catalog Pre-1956 Imprints provided no entries for Martin (The American Library Association 1976).

(1796). The American atlas. Philadelphia: John Reid.

MATHEWS, JOHN (C.1744-1802)

According to Pattison (1957), Mathews was one of three men who were engaged in surveying the Seven Ranges, inaugurating the division of the Ohio Company's lands. He later returned to federal surveying after the Land Act of 1796 was passed by Congress (p.179). Mathews was one of the deputy surveyors appointed under the Land Act of 1800 to work under the direction of Rufus Putnam (Pattison 1957, p.207).

Brown (1897) identifies a John Mathews who was a "delegate" born in Charleston, South Carolina in 1744 and died there in 1802. That Mathews was "active in Revolutionary movements" and became a member of the South Carolina House of Representatives, was a delegate to the Continental Congress (1778-1782), signed the Articles of Confederation (1791), and became Governor of South Carolina. There is no note in Brown or in White regarding surveying or other geography related work (V.5, p.404; White 1904, V.XII, p.160). The National Union Catalog Pre-1956 Imprints provided no entries for Mathews (The American Library Association 1976).

MAVERICK, PETER (1780-1831)

Guthron (1984) noted that Maverick did the engraving for the Cahoone and Fosdick chart of Long Island Sound which was published in 1805 by Blunt. Maverick engraved the map surveyed by Casimir Goerck and Joseph Mangin, who were engaged in 1797 by New York City officials to survey and map the city (Ristow 1985).

Maverick was an engraver born in New York city, the son of Peter R. He was a skilful engraver who worked for book publishers, bank note companies, and taught the craft. He was one of the founders of the National Academy of Design. He is most connected to the engravings of such people of Henry Clay, General Jackson, and other notable people (Brown 1903, V.5, p.413).

(1811). [Commissioners'] Map of New York City. By John Randel, Jr., engraved by Peter Maverick. New York: T. & J. Swords.

(1805). Blunt's new and correct chart of Long Island Sound from Montauk Point to Frogs Point, including Fisher's Island Sound, and Watch Hill Reef. Surveyed in 1805 by John Cahoone, Newport, and N.(icoll) Fosdick, New London. New Edition. Engraved by Peter Maverick. New York: E. M. Blunt.

(1822). Explanations, of the ground plan of the University of Virginia. Charlottesville: (np). Broadside 21x16.5cm.

(1816). General atlas. New York: P. Maverick, Durand & Co. 5 maps. 24°.
(1800). Map of the middle states of North America, shewing [sic] the position of the Geneseo Country, comprehending the counties of Ontario & Steuben, as laid off in townships of six miles square each. New York: (np). $14^{1/2}$x$15^{3/4}$inches.
(1803). A plan and regulation of the city of New York. New York: (np).
(1810). Six maps. Engraved for Whiting & Wason's edition of Dr. Scott's Bible. Newark, NJ: P. Maverick.

Other

Damerum, William. (nd). Map of the southern part of New York. (np).

MELISH, JOHN (1771 - 1822)

Melish was a geographer and map publisher. Ristow (1985) provides a lengthy treatment of Melish and his work (pp.179-190). In that material we find that Melish was born in Glasgow, Scotland. He was apprenticed to a textile manufacturer in Glasgow and attended the University of Glasgow. His education and enterprise led to a partnership in the textile business.

He visited the United States around 1798 and moved here in 1809. In 1811 Melish decided to permanently stay in the United States. He engaged in farming and traveled broadly on the North American continent, ultimately engaging in the writing and publishing of books regarding his findings.

(1815). Account of a society at Harmony, (twenty-five miles from Pittsburgh) Pennsylvania, United States of America. Taken from "Travels in the United States of America, in the years 1806 and 1807, and 1809, and 1811"... London: Printed by R. and A. Taylor. 12p. 21cm.
(nd). Atlas of ancient and modern maps. (np).
(1822). A catalogue of maps and geographical works, published and for sale by John Melish. Philadelphia: author. 24p.
(1819). Catalogue of maps, charts, and geographical works, published and for sale... Philadelphia: (np).
(1814). Description of the roads in the United States. Philadelphia: author.
(1813). A description of East and West Florida and the Bahama Islands, &c. &c. Philadelphia: Printed by G. Palmer. 3-18p. map. 22cm.
(c.1813). A description of the British possessions in North America, &c. &c. Philadelphia: (np). 28p. map.
(1812 & 1813). A description of the British possessions in North America, and of the most important places along the lines in the United States; intended as an accompaniment to Melish's map of the Seat of War in North America. Philadelphia: Palmer. 18p. map. 23.5 cm.
(1814 & 1815). A description of the roads in the United States. Comp. from the most authentic materials. Philadelphia: Printed by G. Palmer. iv4-11p. 82col. 17.5cm.
(1813 & 1815). Description of the Seat of War in North America. Philadelphia: Printed by G. Palmer. 3-34p. 4 maps. 22cm.

(nd). East end of Lake Ontario and River St. Lawrence from Kingston to French Mills, reduced from an original drawing in the Naval Department. H. S. Tanner, sc. (np). map. $18^{3/8}$x$22^{1/2}$inches.
(1815). Falls of Ohio. Philadelphia: (np). map. 16x10cm.
(1818). A general atlas and geography. Philadelphia: author.
(1815). Geographical description of the United States. Philadelphia: Printed for J. Melish, G. Palmer, printer. 32p. maps.
(1816). Geographical description of the United States. Philadelphia: author. 92x145cm. Scale 1:3,800,000.
(1816). Geographical description of the United States, with the contiguous British and Spanish possessions, intended as an accompaniment to Melish's map of these countries. Philadelphia: author.
(1818). As above- 3d ed.
(1816). Geographical description of the United States, with the contiguous countries, including Mexico and the West Indies; intended as an accompaniment of Melish's map of these countries... A new ed. greatly improved. Philadelphia: author. viii9-491,15p. maps. 21.5cm.
(1826). As above- New York: A. T. Goodrich.
(1818). A geographical description of the world, intended as an accompaniment to the map of the world on Mercator's projection. Philadelphia: John Melish and Samuel Harrison. 7ix-xiii2,9-280p. maps. 21.5cm.
(1822). A Geographical Description of the World, intended as an accompaniment to the map of the world. A new ed., greatly improved. Philadelphia: John Melish and Samuel Harrison. 289p. maps. 22cm.
(1815). Geographical establishment. No. 49, South Third Street, Philadelphia. John Melish, geographer and map publisher, has lately published the following works... Philadelphia: author. 8p. 23.5cm.
(1816 & 1818). Geographical intelligence. Philadelphia: author. 12p. map.
(1818). Geographical intelligence including a prospectus of the map and description of the world, map and description of the United States, and universal school atlas and geography with directions for their use. Philadelphia: Melish. 28p. 21.5cm.
(1819). Information and advice to emigrants to the United States: and from the eastern to the western states: illustrated by a map of the United States and a chart of the Atlantic Ocean. Philadelphia: author. viii,144p. maps. 15cm.
(1818). Map of Alabama. Philadelphia: author.
(1813). Map of the American coast from Lynhaven Bay to Narraganset Bay. Engraved by Benjamin Tanner. Philadelphia: author.
(c.1818). Map of Illinois constructed form the surveys in the General Land Office and other documents... Philadelphia: author. 62.5x62.5cm.
(1817). Map of Indiana. Philadelphia: Melish & Harrison. $16^{7/8}$x$12^{1/2}$inches. Scale: 17 miles to 1 inch.
(1819). Map of Mississippi constructed from the surveys in the General Land Office and other documents. Philadelphia: author. $13^{1/2}$x9inches.

(1822, 1824 & 1826). Map of Pennsylvania; conducted by the county surveys authorized by the state. Engraved by B. Tanner. 135x196cm. Scale: 25 statute mi. to 1760 yds.

(1813). Map of the Seat of War in North America. Philadelphia: author. Accompanied by : Description of the Seat of War in North America.

(1813). Map of the Southern Section of the United States. Engraved by Benjamin Tanner. Philadelphia: author. Accompanied by : Description of the Seat of War in Southern Section.

(1813). Map of the Southern Section of the United States, including the Floridas & Bahama Islands shewing the Seat of War, in that department. Engraved by H. S. Tanner. Philadelphia: author.

(1818). Map of Tennessee. Philadelphia: Melish & Harrison. $14^{1/2}$x$23^{3/8}$inches. Scale: $18^{1/3}$miles to 1 inch.

(1813). Map of the United States. H. S. Tanner, sculp. Philadelphia: author. 12x$17^{1/4}$inches.

(c.1816). Map of the United States of America to west lat. 96.75º only. Philadelphia: author. 45x33cm.

(c.1816). Map of the United States with the contiguous British & Spanish possessions. Compiled from the latest & best authorities. Engraved by J. Vallance & H. S. Tanner. Philadelphia: author. 262x88cm. Scale 1:3,360,000. Inset map of the West Indies.

(c.1818). As above- Improved to the 1st of January 1818. 88x145cm. Scale 1:3,900,000.

(1820). As above- Improved to the 1st of January 1818. 94x149cm.

(nd). Melish map, 1819 [Spanish treaty]. (np). No scale. 60x41cm.

(1816). Melish's universal school atlas. Philadelphia: author. 9 maps. 11in.

(1818). Melish's universal school geography, being a companion for his Universal school atlas.... Philadelphia: author. 18cm.

(1812). A military and topographical atlas of the United States; including the British possessions & Florida, comprehending: 1. Map and description of the Seat of War...and a map and plan of Montreal. 2. Southern section of U.S. 3. Am. coast. 4. Detroit River. 5. Queec; to which is added a list of the military districts, a register of the army, and a list of the navy of the U.S... Philadelphia: G. Palmer, printer.

(1813 & 1815). As above- 6,3-34,3-18,3-29,44p. 8 maps. 22cm.

(1813-1814). A new juvenile atlas and familiar introduction to the use of maps: With a comprehensive view of the present state of the Earth. By John Melish, Vallance, and Henry Tanner. Philadelphia: Printed by G. Palmer.

(1816). Northern section of the United States including Canada &c. (np). 38x52cm. Scale: 1 inch to 70 miles.

(c.1820). Ohio. J. Vallance, sc. (1812). Improved to 1820. Philadelphia: author. 9x9inches.

(1819). Reisen durch die Vereinten-Staaten von America, in den Jahren 1806, 1807, 1809, 1811, und 1811. Aus dem Englishchen, und mit Amerkungen begleitet von J. Ernst Ludw. Brauns. Weimar: im Verlage des Gr. H. S. pr. Landes-Industrie-Comptoirs, 1819. 378p. maps. 20cm. [Accents omitted]

(1814). The sine qua non: a map of the United States, shewing [sic] the boundary line proposed by the British commissioners at Ghent, with the documents relative to the negotiations. Philadelphia: author, printed by G. Palmer. 63p. map. 18.5cm.

(1815). As above. 2d edition, with additional documents. iv,3-120p. map. 12°.

(1816). Southern section of the United States including Florida, &c.... Philadelphia: author. 41x53.5cm. Scale: 4.1cm. to 100 miles.

(1813). Statistical account of the United States, with topographical tables of the counties, towns, population, &c. From the census of 1810. Intended as an accompaniment to the portable map of the United States. Philadelphia: Melish & Tanner. 35 (i.e. 37), 1p. 15cm.

(1822). As above- 45p. 16.5cm.

(1815 & 1816). Traveller's directory through the United States: consisting of a geographic description of the United States, with topographical tables of the counties, towns, population, &c., and a description of the roads, compiled from the most authentic materials. Philadelphia: Melish & G. Palmer. 2 parts in 1v. maps. 24°.

(1818). As above- Melish & S. Harrison, geographers.

(1819). As above- 5th edition. 102p. maps. 16°.

(1819). As above- 134p. 16cm.

(1820). As above- 6th edition.

(1822). As above- xix,183p.

(1825). As above- New York: A. T. Goodrich. xix,194p. maps. 18.5cm.

(1831). The travellers manual; and description of the United States: comprising geographical, historical, and statistical details of the union, and of each state... New York: A. T. Goodrich. 9-497,15p. 12 maps. 23.5cm.

(1812 & 1815). Travels in the United States of America in the years 1806 & 1807, and 1809, 1810, & 1811; including an account of passages betwixt [sic] America and Britain, and travels through various parts of Great Britain, Ireland, and Upper Canada. Illustrated by eight maps. Philadelphia: author, printed by Thomas & George Palmer. 2v. maps. 22cm.

(1818). As above- xxiii25-648p. maps. 23.5cm.

(1818). As above- London: G. Cowie & co. xxiii25-648p. maps. 23.5cm.

(1819). United States of America compiled from the latest and best authorities. Philadelphia: author. 52.2x43.8cm.

(1818). Universal school geography. Philadelphia: author.

(1818). Universal school atlas. Philadelphia: author.

(1815). View of the country round Pittsburgh. Philadelphia: author. 16x11cm.

(1822). Views on political economy, from the description of the United States. Philadelphia: author. 27p. tables. 24cm.

(1818 & 1820). The world of Mercator's Projection, revised and improved to 1818 by John Melish. Philadelphia: author.

Other

Berkbeck, Morris. (1818). Letters from Illinois. Illustrated by a map of the United States, shewing [sic] Mr. Birkbeck's journey from Norfolk to Illinois, and a map of the English prairie and the adjacent country, by John Melish... Philadelphia: M. Carey & Son.

Lavoisne, C. (nd). A complete genealogical, historical, chronological, and geographical atlas... (np).

(nd). The North American tourist. (Melish supposed author).

Mayo, Robert. (nd). An atlas of ten select maps of ancient geography, both sacred and profane; with a chronological table of universal history & biography. (np).

MITCHILL, SAMUEL LATHAM (1764-1831)

Mitchell was a geologist and author born in North Hampstead, Long Island, New York and died in New York city. He studied medicine and graduated from the University of Edinburgh, Maryland in 1786. He was appointed U.S. Commissioner to make a treaty with the Iroquois Indians. Between 1801 and 1804 Mitchill was elected Democratic representative in the 7th and 8th congresses.

While a professor of botany in Columbia College (1792-1799) he helped establish the Society for the Promotion of Agriculture, Manufactures and Useful Arts. During that same time he made a geological survey of the state. He was also known as "Nestor of American Sciences" (Brown 1897, V.5, pp.511-512).

White (1897) additionally asserts that Mitchill was much respected in his time. Thomas Jefferson called him the "Congressional Dictionary" because of his political knowledge. He was the champion of Fulton and Livingston as they tried to introduce navigation by steam. His scientific writings were numerous and stimulated research in America and Europe. His pamphlet, "A Picture of New York," was the basis of Washington Iring's Knickerbocker's History of New York (V.IV. p.409).

(1811). Account of the remains of marine animals in a fossil state, in New Jersey. American Mineralogical Journal, 1(2), art. 20, pp.95-96.

(1801). Additional articles on my report to the Agricultural Society on the mineralogy of New York... The Medical Repository, New York City, 5(2), pp.212-214.

(1825). Address on the completion of the Erie Canal. (np).

(1812). An amendment proposed to the geological chart of the United States, as respects the character of the north side of Long Island, which is shown to be alluvial and not primitive, as therein stated... American Mineralogical Journal, 1(3), art. 23, pp.129-133.

(1810). Descriptive catalogue accompanying a suite of mineral specimens presented to the editor... by his colleague... collected on a tour to Niagara... American Mineralogical Journal, 1(1), art. 1, pp.1-5.

(1810). Description of the Fezzan ram, lately imported into New York... American Medical and Philosophical Register, 1(2), pp.245-252.

(1814). A detailed narrative of the earthquakes which occurred on the 16th of December, 1811, and agitated the parts of North America that lie between the Atlantic Ocean and Louisiana; and also particular account of the other quakings of the Earth occasionally felt from that time to the 23d and 30th of January, and the 7th and 16th of February, 1812, and subsequently to the 18th of December, 1813, and which shook the country from Detroit and the Lakes of New Orleans and the Gulf of Mexico; compiled chiefly at Washington, DC. Literary and Philosophical Society of New York, Papers on Natural History, 1(3), pp.281-331.

(1801). Experiments, facts and observations in natural history made during the autumn of 1801... The Medical Repository, New York City, 5(2), pp.205-216.

(1800). Facts in natural history noted during the summer and autumn of 1799. The Medical Repository, New York City, 3(4), pp.420-422.

(1814). The fishes of New York, described and arranged... Literary and Philosophical Society of New York, Papers on Natural History, 1(5), pp.355-492.

(1812). Geological observations on the United States. American Medical and Philosophical Register, 2(3), art. 3, pp.253-256.

(1818). The hedge-hog ray, a species of fish taken occasionally near New York, in the Atlantic Ocean, and, now, as is believed for the first time described. American Journal of Science, 9(2), pp.290-93.

(1810). Information concerning the Canada thistle, a plant which exceedingly annoys the agriculture of the northern and western counties of New York... American Medical and Philosophical Register, 1(2), art. 10, pp.211-216.

(1801). Luminous appearance of ocean water caused by animals... The Medical Repository, New York City, avr. 11, pp.375-383.

(1815). A memoir on the fossil remains of organized beings, more especially of animals around New York. Literary and Philosophical Society of New York, Papers on Natural History, 1(2), p.125.

(1798). A new variety of iron-ore of the argillaceous kind, and resembling basaltes. Medical Repository, New York City, 2(2), pp.219-220.

(1799). Observations on cankerworms, and the means of preventing their effects. Transactions the Society for the Promotion of Agriculture, Arts and Manufactures, New York City and Albany, pt. 4, pp.33-41; 2d ed. p.314-321.

(1822). Observations and facts concerning the Siren Lacertina of Georgia and Carolina... New York Medical and Physical Journal, 3(2), p.184-198.

(1800). Observations in natural history, made in the summer and autumn of the year 1800... The Medical Repository, New York City, 4(2), pp.196-200.

(1805). On the ice islands... Proceedings of the American Philosophical Society, 22, appx. p.380.
(1811). The physical geography of the first range of mountains extending across New-Jersey, from the Hudson to the Delaware; with some experiments on the Chalybeate spring at Schooley's Mountain. American Mineralogical Journal, 1(2), art. 13, pp.70-79.
(1799). A review of B. S. Barton's "fragments of the natural history of Pennsylvania, part. 1" The Medical Repository, New York City, 3(2), art. 2, pp.177-189.
(1800). A sketch of the mineralogical and geological history of New York... The Medical Repository, New York City, 3(4), art. 1, pp.325-335.
(1798). A sketch of the mineralogical and geological history of the state of New York... The Medical Repository, New York City, 1(3), art. 1, pp.293-314.
(1799). A sketch of the mineralogical history of the state of New York, by...the commissioner appointed to make a tour through the state of New York, in the vicinity of Hudson's River... Transactions the Society for the Promotion of Agriculture, Arts and Manufactures, New York City and Albany, pt.4, pp.124-152.
(1814). A sketch of the scenery in the region around Harper's Ferry, where the ridge of Blue Mountains is penetrated by the joint waters of the Potomac and Shenandoah Rivers... American Minerlogical Journal, 1(4), art. 34, pp.211-218.
(1819). Supplement to Mitchill's "Ichthyology of New York." Literary and Philosophical Society of New York, Papers on Natural History, 4(4), pp.284-85.

MOORE, SAMUEL S. (C.1774-1861)

Brown (1897) identifies a Samuel Moore born in Deerfield, New Jersey, the son of Col. David and Lydia (Richman) Moore. That Moore died in Philadelphia in 1861. He graduated from the University of Pennsylvania, became a tutor there, and practiced medicine for a short time. Poor health forced him to abandon his profession and he entered the East India trade, making trips to China and India. In 1824 he was appointed Director of the U.S. Mint at Philadelphia. There he stayed until 1835 when he engaged in mining. Moore was a member of the American Philosophical Society (V.5, p.539; White 1906, V.XII, p.421).

(1802). The traveller's directory or a pocket companion, shewing [sic] the course of the main road from Philadelphia to New York and from Philadelphia to Washington. Philadelphia: Mathew Carey. T. W. Jones, co-author. 52p. 38 maps. 21.5cm.
(1804). As above- 2d ed. 37, 19p. 22cm.

MORRIS, WILLIAM W. (C.1785)

Morris was one of eight surveyors who reported for duty in 1785 to the newly appointed Geographer of the United States, Thomas Hutchins beginning work on the surveying of America. He was not interested in land speculation. Morris was not sent, like some of his fellow surveyors, as a spy to seek out new investments. He was a skilled surveyor and the equal of Hutchins. "He gained Hutchins' special commendation for his assistance in 1785" (Pattison 1957, p.127).

The boundary line between Pennsylvania and Maryland was completed by Morris, Rittenhouse, Hardenberg and Ellicott in 1787 (Barton 1813, p.246). No other information regarding Morris has been found. The National Union Catalog Pre-1956 Imprints provided no entries for Morris (The American Library Association 1976).

MORSE, JEDIDIAH (1761-1826)

Brown (1941) identifies Morse as, "The Father of American Geography" (Brown 1941, p. 7; The Biographical Dictionary 2001, Morse, Jedidiah; Encyclopedia Britannica 2001. Morse, Jedidiah). Morse, on August 23, 1761, was born in Woodstock, Connecticut. He attended and graduated from Yale University in 1783. During that year he opened a young ladies school in New Haven, Connecticut. He later studied for the ministry and was ordained in 1786. His activities included a deep interest in "civilizing the American Indians," which prompted him to become appointed by the Secretary of War to visit and observe Indian tribes with directions to devise suitable means for improving their situation. Morse was also involved in religious controversy by upholding the faith of the New England Church. His interests and travel fostered an interest that prompted him to become involved in writing geography books.

On May 4, 1789, Morse married Elizabeth A. Breese at Shrewsbury, New Jersey. That marriage produced ten children, three of them growing to maturity. Samuel F. B. Morse became a noted portrait painter and the inventor of the telegraph. Sidney Edwards Morse was a prominent journalist in New York. Along with his brother, Richard Cary Morse he founded the New York Observer, the oldest religious newspaper in the state. Sidney worked with his father in the preparation of geographies, and in 1839, in collaboration with Henry A. Munson, he developed a method of printing on the printing press, maps in color. He and Samuel constructed and patented a flexible piston pump and later invented an instrument for exploring the depths of the sea, the bathyometer (Johnson & Malone 1928, V.13, p.245; White 1903, V.XIII, p.353).

Ralph Brown (1941) published an article, "The American Geographies of Jedidiah Morse" in the Annals of the Association of American Geographers. In his article Brown notes that, "By the year 1800, Morse's name was familiar to the reading public and some of his contemporaries were hailing him as 'the father of American geography,' or as Judge Harris of Rhode Island said in

1794, 'our first American Geographer,' (p. 7). Morse has been accepted by most contemporary American geographers as he was identified by Brown. That may be in question.

Morse, nonetheless, was the producer of what appears to be the first of American geography books. In his geographies he was addressing what he saw as the needs of citizens in the new United States from the point of view of an American citizen. He was informing and persuading them that this is a wonderful bountiful country.

(1797-1798). An abridgment of the American gazatteer exhibiting, in alphabetical order, a compendious account of the states, provinces...on the American continent, and its appendant islands; particularly the Wet Indies...to which is annexed, an accurate table of all the post-offices in the United States ...illustrated with a map of North America...Boston: Printed by Thomas and Andrews.

(1796). The American gazatteer, exhibiting, in alphabetical order, a much more full and accurate account, than has been given, of the states, provinces, counties, cities, towns... on the American continent, also of the West-India islands... with a particular description of the Georgia western territory... Collected and compiled... by, and under the direction of, Jedidiah Morse... Illustrated with seven maps. Boston, New England; and London: Reprinted for J. Stockdale 619p. 7 maps (6 fold.). 22cm.

(1797). As above- Boston: Printed by S. Hall, and Thomas and Andrews, and sold by E. Larkin; and New York: Ten Eyck, and S. Campbell.

(1798). As above- 2d ed. Boston, New England; and London: Reprinted for J. Stockdale. 624p. 22cm.

(1804). As above- Charlestown, Mass.: Printed by and for Samuel Etheridge; and Boston: Thomas Andrews.

(1810). As above- 3d ed. Boston: Thomas & Andrews, J. T. Buckingham, printer. 800p. 22.5cm.

(1789). The American geography; or a view of the present situation of the United States of America. Containing astronomical geography. Geographical definitions. Discovery, and general description of America...Particular descriptions of the thirteen United States, and of Kentucky, the western territory and Vermont... Illustrated with two sheet maps... To which is added, a concise abridgment of the geography of the British, Spanish, French, and Dutch dominions in America, and the West Indies—of Europe, Asia and Africa. Elizabeth Town, NJ: Printed by Shepard Kollock. 534p. maps. 20.5cm.

(1792). As above- 2d ed. London: Printed for J. Stockdale. 536p. 22cm.

(1792). As above- 3d ed. Dublin: Printed for J. Jones.

(1794). As above- London: Printed for J. Stockdale. 715p. 24 maps (7 fold fold), 28x22cm.

(1789). The American school geography; or a view of the present situation of the United States of America...illustrated with two sheet maps...to which is added, a concise abridgment of the geography of the British, Spanish, French and Dutch dominions in America, and the West Indies - of Europe, Asia and Africa... Elisabethtown, NJ: Printed by Shepard Kollock.

(1793). The American universal geography, or, a view of the present state of all the empires, kingdoms, states, and republics in the known world, and of the United States of America in particular... Illustrated with maps. Boston: Isaiah Thomas and Ebenezer T. Andrews; Worcester: Thomas; New York: Berry, Rogers, and Berry; Philadelphia: H. & P. Rice; and Charleston: W. P. Young. 2v. 22cm.

(1796). As above- 2d ed. Boston: Isaiah Thomas and Ebenezer T. Andrews.

(1796). As above- 3d ed. Boston: Isaiah Thomas and Ebenezer T. Andrews; and Worcester: Thomas.

(1805, 12). As above- Boston: Printed by J. T. Backingham for Thomas and Andrews.

(1807). As above- Boston:

(1819). As above- 7th ed. Boston: Lincoln & Edmands; and Charlestown: Lincoln & Edmands.

(1798). An abridgment of the American Gazatteer. Exhibiting in alphabetical order, a compendious account of the states, provinces, counties, cities... rivers, bays, harbours... on the American continent, and its appendant islands: particularly the West-Indies... To which is annexed an accurate table of the post-offices in the United States. Boston: Thomas and Andrews; Worcester: I. Thomas; Albany: Thomas, Andrews & Penniman; and Baltimore: Thomas, Andrews & Butler. 398p. 17.5cm.

(1822). Ancient atlas adapted to Morse's new school geography. Boston: Printed by J. H. A. Frost. 5 col. maps (1 double). 50cm.

(182?). Atlas to Morse's geography. (np). 8 fold. maps (1 double). 30cm.

(1804). A compendious history of New England, designed for schools and private families. By Jedidiah Morse, and Rev. Elijah Parish... with... a map. Charlestown, Mass.: Printed and sold by Samuel Etheridge. 336p. 17.5cm.

(1809). As above- 2d ed. Newburyport: Thomas & Whipple. 336p. 17.5cm.

(1808). A compendious history of New England, exhibiting an interesting view of the first settlers of that country, their character, their sufferings, and their ultimate prosperity. Collected and arranged from authentic sources of information, by Jedidiah Morse and Rev. Elijah Parish. London: C. Taylor. 207p. 19cm.

(1820). As above- ...To which is added, a short abstract of the history of New York, and New Jersey. Charlestown, Mass.: Printed by S. Etheridge. 514p. 18cm..

(1814). A compendious and complete system of modern geography. Or a view of the present state of the world. Being a faithful abridgement of The American universal geography, (edition of 1812). with corrections and additions made from information since received..Illustrated by a representation of the solar system, and six maps of the principal divisions of the globe. Boston: Thomas and Andrews; New York: Eastburn, Kirk, and Co.; and Philadelphia: M. Carey. 670p. 6 folded maps. 22cm.

(1797). A description of the soil, productions, commercial, agricultural and local advantages of the Georgia western territory: together with a summary and impartial view of the claims of Georgia and of the United States to this territory, and of the principal arguments adduced by the purchasers against these claims. Collected and stated from various authentic documents. Extracted and publ. in this form, (by permission) from Rev. Mr. Morse's American Gazatteer... Boston: Printed by Thomas and Andrews. 24p. folded map. 19.5cm.

(1795 & 1798). Elements of geography: containing a concise and comprehensive view of that useful science, as divided into, 1. astronomical 2. physical or natural 3. political geography. On a new plan. Adapted to the capacities of children and youth; and designed, from its cheapness, for a reading and classical book in common schools, and as useful winter evening's entertainment for young people in private families. Illustrated with a neat map of the United States, and a beautiful chart of the whole world. 2d. ed., cor. Boston: E. T. Andrews; Worcester: I. Thomas; New York: S. Campbell; Philadelphia: M. Carey; and Baltimore: Thomas, Andrews and Butler. 143p. 2 folded maps. 13.5cm.

(1801). As above- 4th ed. Boston: I. Thomas and E. T. Andrews.

(1804). As above- 5th edition.

(1825). As above- 6th ed., rev. and cor. New Haven: H. Howe. 162p. 15cm..

(1787). To the friends of science. It has long been the universal wish, that the public might be furnished with an accurate geographical... grammar of the United States... Philadelphia: Printed by Robert Aitken & Son. broadside. 27x22cm.

(1811). A geographical and historical view of the world..with notes, correcting and improving the part which relates to the American continent and islands. By Jedidiah Morse...in fine volumes... Boston: Prin. Thomas B. Wait & Co. 21.5x23.5cm.

(1811). A geographical and historical view of the world: Exhibiting a complete delineation of the natural and artificial features of each country; and a succinct narrative of the origin of the different nations, their political revolutions, and progress in arts, sciences, literature, commerce, etc....the whole comprising all that is important in the geography of the globe and the history of mankind. With notes correcting the part which relates to the American continent and islands. By Jedidiah Morse in Five Volumes. Boston: Printed by Thomas B. Wait & Co. 405p. 14x21.6cm.

(nd). Geographical queries... (np). Broadside. 25.5x21cm.

(1790). Geography made easy: being an abridgement of the American geography. Containing, astronomical geography... discovery and general description of America... To which is added, a geographical account of the European settlements in America, and of Europe, Asia and Africa... Calculated particularly for the use and improvement of schools in the United States. 2d ed. Boston: Printed by Isaiah Thomas & Ebenezer T. Andrews. 322p. 7 maps. 12°.

(1791). As above- 3d ed. Boston: Printed by Samuel Hall. 323p. 17.5cm.

(1794). As above- 4th ed. Boston: Printed by I. Thomas and E. T. Andrews. 432p. 17cm.

(1796). As above- 5th ed.

(1798). As above- 6th ed. Boston: Printed by I. Thomas and E. T. Andrews; Worcester: I. Thomas; New York: S. Campbell; Albany: Thomas, Andrews & Penniman; Philadelphia: M. Carey; Baltimore: Thomas, Andrews & Butler.

(1800). As above- 7th ed.

(1802). As above- 8th ed. Boston: I. Thomas and E. T. Andrews.

(1804). As above- 9th ed. Boston: Printed by Joseph T. Buckingham for Thomas and Andrews.

(1806). As above- 10th ed.

(1807). As above- 11th ed.

(1809). Geography made easy: being an abridgment of the American universal geography. To which are prefixed elements of geography. For the use of schools and academies in the United States of America. Boston: Thomas & Andrews. 364p. 18cm..

(1811). As above- 2d ed. Boston: Thomas & Andrews. J. T. Buckingham, printer.

(1812). As above- 3d ed.

(1813). As above- 4th ed.

(1814). As above- 17th ed., and 5th of this new abridgement.

(1816). As above- 18th ed., and 6th of this new abridgement.

(1816). As above- 16th Boston Edition. Boston: Troy, Parker and Bliss.

(1818). As above- 19th ed., and 7th of the new abridgement. Boston: Thomas & Andrews.

(1819). As above- 20th ed. Utica: W. Williams; and Boston: West, Richardson, & Lord.

(1820). As above- 23d ed. Boston: Richardson & Lord.

(1784). Geography made easy. Being a short, but comprehensive system of that very useful and agreeable science...illustrated with two correct and elegant maps, one of the world and the other of the United States, together with a number of newly constructed maps, adapted to the capacities and understanding of children. Calculated particularly for the use and improvement of schools in the United States... Newhaven, Conn.: Printed by Meigs, Bowen and Dana. 214p. 16.5cm.

(1790). The history of America, in two books. Containing, I. A general history of America. II. A concise history of the late revolution. Extracted from the American edition of the encyclopedia. Philadelphia: Thomas Dobson. 360p. 17.5cm.

(1795). As above- 2d ed.
(1798, 1808). As above- 3d ed.
(1808). As above- 4th ed. 455p. 18.5cm.
(1819). As above- Philadelphia: Printed for J. Webster.
(nd). A new and correct edition of the American geography; or, a view of the present situation of the United States of America: containing astronomical geography. - Geographical definitions, discovery and general description of America and the United States... With a particular description of Kentucky, the western territory, and Vermont... To which is added an abridgment of the geography of the British, Spanish, French, and Dutch dominions in America and the West Indies. - Of Europe. (np).
(1802). A new gazatteer of the eastern continent; or, a geographical dictionary: containing, in alphabetical order, a description of all the countries, kingdoms in Europe, Asia, Africa and their adjacent islands. Carefully compiled...by Jedidiah Morse and Rev. Elijah Parish...illustrated with 18 maps. Designed as a second volume, to the American gazetteer. Charleston, Mass.: Printed and sold by Samuel Etheridge. 712p. 22cm.
(1806). As above- 2d ed. Boston: Printed by J. T. Buckingham for Thomas & Andrews. 726p. 3 folded maps.
(1789). A new geographical dictionary. Elizabeth Town: (np).
(1822). A new system of geography, ancient and modern, for the use of schools; accompanied with an atlas... By Jedidiah Morse and Sidney Edwards Morse. 23d ed. Boston: Richardson & Lord. 342p. 18cm.
(1824). As above- 24th ed.
(1826). As above- 25th ed.
(1828). As above- 26th ed. New York: Collins and Hannay.
(1822). A new universal atlas of the world. Comprising twenty maps, carefully prepared, from the last information, and neatly engraved; the world, its several grand divisions, and principal subdivisions. By Jedidiah Morse and Sidney E. Morse. (np). 29x29cm.
(1831). As above- ...2d ed. With numerous additions and improvements by N. & S. S. Jocelyn. New Haven: I. R. Goodno. 21 colored maps (part double). 29cm.
(1821). A new universal gazatteer, or geographical dictionary... Accompanied with an atlas. By Jedidiah Morse and Richard C. Morse. 3d ed. New Haven: S. Converse. 856p. 24.5cm.
(1823). As above- 3d ed.
(1797). Proposals for printing by subscription a gazetteer of America, including the West-India Islands and others adjacent to the continent. The whole collected, compiled and arranged by Jedidiah Morse. Boston: (np).
(1822). A report to the Secretary of War of the United States, on Indian Affairs, comprising a narrative of a tour performed in the summer of 1820, under the commission from the President of the United States, for the purpose of ascertaining, for the use of the government, the actual state of the Indian

tribes in our country. Washington: Davis and Force; printed by S. Converse, New Haven. 400p. folded map. 22.5cm.

(1823). The traveller's guide: or pocket gazatteer of the United States; extracted from the latest edition of Morse's Universal Gazetteer. With an appendix. By Jedidiah Morse, D. D., and Richard C. Morse, A. M. New Haven: N. Whiting. 323p. 16cm.

(1826). As above- New Haven: S. Wadsworth.

Other

Belknap, Jeremy. (1955). Report on the Onaids, Stockbridge, and Brotherton Indians, 1796, by Jeremy Belknap and Jedidiah Morse. New York: Museum of the American Indians, Haye Foundation.

Bigland, John. (1814). Letters on the study and use of ancient and modern history: containing observations and reflections on the causes and consequences of those events which have produced conspicuous changes in the aspect of the world, and the general state of human affairs. By John Bigland... With notes by the Rev. Jedidiah Morse... Philadelphia: Printed for W. W. Woodward.

Morse, Sidney Edwards. (1822). A new system of modern geography, or a view of the present state of the world... Accompanied with an atlas. Boston: G. Clark; and New Haven: Howe & Spalding.

NANCARROW, JOHN (C.1791)

Rittenhouse, Ewing, and John Nancarrow were appointed on October 10, 1791, by Governor Mifflin of Pennsylvania, to "view and mark out a road" from Schuylkill to Lancaster and from Philadelphia to that same borough, and to provide topographical observations (Barton 1813, p.361-362). No other information regarding Nancarrow has been located. The National Union Catalog Pre-1956 Imprints provided no entries for Nancarrow (The American Library Association 1976).

NAU, ANTOINE (C.1800)

The First Part of Captn. Pike's Chart of the Internal Part of Louisiana and A Chart of the Internal Part of Louisiana, including all the Hitherto Unexplored Countries, Lying between the River LaPlatte of the Missouri on the N: and the Red River on the S: the Mississippi East and the Mountains of Mexico West: With a part of New Mexico and the Province of Texas was compiled by Antoine Nau, a French draftsman serving under General James Wilkinson in St. Louis. According to Schwartz and Ehrenberg (1980), "They are the first published maps to depict the geographical knowledge of the entire southwest derived directly from exploring parties" (p.225). The National Union Catalog Pre-1956 Imprints revealed no entries for Nau (The American Library Association 1976).

NEVILLE, JOSEPH (C.1785)

Barton (1813) identifies several people engaged in establishing the boundary line between Virginia and Pennsylvania:

> This boundary-line was begun in May, 1785, by Messrs. Rittenhouse, Ellicott, Porter, and Nevill; assisted by the present Dr. Benjamin Smith Barton, then a youth about nineteen years of age, whose medical and other scientific acquirements rendered him, even at that early period of life, an useful associate of the commissioners. Mr. Nevill (who was employed on the part of Virginia) left the other commissioners late in August. (pp.306-307)

No other information about Neville has been found. The National Union Catalog Pre-1956 Imprints did not list any publications for Neville (The American Library Association 1976).

NICHOLS, FRANCIS (1737-1812)

Nichols is identified by Brown (1903) as a soldier born in Crieve Hill, Enniskillen, Ireland in 1737 immigrating to America in 1769. He became captain and quartermaster in the American army. He settled in Cumberland County, Pennsylvania, enlisted in the patriot army, was captured at the Battle of Quebec. Later released and his sword returned to him. Nichols rose to the rank of brigadier-general. Following the war he became the first U.S. Marshall of the Eastern District of Pennsylvania. He died in Pottstown, Pennsylvania (V.VI, p.6).

(1809). A compend of geography, containing a concise description of the different countries of the Earth. For the use of students. Philadelphia: Printed for author by T & G Palmer. 190p. 11 plates. 17.5cm.

(1813 & 1818). An elementary treatise of geography. Containing a concise description of the different countries of the world. Compiled from the best modern travels. Also a compendious view of the solar system. For the use of schools...A new edition corrected and improved. Philadelphia: Printed for Francis Nichols by William Fry. 162p. 15cm.

(1811). A new atlas, adapted to the use of students of geography and history, and intended to illustrate the following works published by F. Nichols. I. A compend of geography...By F. Nichols. II. Geographical delineations...By J. Aiken. III. Elements of general history...By A. F. Tytler. Philadelphia: F. Nichols. 11 maps. 4°.

Other

(1908). The Penn. magazine of history and biography. Published quarterly by Historical Society of Pennsylvania, pp.108-10, 119-120, Philadelphia.

NORMAN, I. (C.1791)

A map by Norman is found in Schwartz and Ehrenberg (1980):

> Shows Phelps-Gorham land sale in western New York east of Genesee River, the Seven Ranges west of the Ohio River, and new

townships between the Penobscot River and the Scioto River (Schodic on map)-all for first time on an inclusive map of United States. 'A Map of the Lakes and Rivers Between Lake Superior and the North Sea' included as inset. (p.206)

(1791). The United States of America laid down from the best authorities agreeable to the Peace of 1783. Boston: I. Norman.

NORMAN, JOHN (1748-1817)

Norman was born in England but worked in Philadelphia and Boston (Guthorn, 1984) . Two of his charts appeared in Clark's (1790) A Complete Set of Charts of the Coast of America, Extending from Cape Breton on the East, into the Gulph of Mexico, with Nautical Directions. In 1790 he advertised a new set of charts of the West Indies and that he was then engraving a set of charts of the American coast on a large scale. The map, "The United States of America Laid Down From the Best Authorities Agreeable to the Peace of 1783," published by I. Norman was likely related to John or William Norman (Schwartz & Ehrenberg 1980). Ristow (1985) tells us that Norman was an engraver who worked with Osgood Carleton (pp.66, 91, 224).

(1798). An accurate map of the Commonwealth of Massachusetts exclusive of the District of Maine compiled pursuant to an act of the General Court from actual surveys of the several towns &c. Taken by their order... By Osgood Carleton. Boston: Published and Sold by O. Carleton and J. Norman. Sold also by W. Norman. 89x119cm. Scale 1:253,440.

(1791, 1792 & 1794). The American pilot, containing, the navigation of the sea coasts of North America, from the Streights [sic] of Beel Isle to Cayenne, including the island and banks of Newfoundland, the West India Islands, and all the islands of the coast. With, particular directions for sailing to, and entering the principal harbours, rivers, &c. Likewise, necessary directions for those who are not fully acquainted with the use of charts. Certificate. At the request of the publisher, I have carefully examined the charts...and find them to be as accurate as any of the kind hitherto published. Osgood Carleton... Boston: Printed and sold by John Norman. Boston: Published and Sold by O. Carleton and J. Norman. 13 maps. 55cm.

(1790). The Atlantic Neptune. Boston: Published by Matthew Clark.

(1789). The Boston directory... Illustrated with a plan of the town of Boston. Boston: author.

(1789). Chart of the coast of America from George's Bank to Rhode Island including Nantucket. Matthew Clark Sc. Boston: J. Norman.

(1795). Map of the District of Maine, drawn from the latest surveys and other best authorities. Boston: Published and Sold by O. Carleton and J. Norman.

(1786). The town and country builder's assistant...explaining short and easy rules...for drawing and working, the five orders of columns entire: or any part of an order, without regard to the module or diameter. And to enrich them with their rusticks, flutings, cablings, dentules, modillions, &c. Also to proportion their doors, windows, intercolumnations, portico's and arcades. Together with a variety of raking, circular, scrolled, compound, and contracted pendements...The whole illustrated by upwards of 200 examples, engraved on 60 folio copper plates. By a lover of architect. Boston: Engraved, printed, and sold by J. Norman. 31.5cm.

(1791). The United States of America laid down from the best authorities agreeable to the Peace of 1783. By Osgood Carleton, engraved by John Norman. Boston: J. Norman.

U.S. Senate. 25th Congress, 3rd Session. John A. L. Norman, praying donation of land in aid of Dade Institute of Florida, (S.doc.116). Washington: Government Printing Office, 1838-1839. (Serial Set 339).

Other

Swan, Abraham. (1775). The British architect: or, The builder's treasury of staircases. Containing, I. An easier, more intelligible, and expeditious method of drawing the five orders, than has hitherto been published... II. Likewise stair cases... III. Designs of arches, doors, and windows. IV. A great variety of new and curious chimney-pieces... V. Corbels, shields, and other beautiful decorations. VI. Several useful and necessary rules of carpentry... Philadelphia: Printed by R. Bell for J. Norman.

NORMAN, WILLIAM (C.1807)

William was the son of John Norman (Guthorn 1984). Guthron goes on to provide some biographical information. William was a printer, bookseller, engraver and stationer in Boston who engraved a "Plan of the town of Boston" in 1782. In 1798 he published The American Pilot (pp.7-8).

(1803). American atlas containing the navigation of the sea coast of North-America...describing also the capes, headlands, rivers... Boston: author. 5 lp. 15 folded charts. 54x41cm.

(1798 & 1803). The American pilot containing the navigation of the sea coast of North America, from the streights of Belle-Isle to Essequebo, including the island and banks of Newfoundland, the West-India islands, and all the islands on the coast. With particular directions for sailing to, and entering the principal harbours, rivers, &c. Describing also the capes, head lands, rivers, bays, roads, havens, harbours, straits, rocks, sands, shoals, banks, depths of water and anchorage. Shewing the courses and distances from one place to another, the ebbing of the sea, the setting of the tides and currents, &c. With many other things necessary to be known in navigation. Likewise necessary directions for those who are not fully acquainted with the use of charts. By William Norman. Boston: W. Norman. 3 p.l. 9 maps

(c.1816). As above- 11 maps (2 dated 1816).

(c.1791). A new and accurate chart of the bay of Chesapeake including Delaware Bay...as far as the navigable part of the rivers Patowmack [sic]

Patapsco, N East drawn from several draughts...Chiefly from those of Anthony Smith Pilot of St. Marys...Boston: Printed & sold by W. Norman.

(1798). A new chart of the N.W. coast of America with part of the coast of California including the Sandwich Islands and the principal harbours As surveyed by George Vancouver Esqr. Revised and corrected by Osgood Carleton Esqr. Teacher of mathematics Boston. Boston: W. Norman.

(1782). A plan of the town of Boston. Boston: W. Norman.

(1806). As above- $35^{1/2}$x$18^{7/8}$inches. Scale: 346.5 feet to 1 inch.

O'NEILL, JOHN (C.1808)

Our sources reveal no information about O'Neill or his work beyond the publications noted. The National Union Catalog Pre-1956 Imprints provided no entries for O'Neill (The American Library Association 1976).

(1808). A new and easy system of geography and popular astronomy; or an introduction to universal geography, ... the whole arranged in catechetical from designed as well for the use of schools, as the improvements by sundry teachers. Baltimore: Printed by G. Dobbin & Murphy. 504p. folded map. 18.5cm.

(1812). As above- ... Illustrated with a map of the world, a map of North America and a plan of the solar system. 2d ed., with considerable additions by Joseph James. Baltimore: Fielding Lucas, Jr. and Edward J. Coale. Fry and Kammerer, printers. 450p. 2 folded maps.

(1814). As above- 3d ed.

(1816). As above- 4th ed., Baltimore: F. Lucas, Jr. 359p. 17cm.

(1819). As above- ...5th ed., revised and corrected by James Bain. Baltimore: Published by Fielding Lucas, Jr. and J. Robinson. 389p. 3 folded maps. 18cm.

PAGE, JOHN (1744-1808)

Page engaged in celestial observations that resulted in published findings in the American Philosophical Society's Transactions in April 2, 1795 (Barton, 1813) . Along with Page, the names of Rittenhouse, Lukens, and Andrews are noted (p.xli).

White (1891) notes that Page was Lieutenant Governor of Virginia 1776-1779 and Governor of Virginia, born at Roswell, Bloucester county, Virginia. His great-great-grandfather was an English merchant who moved to this country settling in Virginia. Page studied at William and Mary college, from which he graduated in 1763. He was a friend and follower of Thomas Jefferson. Jefferson appointed him U. S. Commissioner of Loans for Virginia which Page continued to hold until his death. He is described as a man of learning, a statesman, an admirable soldier, and a theologist (The Biographical Dictionary 2001. Page, John; White 1891, V.III, p.219).

(1786). A letter from David Rittenhouse, Esq. to John Page, Esq. in answer to one from Mr. Page: concerning a remarkable meteor, seen in Virginia and

Pennsylvania, on the 31st of October 1779. (Mr. Rittenhouse's letter is dated January 16, 1780). American Philosophical Society, Transactions, II, May 2, 1783.

PARISH, ELIJAH (1762-1825)

Parish, a geography textbook author, was born on November 7, 1762 in Lebanon, Connecticut and died on October 15, 1825 in Byfield, Massachusetts. He attended Plainfield Adacemy, then entered Dartmouth College. He was admitted to Phi Beta Kappa chapter of Hanover in 1788. Parish studied theology under the Rev. Ephraim Judson and was installed as pastor of the Congregational Church at Byfield, Massachusetts on December 20, 1787. His theological views were strict Congregational orthodoxy. Johnson and Malone (1928) attest to his success as a pastor:

> As a pastor, he was unusually successful and at his death there was not a more united parish in the state. Indeed, he appears to have quite dominated the life of his people, for "it is in no respect an exaggeration to say that any opinion expressed in opposition to their pastor, political, religious, or regarding measures of policy, would have had little chance of finding favour [sic] among his people. (V.14, p.204)

Typical of New England ministers, Parish was a Federalist. In 1810 he was asked by the legislature to preach the annual election sermon. He delivered an attack on the national administration so acrimoniously that the legislature, refused the usual compliment of requesting a copy of the sermon for publication. It was, however, published by subscription and widely read. Parish was also critical of the War of 1812 and published sermons denouncing it (Marquis 1963; Historical Volume, p. 393; White 1891, V.IV, p.526).

(1809). A compendious history of New England, designed for schools and private families. By Jedidiah Morse, D. D. and Elijah Parish, D. D. 2d ed., with improvements by the authors. Newburyport: Thomas & Whipple.

(1807). A compendious system of universal geography, designed for schools. Compiled from the latest and most distinguished European and American travellers, voyagers and geographers. Newburyport, MA: Thomas & Whipple. 213 p. 17.5cm.

(1808, 1809 & 1810). As above- ...2d ed., with many improvements.

(1810). A new system of modern geography; or, a general description of all the considerable countries in the world. Compiled from the latest European and American geographies, voyages and travels. Designed for schools and academies. Ornamented with maps. Newburyport, MA: Thomas & Whipple. 370p. 2 folded maps. 18.5cm.

(1812 & 1814). As above- Newburyport, MA: E. Little & Co. 366p. 18cm..

(1813). Sacred geography: or, a gazatteer of the Bible. Containing, in alphabetical order, a geographical description of all the countries, kingdoms, nations and tribes of men, with all the villages, towns, cities, provinces, hills, mountains, rivers, lakes, seas, and islands, mentioned in the Sacred Scriptures, or Apocryphs, including an account of the religion, government, population, fulfillment of prophecies, and present condition of the most important places. Embellished with a new map of the principal countries mentioned in the Sacred Scriptures. Boston: Samuel T. Armstrong. 522p. folded map. 22cm.

Other

Morse, J., & Parish, E. (1804). A compendious history of New England. (np).

Morse, J., & Parish, E. (1802). A new gazetteer of the Eastern Continent. (np).

Note. (Johnson & Malone 1928, V.14, p.204).

Morse, J. (1826). Sermons, practical and doctrinal, by the late Elijah Parish...With a biographical sketch of the author. Boston: Crocker & Brewster. 421p. 24cm.

PARKER, ALEXANDER (C.1785)

Parker was one of eight surveyors who reported for duty in 1785 to the newly appointed Geographer of the United States, Thomas Hutchins, beginning work on the surveying of America. He viewed himself as a servant of the federal government, not that of the state of Virginia. Parker was a "woodswise man" who worked as a Virginia county surveyor capable of working with Hutchins as an equal (Pattison 1957, p.127). The National Union Catalog Pre-1956 Imprints revealed no entries for Parker (The American Library Association 1976).

PATTILLO, HENRY (1726-1801)

Patillo authored a variety of books, including geographies. He is identified as a clergyman, born in Scotland in 1726. On July 12, 1758 he married Mary Anderson. He was licensed in 1757, by the Presbytery of Hanover, Virginia, to preach. Patillo was pastor of churches at Willis Creek, Byrd, and Buck Island, Virginia from 1758 through 1762. From 1764 to 1774, he pastored churches at Hawkfields and Little River, North Carolina. His last church was at Grassy Creek, North Carolina from 1774 until 1801. Patillo died in Dinwiddie County, Virginia in 1801 (Marquis Company 1963; White 1897, V.VII, p.301).

(1796). A geographical catechism, to assist those who have neither maps nor gazetteers... with as much of the science of astronomy, and the doctrine of the air, as is judged sufficient for the farmer... Halifax, NC: Printed by Abraham Hodge. 62p. 12x21.5cm.

(1909). As above- N. W. Walker and M. C. S. Noble, (Eds.). Chapel Hill, NC: The University Press. 22cm.

PAYNE, JOHN (C.1794)

Payne's contribution to geographic knowledge was his writing. Information about the life of Payne has not been located.

(1794, 95). An epitome of history; or, a concise view of the most important revolutions and events, which are recorded in the histories of the principal empires, kingdoms, states, and republics, now subsisting in the world: also their forms of government: accompanied with short accounts of the different religions... London: Printed for J. Johnson, and New York: Berry, Rogers, and Berry. 2v. 373p. 21.5cm.

(1796). Geographical extracts forming a general view of Earth and nature...Illustrated with maps. London: G. G. and J. Robinson. 530p. 2 fold. maps. 22cm. (tr.).

(1793). The naval, commercial, and general history of Great Britain, from the earliest times to the rupture with Spain, in the year 1779...with an illustrative and political preface to each volume, and a copious general index to the whole...The 2d ed., corrected and much improved. London: Printed for the Proprietors and sold by J. Parsons. 5v. plates. 1 folded map.

(1810). A new and complete history of Europe, from the earliest ages, down to the present time; describing its subdivisions of republics, states, empires, and kingdoms: extent, boundaries and remarkable appearance of the country; cities, towns and curiosities of nature and art... 1st American ed. from the last English ed. New York: Reprinted for the author. 24cm.

(1798, 1799 & 1800). A new and complete system of universal geography...to which is added, a view of astronomy, as connected with geography...with a copious index annexed to each volume. Being a large and comprehensive abridgment of universal geography...with additions, corrections and improvements...by James Hardie...in four volumes. New York: Printed for and sold by John Low.

(1798). A new and complete system of universal geography: Describing Asia, Africa, Europe and America; with their subdivisions of republics, states, empires and kingdoms: The extent, boundaries, and remarkable appearances of each country; cities, towns and curiosities of nature and art. Also giving a general account of the fossil and vegetable productions of the earth. The history of man, In all climates, regions, and conditions; customs, manners, laws, governments, and religions: The state of arts, sciences, commerce, manufactures, and knowledge. Sketches of the ancient and modern history of each nation and people to the present time. To which is added, A view of astronomy, as connected with geography: and of the universe in general. With a copious index annexed to each volume. Being a large and comprehensive abridgment of universal geography. New York: Printed for and sold by John Low, Book Seller.

(1791 & 1792). Universal geography, formed into a new and entire system; describing Asia, Africa, Europe, and America; with their subdivisions of empires, kingdoms, states, and republics: the extent, boundaries, and remarkable appearances of each country; cities, towns, and curiosities of nature and art. Also giving a general account of the fossil and vegetable productions of the earth, and of every species of animal; the history of man, in all climates, regions, and conditions; customs, manners, laws, governments, and religions; the state of the arts, sciences, commerce, manufactures, and knowledge. Sketches of the ancient and modern history of each nation and people. To which is added, a short view of astronomy, as connected with geography; of the planetary system to which the Earth belongs; and of the universe in general. With a set of maps, drawn from the best materials, every one of which is very neatly coloured; and a great variety of copperplates. London: Printed for the author, and sold by J. Johnson and C. Stalker. 2v. 36cm.

(1793 & 1794). As above- Dublin: Printed by Z. Jackson.

(1809). As above- 3v. 31 plates. 9 maps. 29.5cm.

(1798 & 1800). As above- New York: Printed for and sold by John Low. 4v. 22.5cm.

PEASE, SETH (1764-1819)

Pease, in 1798, produced a map of the Connecticut Western Reserve, engraved by Amos Doolittle (Schwartz & Ehrenberg 1980, p.216). It showed east and west surveys of the Cuyahoga River to the Vermilion River with thirty-two townships. A note by Ristow (1985) indicates that Pease was a "former Surveyor General South of the State of Tennessee" worked with Isaac Briggs and Thomas Freeman to produce map records of the townships and ranges on the west side of the Mississippi River (p.145).

(1914). Journals of Seth Pease to and from New Connecticut, 1796-1798. In Western Reserve Historical Society. Tracts. No. 94, p.27-132. Cleveland, OH: Western Reserve Historical Society.

(1798). A map of the Connecticut Western Reserve from actual survey. Newhaven: (np).

(c.1817). A map of the state of Louisiana with part of the state of Mississippi and Alabama Territory by Maxfield Ludlow Chief Clerk Surveyor Genls. Office south of Tennessee engraved by Charles and J. G. Warnicke. Eastern district of Pennsylvania: (np).

PEASLEY, A. M. (C.1796)

Guthron (1984) places Peasley among engravers listed as being employed by the Blunts (p.10). Peasley's engravings appeared in Blunt's American Coastal Pilot.

(1796). American coast pilot: Containing directions for the principal harbors, capes, and headlands on the coast of North and part of South America... with the prevailing winds, setting of the currents, &c., and the latitudes and longitudes of the principal harbors and capes; together with tide tables and variation. (np).

Note. [Brown 1897, V.1, p.333].

(1804 & 1806). The bay of Chesapeake from its entrance to Baltimore. Compiled by Edmund M. Blunt, engraved by A. M. Peasley. (np).

(1804). Charleston harbour [sic]. Engraved by A. Peasley. Newburyport, MA: Edmund M. Blunt.

PERRY, WILLIAM L. (OF KELSO, SCOTLAND) (C.1788)

Perry's contribution to American geographic knowledge was through his publication. No biographical information about Perry has been located.

(1788). Appendix to the Royal standard English dictionary; containing the proper names of persons and places in the Old and New Testaments; names of the principal cities, rivers, mountains, and c. in the known world; the ancient and modern poets, philosophies, patriots, statesmen, and c... Worcester, Mass: Isaiah Thomas.

PETRIE, EDMUND (C.1790)

Petrie produced one of the earliest post revolutionary maps in the new United States. It was perhaps the earliest fire insurance map published. The map was published, in London -1790, for the Phoenix Assurance Company, Ltd.:

> The map lists and locates public buildings and wharves, gives the width of streets and lanes, identifies ninety-nine private and commercial properties, and locates the fire station and public wells. (Ristow 1985, p.244-247)

(1790). Ichonography of Charleston, South Carolina. London: Phoenix Fire-Company of London.

PICKERING, OCTAVIUS (C.1780)

Pattison (1957) reports that Pickering as a man, prior to the 1780s, who was engaged in the mercantile business of Philadelphia. He became quartermaster general in 1780 and remained so until the end of the revolutionary war. "After the adoption of the Constitution, Pickering held three cabinet posts under President Washington, and later served as United States Senator from Massachusetts, 1803-1811" (p.87).

The American Library Association (1978) lists several publications by an Octavius Pickering who lived from 1791-1868. Although there may be a relationship, this does not appear to be our man (V.457, p.272).

(1827). On the geographical distribution of plants. The American Philosophical Society, 3(5), pp.274-84.

Other

Bowman. (1995). Octavius Pickering, Dictionary of American Biography, XIV. (np).

PIKE, ZEBULON MONTGOMERY (1779-1813)

Pike was primarily a soldier and explorer. He was born in Lamberton, now a part of Trenton, New Jersey on January 5, 1779 and was killed on April 27, 1813. His death was the result of an exploding enemy powder magazine as he led his American troops in an attack on the British position at York (now Toronto), Canada during the War of 1812. Pike had risen from the rank of first lieutenant in 1805 to the rank of brigadier-general in 1813 (The Biographical Dictionary 2001. Pike, Zebulon; Encarta Encyclopedia 2001. Pike, Zebulon; Encyclopedia Britannica 2001. Pike, Zebulon).

As an explorer, Pike began his career as head of a company assigned to discover the head of the Mississippi. He later explored the headwaters of the Arkansas and Red rivers. He also explored numerous other places in the west. His Papers of Zebulon M. Pike, 1806-1807, were taken by Spanish officials and not recovered from Mexican archives until 1908 (Johnson & Malone 1928, V.7, pp.599-600).

(1810). An account of expeditions to the sources of the Mississippi, and through the western parts of Louisiana, to the sources of the Arkansaw, Kans, La Platte, and Pierre Juan Rivers; performed by order of the government of the United States during the years 1805, 1806, and 1807. And a tour through the interior parts of New Spain, when conducted through these provinces by order of the captain-general, in the year 1807... illustrated by maps and charts. Philadelphia: C. & A. Conrad & Co.; Baltimore: Fielding Lucas, Jr.; John Binns, printer. 5+105+9+[107]-277+[2]+65+(1)+53+87p. 2 charts. 4 maps. 1 port. 2 tab.

(1807). An account of a voyage up the Mississippi River, from St. Louis to its source; made under the orders of the War Department, by Lieut. Pike, of the United States Army, in the years 1805 and 1806. Compiled from Mr. Pike's journal. Washington: (np). 68p. 22.5cm.

Note. This appears to be a paraphrase of Pike's journal by Nicholas King.

(1885). As above- Washington, DC: American Register.

(1960). As above- Louisville, KY: Lost Cause Press.

(nd). Carte de l'interieur de la Louisiane comprenant tous les pays jusqu'a alors inconnus, entre la riviere Plate au N., La riviere Rouge au S., le Mississippi a l' E., et les montagnes du Mexique a l'O, avec une partie du Nouveau Mexique et de la province de Texas, par le Major Z. M. Pike. (np). 50x43cm. [Accent omitted]

Compensation to persons engaged in several exploring expeditions under Captain Zebulon Pike, American State Papers: Miscellaneous, V.XXXVII (Doc. 248 &259).

(1906). ...The discovery of Pike's Peak. 1806. From the diary of an expedition made under the orders of the War Department, by Captain Z. M. Pike, in the years 1806 and 1807, to explore the internal parts of Louisiana. Boston: Directors of the Old South work, 24p., 20cm. (Old South leaflets. General series. 7(174).

(1895). The expeditions of Zebulon Montgomery Pike, to headwaters of the Mississippi River through Louisiana Territory, and in New Spain, during the years 1805-6-7. A new ed., now first reprinted in full from the original

1810, with copious critical commentary, memoir of Pike... and complete index by Elliott Coues... New York: F. P. Harper. 3v. 7 maps (6 folded). 24.5cm.

(1811). Exploratory travels through the western territories of North America: comprising a voyage from St. Louis, on the Mississippi, to the source of that river, and a journey through the interior of Louisiana, and the north-eastern provinces of New Spain. Performed in the years 1805, 1806, and 1807, by order of the government of the United States... London: Printed for Longman, Hurst, Rees, Orme. 436p. 2 maps. 28x23cm.

(1889). As above- London & Denver: W. H. Lawrence & Co. 394p. 4 maps (2 double). Introduction signed William M. Maguire.

(nd). Lt. Zebulon Pike's notebook of maps, traverse tables, and meteorological observations, 1805-1807. Washington, DC: National archives and Records Service.

(1908). Papers of Zebulon M. Pike, 1806-1807. Contributed by Herbert E. Bolton... New York: The Macmillan Co. 827p. 27cm.

(1872). Pike's explorations in Minnesota, 1805-6. In Minnesota Historical Society. Collections. St. Paul: (np). V.1, pp.368-415. 23.5cm.

(nd). Premiere partie de la carte de l'interieur de la Louisiane par Z. M. Pike. Reduit... Par Antoine nau. 54.5x46cm. [Accent omitted]

(1813). Reise durch die westlichen gebiete von Nord-America, enthaltend die tagebucher der reisen von St. Louis, den Mississippi hinauf bus an die quellen dieses flusses; durch das innere von Louisiana, und durch die nordostlichen provinzen von Neu-Spanien, gemacht in den jahren 1805, 1806, u. 1807, auf befhl der regierung der Vereinigten Staaten. Von Zebulon Montgomery Pike... Aus dem englischen ubersetzt von Ph. Ch. Weyland. Weimar: Landes-industrie-comptoir. 556p. folded map. 21.5cm. [Accent omitted]

(1813). Reize naar Nieuw-Mexico en de binnenlanden van Louisiana, voorgegaan door eenen togt naar de bronnen der Mississippi, gadaan op last van het gouvernement der Vereenigde staten in de jaren 1805, 1806, 1807, door den majoor Z. M. Pike. Uit het Engelsch vertaald... Amsterdam: C. Timmer. 2v. folded maps. 23cm. [Accent omitted]

(1810). Sources of the Mississippi. Philadelphia: Conrad.

(1812). Voyage au Nouveau-Mexique, a la suite d'une expedition ordonnee par le governement des Etats-Unis, pour reconnoitre les sources des rivieres Arkansas, Kanses, La Plate, et Pierre-Jaune, dands l'interieur de la Louisiane occidentale. Precede d'une excursion aux sources du Mississippi, pendante les annees 1805, 1806, et1807. Par le major Z. M. Pike. Traduit de l'anglais par M. Breton... Orne d'une nouvelle carte de la Louisiane, en trois parties. Paris: D'Hautel. 2v. 3 folded maps. 20.5cm.[Accent omitted]

Other

DeWitt, Donald. (1906). Pike and Pike's Peak; a brief life of Zebulon Montgomery Pike and extracts from his journal of exploration. (np). 52p. plates.

Hart, Stephen Harding, & Hulbert, Archer Butler. (Eds.). (1932). ...Zebulon Pike's Arkansaw journal: in search of the southern Louisiana purchase boundary line (interpreted by his newly recovered maps). Edited, with bibliographical resume, 1800-1810, by Stephen Harding Hart and Archer Butler Hulbert... Colorado Springs, CO: The Stewart commission of Colorado College; and Denver: The Denver Public Library. 200p. maps. 24.5cm.

Humphreys, Mary Gay. (Ed.). (1911). The boy's story of Zebulon M. Pike. New York: C. Scribner's Sons. 377p. 20.5cm. Note. Based on his diary and reports.

Kneen, Orville Hayter. (1929). Young pioneers on western trails, young explorers' own stories of adventure in the unknown West, by Orville H. Kneen, with a frontispiece and a map. New York: Frederick A. Stokes Co.

Quaife, Milo Milton. (Ed.). (1925). ... The Southwestern expedition of Zebulon M. Pike. Chicago: R. R. Donnelley & Sons Co. 289p. folded map. 17.5cm.

PINKHAM, PAUL, CAPT. (C.1796)

Pinkham's Nantucket Shoals was a chart included in Blunt's American Coast Pilot... (Guthron 1984, p.8). Ristow (1985) reports that Pinkham was a surveyor of George's Bank (p.228).

(1796). American coast pilot: Containing directions for the principal harbors, capes, and headlands on the coast of north and part of South America... with the prevailing winds, setting of the currents, &c., and the latitudes and longitudes of the principal harbors and capes; together with tide tables and variation. (np).

(1797). A chart of George's Bank, including Cape Cod, Nantucket and the Shoals lying on their coast, with directions for sailing over the same &c. Surveyed by Capt. Paul Pinkham. Engraved and Printed for Edmund M. Blunt Proprietor of American Coast Pilot.

(1791). A chart of Nantucket Shoals surveyed by Capt. Paul Pinkham. Boston: John Norman. $29^{3/4}$x19inches. Scale 1:145,000.

(nd). Directions to and from the light-house, on the north-east point of Nantucket... (np).

POPPLETON, THOMAS H. (C.1811)

Poppleton advertised himself as a "Practical land Surveyor and Draughtsman" in the Federal Gazette and Baltimore Daily Advertiser, in 1811. That may place him among either the first or second generation of American geographers. Surveys from that advertisement took place in 1811 or 1812, but the map was not published until 1817 (Ristow 1985, p.252-254).

(1817). Plan of the city of New York the greater part from actual survey made expressly for the purpose (the rest from authentic documents) by Thos. H. Poppleton, city surveyor. Engraved by Peter Maverick. New York: Printed by Samuel Maverick.

(1823). This plan of the city of Baltimore as enlarged & laid out under the direction of the commissioners appointed by the General Assembly of Maryland in Feb. 1816 Is respectfully dedicated to the citizens thereof by their Obt. Servt. T. H. Poppleton surveyor to the Board C. P. Harrison, Script. Sculpt. New York: (np).136x110cm.

PORTER, ANDREW (1743-1813)

Porter was a member of the Pennsylvania Boundary Commission that worked on the completion of the western end of the Mason and Dixon Line. He rose to the rank of brigadier-general of the Pennsylvania militia and major-general and surveyor-general. Porter died in Harrisburg, Pennsylvania

Identified by Brown (1897) as a soldier, Porter was born in Worcester, Montgomery County, Pennsylvania. He was the son of Robert who emigrated from Londonderry, Ireland. Andrew was a school teacher who became a soldier and fought in the battles of Trenton, Princeton, Brandywine, and Germantown. He received personal commendation by General Washington. (V.6, 304).

Barton (1813) names several people engaged in establishing the boundary line between Virginia and Pennsylvania:

> This boundary-line was begun in May, 1785, by Messrs. Rittenhouse, Ellicott, Porter, and Nevill; assisted by the present Dr. Benjamin Smith Barton, then a youth about nineteen years of age, whose medical and other scientific acquirements rendered him, even at that early period of life, an useful associate of the commissioners. (pp.306-307)

Porter was also working with Archibald McLean, who was among the men who completed surveying the boundary of Pennsylvania to the Ohio River (Pattison 1957).

(nd). Pennsylvania line. Artillery regiment account; or other accounts: 1773-1808. (np).

Other

(1880). A sketch of the life of General Andrew Porter, Pennsylvania Magazine of History and Biography, IV, p.261-301.

POUPARD, JAMES (C.1786)

Poupard was a Philadelphia jeweler, goldsmith and engraver. He engraved the plan of Baltimore surveyed by Folie and Varle' (Ristow 1985, p.252).

(1786). A chart of the Gulf Stream. James Poupard Sculp. Inset: (The Atlantic Ocean and the two coasts). In The Proceedings of the American Philosophical Society, II. $8^{1/2}$x10inches.

(1792). Plan of the town of Baltimore and its environs dedicated to the citizens of Baltimore taken upon the spot by their most humble servant A. P. Folie, French Geographer. (np): A. Folie.

PRENTIS, WILLIAM (C.1807)

Virginia's first state map was prepared in 1807 by William Prentis and William Davis, in collaboration with Bishop Madison, from information provided by local surveyors (Schwartz & Ehrenberg 1980, p. 224). Between them they produced and presented their map to the Virginia General Assembly, 1805 (Ristow 1985, p.121).

(1807). A map of Virginia formed from actual surveys, and the latest as well as most accurate observations. Drafted by William Davis. Richmond: Dedicated to the General Assembly of Virginia...by their Fellow Citizens, James Madison, William Prentis, William Davis, Proprietors. 114x175cm. Scale 1:440,000.

PRICE, JONATHAN (C.1790)

A Chart of the Coast of North Carolina between Cape Hatteras & Cape Fear was prepared in 1806 by Jonathan Price and Thomas Coles (Schwartz & Ehrenberg 1980). The survey was requested by President Thomas Jefferson and the U. S. Congress because of the hazards that the Outer Banks presented to U. S. shipping interests (p. 221).

Guthron (1984) goes on to relate that Price and Thomas Coles were appointed by Congress to survey 70 leagues along the east coast that was particularly hazardous to shipping. The survey was completed of Cape Fear River and the coast from Cape Roman to Cape Henry. That expedition proved so valuable that the Federal Government became the principal supporter of a national survey of the entire coastline (p. 5). Ristow (1985) reveals that Price worked with John Strothers to produce several maps (pp.123-125).

(1808). First actual survey of the state of North Carolina. By Jonathan Price and John Strother. Engraved by William Harrison Jr. Philadelphia: Printed by C. P. Harrison. 74x152cm. Scale 1:500,000.

(1808). A chart of the coast of North Carolina between Cape Hatteras & Cape Fear from a survey taken in the year 1806 by Thomas Coles and Jonathan Price pursuant to an Act of Congress. Engraved by C. P. Harrison. (np).

(1795 & 1796). A description of Occacock [sic] Inlet, and of its coasts, islands, shoals and anchorage, with the courses and distances to and from the most remarkable places and directions to sail over the bar and through the channels, adorned with a map taken by actual survey. Engraved by W. Johnston. New Bern, NC: F. X. Martin.

(c.1800). A map of Cape Fear River and its vicinity from The Frying Pan Shoals to Wilmington by actual survey addressed to the commissioners of navigation of Port Wilmington by Price & Strother. Engraved by W. Barker. Philadelphia: Price & Strother.

(1796 & 1798). To navigators this chart being an actual survey of the sea coast and inland navigation from Cape Henry to Cape Roman is most respectfully inscribed by Price & Strother. Engraved by W. Johnston. New Bern, NC: Price & Strother. 38x13$^{3/4}$inches. Scale 1:490,00.

(1838). Report of Jonathan Prie and Woodson Clemons, on sundry surveys made by them between the Roanoke and Pamtico, and the Pamtico and Neuse. (np). 35p. 24cm.

(nd). Report of the commissioners for surveying the coast of N.C. to the Hon. Albert Gallatin, Secretary of the Treasury of the U.S. (see U.S. Commissioners for Surveying the coast of North Carolina). (np).

PURCELL, JOSEPH (C.1754)

Purcell was one of the many, "British cartographers active in America during the Revolution who became American citizens and continued their work in the United States" (Schwartz & Ehrenberg 1980, p. 209).

Ristow (1985) notes that a Purcell and a Doolittle map were folded in the 1789 edition of Jedidiah Morse's The American Geography, (p. 71).

(c.1776). A map of the Southern Indian District of North America compiled under the citizen of the United States. Charleston, SC: (np). 8p. 19cm.

(1787). Histoire de la revolution d'Amerique, par rapport a la Caroline Meridionale; par M. David Ramsay... traduite de l'anglois... A Londres et se trouve a Paris: chez Froulle', 1787. [Accent omitted]

(1796). Histoire de la revolution d'Amerique, par rapport a la Caroline Meridionale; par M. David Ramsay... Traduite de l'anglais. Ornee de cartes et de plans, divisee en cinq volumes; le cinquieme est d'augmentation, & contient des observations sur le commerce des Etats-Unis d'Amerique. Par Jean Lord Sheffield; traduites de l'anglais par Mirabear... A Paris: Chez Moutardier. 5v. folded maps. 8°. [Accent omitted]

(1789). History of the American Revolution. Philadelphia: Printed and sold by R. Aitken & Son.

(1791). As above- London: Sold by J. Johnson. 2v. in 1. 22.5cm.

(1793 & 1795). As above- Dublin: Printed for W. Jones.

(1793). As above- London: Printed for J. Stockdale.

(1811). As above- Trenton: Printed and sold by James J. Wilson.

(1815). As above- Lexington, KY: Downing and Phillips. 2v. 22.5cm.

(1785). The history of the revolution of South Carolina, from a British province to an independent state. Trenton: Printed by Isaac Collins. 2v. maps. 20cm.

(1809). History of South Carolina from its settlement in 1670 to the year 1808. Charleston: David Longworth. 2v. map. plan. 21cm.

(1835 or 1839). The history of the United States. In Samuel L. Knapp. Library of American history. New York: (np). V.1, pp.332-430; V.2, pp.4-92.

(1813, 1816, 1817 & 1818). History of the United States, 1607-1808; or, the thirty-third of their sovereignty and independence, by David Ramsay.

Continued to the treaty of Ghent by S. S. Smith and other literary gentlemen. Second ed. rev. and corrected. Philadelphia: M. Carey. 3v.

(1790). On the means of preserving health in Charleston and its vicinity. (np)

(1804). An oration, on the cession of Louisiana, to the United States, delivered on the 12th May, 1804, in St. Michael's Church, Charleston, South Carolina... Charleston: Printed by W. P. Young. 27p. 22cm.

(1796). A sketch of the soil, climate, weather, and diseases of South-Carolina, Read before the Medical Society of that state. Charleston: Printed by W.P. Young. [4],30p. 2 fold. tbls.

(1819). Universal history Americanised; or, an historical view of the world, from the earliest records to the year 1808. With a particular reference to the state of society, literature, religion, and form of government, in the United States of America. To which is annexed, a supplement containing a brief view of history, from the year 1808 to the battle of Waterloo... In twelve volumes. Philadelphia: M. Carey & Son.

Other

Ford, Timothy. (1792). An enquiry into the constitutional authority of the Supreme Federal Court, over the several states, in their political capacity. Being an answer to observations upon the government of the United States of America: by James Sullivan, esq., Attorney General of the state of Massachusetts. By a citizen of South-Carolina. Charleston: Printed by W. P. Young.

Note. Ramsay supposed author.

Ramsay, Mrs. Martha (Laurens). (1827). Memoirs of the life of Martha Laurens Ramsay... With an appendix, containing extracts from her diary, letters, and other private papers. And also from letters written to her, by her father, Henry Laurens, 1771-1776. By David Ramsay... 3d ed. Boston: Crocker and Brewster; New York: J. Leavitt.

PUTNAM, RUFUS (1738-1824)

Putnam was a millwright, a farmer, a soldier, and a surveyor. He fought in the French and Indian War, 1757-1760. He joined the Continental Army in 1775, entering as a lieutenant colonel. During his time in the army he organized batteries and fortifications in Boston and in New York City. He engaged in the Battle of Saratoga. At West Point, he built new fortifications. In 1783 he was promoted to brigadier general (The Biographical Dictionary 2001. Putnam, Rufus; Encarta Encyclopedia 2001. Putnam, Rufus; Encyclopedia Britannica 2001. Putnam, Rufus)

Brown (1897) discloses that Putnam was a soldier born in Sutton, Massachusetts the son of Elisha and Susanna (Fuller) Putnam. He had little formal education, enlisting in the army in 1757 and becoming a scout. With Col. Israel Putnam and others in 1773, he explored and surveyed lands in East Florida.

As the Revolutionary War materialized, he joined an American regiment and set about engineering and planning fortifications at Prospect Hill and Dorchester Heights near Boston. Though his engineering skills were questioned by Congress, following the fall of New York, General Washington addressed him as "the best engineer in the army." In 1783 he was commissioned brigadier-general and requested by General Washington to compile plans to fortify the whole country. Putnam was appointed by congress as Surveyor of the Northwest Territory. Putnam founded and died in Marietta, Ohio (V.6, pp.377-379).

Putnam took control of federal surveying after the Land Act of 1796 (Pattison 1957, p. 179). He became the leader of the Ohio Company Colony. "As Superintendent of Surveys for the Ohio Company, he was the first administrator of the contract system, a financial arrangement which he later brought over into federal public land surveying" (p. 179). Lots for French settlers at Gallipolis were surveyed under the supervision of Putnam. In 1796 he was superintendent of the survey of Zane's Trace. Washington appointed him surveyor general for the United States where he served until 1803 (Pattison 1957, pp.200-201).

Other

Buell, Rowena (Ed.). (1903). The memoirs of Rufus Putnam and certain official papers and correspondence. Boston: Houghton, Mifflin and Co.

PUTNAM, WILLIAM R. (C.1800)

William R. Putnam was one of the deputy surveyors appointed in the Land Act of 1800 to work under the direction of Rufus Putnam, his father (Pattison 1957, p.207). No other information regarding Putnam has been located. The National Union Catalog Pre-1956 Imprints provided no entries for Putnam (The American Library Association 1976).

RAMSAY, DAVID (1749-1815)

Born in Dunmore, Lancaster county, Pennsylvania, the son of James and Jane (Montgomery) Ramsay. He attended and graduated from the College of New Jersey. Following his schooling, he practiced medicine in Cecil county, Maryland for a short period, but moved to Charleston, South Carolina in 1773. He joined the American army as a surgeon in 1776 and participated in the defence of Savannah. He had a colorful life: Captured by the British, Delegate from South Carolina to the Continental Congresses (1782-1784 and 1785-1786), member of the South Carolina senate and its president for seven years. He was shot by a maniac and died in Charleston, South Carolina (V.6, pp.395-396).

Trenton, New Jersey was the site of the publication, in 1787, of Ramsay's History of the Revolution of South Carolina (Schwartz and Ehrenberg 1980).

It contained five maps of battles in the Carolina region.

(1789). History of the American Revolution. (np)

(1785). History of the revolution of South Carolina from a British province to an independent state. (np)

(1809). History of South Carolina from its settlement in 1670 to the year 1808. (np)

(1816-1817). History of the United States, 1607-1808, continued to the Treaty of Ghent by Samuel S. Smith and others and published posthumously. 3 Vols. Universal History Americanized. (np)

(1790). On the means of preserving health in Charleston and its vicinity. (np)

(1796). A sketch of the soil, climate, weather, and diseases of South-Carolina, read before the Medical Society of that state. Charleston: Printed by W.P. Young. [4], 30p. 2 fold. tbls.

RANDEL, JOHN, JR. (C.1807)

Randel was, in 1807, Secretary and Chief Surveyor of the commission charged by the New York State legislature to lay out streets and roads for the extent of Manhattan Island (Ristow 1985). The result was three manuscript maps completed in 1811, a Map of the City of New York and Island of Manhattan as Laid Out by the Commissioners Appointed by the Legislature April 3d 1807. The commissioners were Governor Morris, Simeon DeWitt, and John Rutherford. The map, engraved by Maverick, was issued in six separate sheets for wall mounting was published by William Bridges, an architect and surveyor of the city. Randel promptly decided to cancel his plans to publish the map (p. 249).

The following publications were located in the National Union Catalog, Pre-1956 Imprints (The American Library Association, 1976). Although the publications noted may not be by our man, the publications cited do reveal interests in and near New York:

(1864). City of New York, north of Canal Street, in 1808 to 1821. In D. T. Valentine. Manual of the corporation of the city of New York. New York: (np). pp.847-856. 18.5cm.

(1822). Description of a direct route for the Erie Canal, at its eastern termination: with estimates of its expense, and comparative advantages. Albany, NY: Printed by G. J. Loomis & Co. 72p. 20.5cm.

(1836). As above- By John Randel, Jun. Printed by Martin and Winne. 50p. 23cm.

(1848). The elevated railway, and its appendages, for Broadway, in the city of N. York, invented by John Randel, jun... A statement of facts, including the opinions of celebrated machinists, and also, "The mechanics' institute of the city of New York." New York: J. M. Ellicott, printer. 19p. 22cm.

(1848). Explanatory remarks and estimates of the cost and income of the elevated railway, and its appendages, &c. for Broadway in the city of New York. Invented by John Randel, jun. New York: G. F. Nesbitt, stationer and printer. 16p. 24cm.

RATZER, BERNARD (C.1769)

During the American Revolution Ratzer was a British citizen, as were most patriots. He was responsible for the original publication and republishing of a map of New York and a map of New Jersey, first published in 1769. Phelps Stokes is quoted as commenting about the map as, "one of the most beautiful, important and accurate plans of New York" (Schwartz & Ehrenberg 1980, p.187). Although some of Ratzer's publications fall outside the dates for the American Geographers: 1784-1812, his contributions likely had influence on the outcome of some skirmishes and perhaps the Revolution.

(1776). Plan of the city of New York, in North America surveyed in the years 1766 and 1767. To His Excellency Sir Henry Moore, Bart.... this plan of New York and its environs surveyed and laid down is most humbly dedicated by ...B. Ratzer... London: Jeffreys and Faden. 91.2x62.2cm.

(1774). Plan of Prince Rupert's Bay in the island of Dominica, surveyed in May 1774. (np).

(nd). Plans and sketches of the forts and batteries in Jamaica. (see under Richard Jones, eng.) (np).

(1777). The Province of New Jersey, Divided into east and west, commonly called The Jerseys. This map drawn by Bernard Ratzer and By Gerard Banker. (np). 80.1x60.2cm.

(1778). As above - Second edition with considerable improvements. by William Faden. (np). 61.8x81.9cm.

(1769). To his excellency Sir Henry Moore, Captain General and Governor in Chief, in & over the province of New York & the territories depending thereon in America, Chancellor & Vice Admiral of the same... This plan of New York... Engraved by Bernard Ratzer. (np).

RAWSON, ANDREW (1773-1835)

Rawson's only contribution to American geography may have been the following text. No publications were found in The National Union Catalog, Pre-1956 Imprints (The American Library Association, 1976). Like several people who contributed to American's knowledge, virtually nothing beyond this single publication is known about Rawson.

(1802). A treasure for children: being a small spelling book...interspersed with moral lessons and fables. To which are added sketches of geography... Worcester, Mass: Printed by Sewell Goodridge for William Parker.

REED, ABNER (1771-1866)

Reed is one of those map engravers who may more appropriately be assigned to the second generation of American geographers, but is included because he was active prior to 1812 (Schwartz & Ehrenberg 1980). Moses Warren and George Gillet prepared a state map for Connecticut that was engraved by Reed and published in Hartford by Hudson and Goodwin in 1812. The second atlas printed in the United States, to which Reed contributed

engravings, was in 1796 published at New York (Schwartz & Ehrenberg 1980).
(1813). Connecticut, from actual survey, Made in 1811; by and under the direction of. Moses Warren and George Gillet; and by them compiled. Published under the authority of the General Assembly, by Hudson & Goodwin. Engraved by Abner Reed, E. Windson. Hartford, CN: Hudson & Goodwin.

REID, JOHN (C.1796)

Schwartz and Ehrenberg (1980) state that Reid published the second atlas printed in the United States (p.215). The National Union Catalog, Pre-1956 Imprints lists the works of two John Reid's: One appears to have been a British surveyor publishing in 1848; The other's dates (1776-1822) make him a possibility, but all publications listed were about medical matters (The American Library Association 1977, V.486, pp.462-463).

(1796). The American atlas. New York: John Reid.

REVERE, PAUL (1735-1818)

It seems fitting to include Revere among those who engraved maps just prior to and during the Revolution. Schwartz and Ehrenberg (1980) name Revere three times in their text: pages 115, 178, and 179. The National Union Catalog, Pre-1956 Imprints lists no publications under the name Paul Revere (The American Library Association, 1977).

Revere was a multi talented person. He was one of America's greatest artists in silver. In addition he made surgical instruments, sold spectacles, replaced missing teeth, and engraved copper plates. He was very active throughout the American Revolution, participating in the Boston Tea Party and was commander of Boston Harbour's principal defense at Castle William. Revere was immortalized by the ballad by Henry Wadsworth Longfellow (Encyclopedia Britannica 2001. Revere, Paul)

(1774). Part of the province of East Florida. Engraved by Paul Revere and Abel Buell. Published with Roman's A Concise Natural History of East and West Florida. New York: author. 62x221cm.

Note. The Florida panhandle.

(1774). As above - New York: author. 147x169cm.

Note. Tampa Bay through Cape Canaveral.

RITTENHOUSE, DAVID (1732-1796)

Rittenhouse was an instrument maker, astronomer and mathematician who may have produced the first telescope in the United States. He was born on April 8, 1732 at Paper Mill Run, near Germantown, Pennsylvania. His family had emigrated from Amsterdam to America in 1688. He died on June 22, 1796 of cholera. His mother was a Welsh Quakeress and his father a farmer. Rittenhouse had little schooling, but appears to have developed mathematical and mechanical ability, perhaps due to a chest of books and tools inherited from his maternal uncle.

At nineteen he opened an instrument shop on his father's farm, mainly for clock making. He is said to have mastered an English translation of Isaac Newton's Principia, developing a sound knowledge of physical science from it. He was interested in telescopes and constructed one in 1756 (Encarta Encyclopedia 2001. Rittenhouse, David; Encyclopedia Britannica 2001. Rittenhouse, David).

Rittenhouse's first public service job was that of surveying the boundary line, in 1763-64, to settle a dispute between Penn and Lord Baltimore. In 1767 he designed his celebrated orrery. The instrument that represents motions of bodies of the solar system and illustrates solar and lunar eclipses and other phenomena "for a period of 5000 years, either forward or backward." In 1785 he invented the collimating telescope and introduced spider threads in the eyepiece of his instruments. He was active in a variety of activities and developments that have been important in furthering our understanding of Earth - Sun relationships (Johnson & Malone 1928, V.8, pp.630-632).

Rittenhouse and Andrew Ellicott directed a team of axemen who marked the boundary, in June 1785, through the forests of the Allegheny Plateau. Pattison (1957) went on to report Ellicott's opinion of Rittenhouse's telescope, "The boundary itself aligned through the use of a Transit Instrument, equipped with what Ellicott termed 'a most excellent telescope'" (p.121). Barton (1813), contributed numerous pages to praising Rittenhouse as an astronomer and philosopher. Rittenhouse's Orrery, a mechanism "adapted to an Armillary Sphere is the only machine that can exhibit a just idea of the true System of the World," received considerable attention in Rittenhouse's Memoirs (p.194).

His other mechanical contributions include contributions to The Norriton Observatory: An equal Altitude Instrument, a Transit Telescope, and an excellent Time-piece (pp.169-171). As a member and President of the American Philosophical Society, Rittenhouse interacted with such personages as Benjamin Franklin, William Smith, Thomas Jefferson, John Ewing, Andrew Porter, Andrew Ellicott, and Thomas Hutchins. Several members of the society participated in the development and spreading of geographic knowledge. Barton (1813) provided a chronological list of papers published by the American Philosophical Society:

(1793). An account of a comet, (first observed by Mr. Rittenhouse, on the 11th of January, 1793:) in a letter from D. Rittenhouse to Mr. Patterson. American Philosophical Society, Transactions, III, February 15, 1793.

(1793). An account of the effects of a stroke of lightning on a house furnished with two conductors; in a letter from Messrs. David Rittenhouse and Francis Hopkinson, to Mr. R. Patterson. American Philosophical Society, Transactions, III, October 15, 1790.

(1770). Account of the Terrestrial measurement of the difference of longitude between the observations of Norriton and Philadelphia; drawn up by the Rev. Dr. Smith, in behalf of Mr. Lukens, Mr. Rittenhouse and himself, the committee appointed by the Am. Philos. Society, for that purpose, agreeably to the request of the Astronomer-Royal of England. American Philosophical Society, Transactions, Aug. 17, 1770.

(1793). An account of several houses, in Philadelphia, struck with lightning on the 7th of June, 1789; by Mr. D. Rittenhouse and Dr. John Jones. American Philosophical Society, Transactions, III, July 17, 1789.

(1771). Account of the Terrestrial measurement of the difference of longitude between the observatories of Norriton and Philadelphia; drawn up by the Rev. Dr. Smith, in behalf of Mr. Lukens, Mr. Rittenhouse and himself, the committee appointed by the American Philosophical Society, for that purpose, agreeably to the request of the Astronomer-Royal of England. American Philosophical Society, Transactions, I, August 17, 1770.

(1771). An account of the transit of Mercury over the Sun, Nov. 9. 1769, as observed at Norriton, in Pennsylvania, by Dr. Smith, and Messrs. Lukens, Rittenhouse, and O. Biddle, the committee appointed for that purpose by the American Philosophical Society: drawn up and communicated, by direction and in behalf of the committee, by Dr. Smith. American Philosophical Society, Transactions, I, June 20, 1769.

(1786). Astronomical observations; communicated by Mr. Rittenhouse. American Philosophical Society, Transactions, II, (nd).

(1793). Astronomical observations made at Philadelphia: viz. of a Lunar eclipse, on the 2d of November, 1789; of the transit of Mercury over the Sun's disk, on the 5th of November, 1789; of an eclipse of the Moon, on the 22d of October, 1790; of an eclipse of the Sun, on the 6th of November, 1790; and of an annular eclipse of the Sun, on the 3d of April, 1791; with an account of corresponding observations of the two first of these phenomena, made at the University of William and Mary in Virginia, by the Rev. dr. Madison; and of the second, alone, made at Washington-College in Maryland, by the Rev. Dr. Smith. American Philosophical Society, Transactions, III, February 4, 1791.

(1813). Astronomical observations, made in the years 1776, 1777, and 1778, at Philadelphia, by the Rev. Dr. W. Smith, and David Rittenhouse, John Lukens, and Owen Biddle, Esquires: copied from a manuscript account of those observations, drawn up by Dr. Smith; never before published. Astronomical Observations, 1776. In Barton, W. Memoirs of the life of David Rittenhouse, LLD.F.R.S. Late President of the American Philosophical Society, &c., pp.581-582.

(1771). Calculation of the transit of Venus over the Sun, as it is to happen, June 3d 1769, in Lat. 40° N. Long. 5h. W. from Greenwich. American Philosophical Society, Transactions, I, June 21, 1768.

(1799). A communication, on the mode of determining the true place of a planet, in an elliptical orbit directly from the mean anomaly by converging series; by David Rittenhouse, LL.D. President of the American Philosophical Society. American Philosophical Society, Transactions, IV, February 5, 1796.

(1780-1785). The continental almanac, for the year of our Lord, 1780-1785. By Anthony Sharp, Philom. [pseud], David Rittenhouse. Philadelphia: Printed and sold by Francis Bailey. 6v. illustrations. 17cm.

(1771). A description of a new orrery; planned, and now nearly finished, by David Rittenhouse, A. M. of Norriton, in the county of Philadelphia: Communicated by Dr. Smith. American Philosophical Society, Transactions, I, March 21, 1768.

(1771). An easy method of deducing the true time of the Sun's passing the meridian, per clock, from a comparison of four equal altitudes, observed on two succeeding days; by David Rittenhouse, A. M. Communicated by William Smith, D. D. Prov. Coll. of Philadelphia. American Philosophical Society, Transactions, I, August 17, 1770.

(1794). On the improvement of time-keepers. by David Rittenhouse, LL.D. Pres. Am. Philos. Society. American Philosophical Society, Transactions, Read, November 7, 1794.

(1786). A letter from David Rittenhouse, Esq. to John Page, Esq. in answer to one from Mr. Page: concerning a remarkable meteor, seen in Virginia and Pennsylvania, on the 31st of October 1779. (Mr. Rittenhouse's letter is dated January 16, 1780). American Philosophical Society, Transactions, II, May 2, 1783.

(1793, August). Meteorological observations, made in Philadelphia. (np). 5p.

(1796). On the mode of determining the true place of a planet, in an elliptical orbit, directly from the mean anomaly by converging series; by David Rittenhouse, Pres. Am. Phil. Society. American Philosophical Society, Transactions, Read February 5, 1796.

(1786). A new method of placing the meridian mark; in a letter to the Rev. Dr. Ewing, Provost of the University. American Philosophical Society, Transactions, II, November 1785.

(1771). Observations on the comet of June and July, 1770; with the elements of its motion and the trajectory of its path; in two letters from David Rittenhouse, M. A. to William Smith, D.D. Prov. Coll. Philadelphia. American Philosophical Society, Transactions, I, August 3, 1770.

(1786). Observations on a comet lately discovered; communicated by David Rittenhouse, Esq. American Philosophical Society, Transactions, II, March 19, 1784.

(1786). An optical problem, proposed by Mr. Hopkinson, and solved by Mr. Rittenhouse. (Hopkinson's letter is dated March 16, 1785: The answer is without date.) American Philosophical Society, Transactions, II, February 17, 1786.

(1799). A paper, On the improvement of time-keepers; by David Rittenhouse, LL.D. President of the American Philosophical Society. American Philosophical Society, Transactions, IV, November 7, 1794.

(1794). A short account of the malignant fever, lately prevalent in Philadelphia: with a statement of the proceedings that took place on the subject, in different parts of the United States. To which are added, accounts of the plague in London and Marseilles; and a list of the dead, from August 1, to the middle of December 1793... 4th ed., Includes the statistical list of burials, David Rittenhouse's meteorological observations (as in previous editions) and an enlarged "List of the names of the persons who have died." Philadelphia: Printed by the author. 160p. 24cm.

(1781-). The town and country almanack... Boston: (np). illustrated. 18cm.

Other

Barton, W. (1812). Letter to John Vaughan concerning the late Dr. Rittenhouse. Ms. communications to the American Philosophical Society, 1, Natural Philosophy. Barton, W.

Buzby, B. G. (1813). Memoirs of the life of David Rittenhouse, LLD. F. R. S., late President of the American Philosophical Society, &c. interspersed with various notices of many distinguished men; with an appendix, containing sundry philosophical and other papers, most of which have not hitherto been published. Philadelphia: E. Parker. 614p. 26cm.

Bailey, (nd). Bailey's Rittenhouse almanac. Philadelphia: author. Calculated by Abraham Shoemaker, etc., for Philadelphia. 18cm. Carey, M.

Dunlap, J. (nd). Der Gantz neue verbesserte Nord- Americanische Calender. (np). [Accent omitted]

Ford, Edward. (1770). Father Abraham's pocket almanac for 1771. Philadelphia: author.

Herbert, S., Jr. (1946). David Rittenhouse, astronomer- patriot, 1732-1796. Philadelphia: University of Pennsylvania Press.

Lessing, G. E. (1774). The Lancaster almanack [sic], (improved) Being a prognosticator, for the year of our Lord, 1775... Lancaster, PA: Printed and sold by author. 10.5x17cm.

Weatherwise, A. (1777). The Lancaster almanack [sic] for the year of our Lord, 1778: being he second after leap year, the third year of American independency... Lancaster: Printed by F. Bailey. 36p. illustrated. 18cm.

ROBERTS, JOHN (C.1789)

Roberts was the engraver of James McComb Jr.s' A New & Accurate Plan of the City of New York in the State of New York in North America , compiled by B. Taylor. It is described as:

> One of the most accurate and beautiful engraved plans of the city, and particularly interesting on account of its tiny bird's-eye views of some of the most important buildings, as well as for the clear idea which it gives of the country lying along the East River in the vicinity of Crown Point" (Ristow 1985, pp.247-249). Also reported by Ristow is the inclusion of one of Roberts' maps in the Matthew Carey atlas. (p.153)

(1796). The American atlas. Philadelphia: Matthew Carey.

(1797). A new & accurate plan of the city of New York in the state of New York in North America. New York: (np).

ROMANS, BERNARD (1720-1784)

Schwartz and Ehrenberg (1980) assert that Romans executed and published the first separate American imprint of the Battle of Bunker Hill, published in Philadelphia in 1775. An interesting note is the portion of the title, "To the Hone. Jno. Hancock Esqre., President of the Continental Congress, This Map of the Seat of Civil War in America is Respectfully Inscribed" (p.185). Romans, however, is chiefly remembered for his 1774 Concise Natural History of East and West Florida. After his personal surveys of Georgia and Florida, he moved his work to New York in search of subscribers (Ristow 1985).

When the Revolution broke out, Romans received a commission as captain in the a Pennsylvania artillery company, 1776-1778. Romans was ordered by Washington to prepare to report on defenses for Fort Constitution (Harley, et.al. 1978). He was actively engaged in geography during that time, attested to by two maps of New England provinces, both of which are rare today and without the name of Romans. In 1777 he published a map of Connecticut with parts of New York, New Jersey, and Rhode Island in the Boston Gazette and Country Journal (May 19, 1777) and in the Connecticut Gazette and the same map was advertised for sale in the Universal Intelligencer (Romans 1985, p. 57).

Romans was born in Holland and died on a return voyage to America following imprisonment by the British. He may have been murdered for a large sum of money he carried with him. In his journal he claims to have been the first surveyor in Florida. He became a member of the American Philosophical Society (White 1897, V.7, p.176-177).

(1778). Annals of the troubles in the Netherlands, from accession of Charles V., Emperor of Germany. A proper and seasonable mirror for the present Americans. Collected and translated from the approved historians in the native tongue. Hartford: Watson and Goodwin. 2v. in 1. 18cm.

(1775). Chart, containing the peninsular of Florida, the Bahama Islands to north side of the island of Cuba, the old straight of Bahama an all the islands, keys, rock, &c. in these seas. New York: author. Six sheets.

(1778). Chorographical map of the country round Philadelphia. New Haven: (np).

(1778). Chorographical map of the Northern Department of North America. New Haven: (np).

(1779). Complete pilot of the Gulf Passage, or, Directions for sailing through the gulf of Florida, or New Bahama Channel, and the neighboring parts. By Capt. Bernard Romans, Capt. W. G. de Brahm, Capt. Hester, Capt. Bishop, etc. London: (np). 4(60)p.

(1789) As above- London: R. Sayer. 74p. 20.5cm.

(1775). A concise natural history of East and West Florida; containing an account of the natural produce of all the southern part of British America, in the three kingdoms of nature, particularly the animal and vegetable.

Likewise, the artificial produce now raised or possible to be raised, and manufactured there, with some commercial and political observations in the part of the world; and a chorographical account of the same. To which is added, by way of appendix, plan and easy directions to navigators over the Bank of Bahama, the coast of the two Floridas, the north of Cuba, and the dangerous Gulph [sic] Passage. Noting also, the hitherto unknown watering places in that part of America, intended principally for the use of such vessels as may be so unfortunate as to be distressed by weather in that difficult part of the world. By Captain Bernard Romans. Illustrated with twelve copper plates, and two whole sheet maps. New York: author. viii,342p. 3 maps. 18.5cm.

(1776). As above- Printed and sold by R. Aitken.

(1775). An exact view of the late Battle at Charlestown, June 17th, 1775. Boston: (np).

(nd). A general map of the southern British colonies in America. (np).

(1772). A general map of West Florida. (np). $53^{1/2}$x$108^{3/16}$inches.

(1775). Map of the Seat of Civil War in America... New York: (np).

(1775). Map of the seat of the war. (np).

(1773). A map of West Florida, part of Et. Florida, Georgia, part of South Carolina...compiled under the direction of Yr. St. John Stuart Esqr. (np).

(1800). A new book of directions to accompany the charts of the Windward Passage, or, Instructions for sailing through the several passages to the eastward of Jamaica, &c. By Captains B. Romans, Bishop, Hester, Lieut. Woodriffe, and other experienced navigators. To which are added, the reports and descriptive instructions of the commanders sent by the French government, to explore the Windward passages. Newly tr. from the French by Monsieur Puysegur. London: R. Laurie and J. Whittle.

(1794, 1779 & 1800). A new and enlarged book of sailing directions for Capt. B. Romans', &c., &c., Gulf and Windward pilot; containing full instructions for sailing through the Gulf of Florida; or, the old and new channels of Bahama: together with directions for the Windward Passage, &c., &c., with the additions of Captains W. G. Debraham, Bishop, Hester, Archibald Dalzel, esq., George Gauld, esq., Lieut Woodriffe, and other experienced navigators. London: R. Laurie and J. Whittle. 74p. 21.5cm.

(July 6, 1778). A new map of the Northern Department, containing the country from Red Hook, on the North River, (the place where the late map of Connecticut left off) to Three Rivers, in Canada, and from the heads of Merrimack River, in New Hampshire, to the heads of Delaware and Susquehannah, westward, including Fort Schuyler, and the Oneyda Lake, with part of Ontario. Boston Gazette and Country Journal. (np).

(1774). Part of the province of East Florida. Engraved by Paul Revere and Abel Buell. Published with Roman's A Concise Natural History of East and West Florida. New York: author. 62x221cm. The Florida panhandle.

(1774). Part of the province of East Florida. Engraved by Paul Revere and Abel Buell. Published with Roman's A Concise Natural History of East

and West Florida. New York: author. 147x169cm. Tampa Bay through Cape Canaveral.
(1794). Plano redusido que contiene la consta de la Florida de Cayo Viscaino el Rio de Sta Maria...por Bernardo Roman...ano de 1786. Colocad...por.. Dn Miguel Josef Serguera... ano de 1794. (np). 30x27cm. [Accents omitted]
(May 19, 1777). Proposals for printing a new map of the state of Connecticut with the parts of New York, New Jersey, and Rhode Island; collected from the best and latest surveys. Boston Gazette and Country Journal.
(nd). The Seat of War, in the Southren British Colonies. (np).
(1892). Sargent's Silva North America, 4, p. 5.

Other

Phillips, P. L. (1924). Notes on the life and works of Bernard Romans. Washington, DC: (np).

ROPES, WILLIAM (C.1804)

Guthron (1984) lists Ropes as an assistant to Bowditch when he surveyed Marblehead, etc. in 1804, 1805, and 1806 (p.34). No publications were found in The National Union Catalog Pre-1956 Imprints (The American Library Association 1976).

(1834). Chart of the harbours of Marblehead, Beverly and Manchester from a survey taken in the years 1804, 5 & 6 by Nathaniel Bowditch AM, AAS, assisted by Geo. Burchmore & Wm. Ropes 3d. Engraved by Hooker and Fairman. Second edition published in 1834, with additions and alterations by Charles M. Endicott and Joseph Perkins, Jr. (np).

ROWSON, MRS. SUSANNA (HASWELL) (1762-1824)

Susanna Haswell Rowson was so multi-talented and her career so multifaceted on both sides of the Atlantic Ocean that a number of identifying occupations and titles appropriately could follow her name. These include English novelist, English poet and lyricist, English actress, American novelist, American actress, American poet and lyricist, pioneer American feminist, American educator, and American geographer. In the history of the arts in America, she is best known as a novelist, but she preferred to think of herself--even while in the process of writing novels--as an educator (Encyclopedia Britannica 2001. Rowson, Susanna).

Susanna Haswell was born in 1762 in Portsmouth, England, to Susanna Musgrave Haswell and William Haswell. The infant's mother died 10 days later and her father, a lieutenant in the Royal Navy, was shipped off to Massachusetts, soon thereafter leaving the little girl in the care of a nurse (Johnson & Malone 1917).

In 1765 William Haswell married an American woman, Rachel Woodward, and two years later returned to England to bring his daughter to Massachusetts. The trans-Atlantic voyage of five-year-old Susanna, accompanied by her father and nurse, was beset by stormy seas, near-starvation, and a virtual shipwreck in Boston Harbor. A quarter-century later Susanna Rowson would incorporate the details of this misadventurous voyage into one of her novels.

The Haswell's settled on the Nantasket Peninsula southeast of Boston Harbor. The British naval base at which Lieutenant Haswell was stationed was at Hull, the largest community on the peninsula. Susanna's childhood in Massachusetts was described by one biographer as "idyllic" (Parker 1986, p.5). Reading the classic literature in her father's library was a childhood pastime that contributed to her education and perhaps later influenced her career choices. During several summers she spent much time with James Otis, a noted lawyer and statesman who was a neighbor and frequent visitor to the Haswell home. "My little scholar," Otis called the child (Nason 1870, p.15). Susanna's stepmother was a pious woman who instilled in the girl an appreciation of biblical authority. Later Susanna Rowson would say, speaking of her education, "I was early accustomed to make the Bible my study and guide; and to settle all questions of morality by The Sermon on the Mount and by the decalogue" (Nason 1870, p.16).

In 1786 Susanna Haswell married William Rowson, a hardware merchant who was a part-time musician with connections to the theatre. Although little is known about Susanna Rowson's early involvement in the theatre, probably both Rowsons were appearing on stage with provincial companies before the end of 1786 (Parker1986, p.11). Susanna Rowson launched her career as a novelist the same year. Securing the patronage of a well-known noblewoman, Rowson published Victoria by subscription. Like her later novels, Victoria sought to teach lessons to girls and young women; in this case the lessons were that serious consequences follow the disobedience of parental instruction and the loss of virginity outside marriage.

In 1793 the Rowsons joined about 50 other English actors as part of an immigrant company enlisted by the owner of the Philadelphia New Theatre. During her first season, Susanna Rowson performed 35 roles, and in her second season she learned 22 additional parts (Parker 1986, p.15). Despite her grueling schedule on stage, she found time to write song lyrics for the theatre. Many of her songs were published and sold as sheet music. Gifted as a lyracist, she would continue writing songs for as long as she lived. She also wrote three plays during her stay in Philadelphia.

Charlotte Temple already had created a sensation in New England by the time the doors to Mrs. Rowson's Young Ladies Academy opened in Boston in 1797. But how, people must have wondered, did being a successful novelist and actress qualify Mrs. Rowson to instruct their children? When the school year began, the school had only one student, but the teacher's reputation for excellence of instruction spread so rapidly that by the end of the academic year she had a hundred pupils and more applications than she could ever hope to accommodate (Nason 1870, p.90). From the academy's first year, therefore,

Rowson required additional staff. During most of the years that she operated the school, several teachers were on staff. For example, in one academic year early in the history of the academy, Rowson employed a professor of music, a French teacher, a penmanship teacher, and a dancing master from France (Nason 1870, p.102-103). In 1799 she introduced a piano, almost unknown in America, which her music teacher used in teaching her students. Rowson had learned to play the instrument in London and often played the one in her school. She taught geography, history, reading, writing, painting, embroidery, and Bible studies.

When Susanna Rowson opened her academy in 1797, her choice of textbooks was restricted by the infancy of textbook publishing in America. For geography, one of her favorite academic subjects, the only textbooks available were Jedidiah Morse's Geography Made Easy (1784), The American School Geography (1789), and Elements of Geography (1795); Robert Davidson's Geography Epitomized (1784); Nathaniel Dwight's A Short But Comprehensive System of the Geography of the World (1795); Charles Smith's Universal Geography Made Easy (1795); and Benjamin Workman's Elements of Geography (1795).

Two atlases, a gazetteer, and several books that included some geography along with other subjects also had been published. Rowson did not consider these books to be appropriate for her students, so she developed her own geography program utilizing information published in the works of Morse and various English writers. After a few years of development, her geography lectures and exercises had become so acclaimed that she decided to publish them in a textbook that could be used in her academy and in other schools. An Abridgment of Universal Geography was published in Boston by John West in 1805, by which time at least three other geography textbooks had been published in the United States: John Payne's A New and Complete System of Universal Geography (1798), W. William's First Principles of Geography (1798), and John Pinkerton's, who was British, Modern Geography (1804).

Her maritime family had instilled in her a keen interest in navigation, exploration, and the landscapes and cultures of foreign lands. Her travels outside the United States had been limited to the British Isles and France, but even so she had seen more of the world than some of the early American writers of geography textbooks.

In her first geography textbook, as in her novels and other writings, Rowson defended the honor and rights of women and described with disdain countries in which women were mistreated. Egypt, Turkey, and Tibet were among the countries in which, she said, women were shown inadequate respect. As a feminist--and the only woman among the early American geographical writers--Rowson was more inclined than her contemporaries to focus on the status of women in her characterizations of cultures.

Patricia Parker, an objective biographer, commented on the writing in An Abridgment of Universal Geography, noting that Rowson had ...her own biases, as a conscientious textbook writer was expected to do. Most of the book is judiciously presented, and even in her most subjective moments she appeared no less objective than her contemporaries....Rowson's style is straightforward, with concise and clear sentences that read easily...her selection of material as well as her style made the information highly readable (Parker 1986, p.106). The book had "some popularity in its day" and was being offered by New England booksellers at least as late as 1815 (Parker 1986, p.107). No second edition of the book is known, although the Library of Congress has a copy that bears an 1806 date.

White (1907, V.IX, p.317) acknowledges Rowson's contributions to America. Her novel Charlotte Temple, was America's first best seller.

(1805). An abridgment of universal geography, together with sketches of history designed for the use of schools and academies in the United States... Boston: Printed for John West...David Carlisle, Printer. 302p. 17.5cm.

(1822). Exercises in history, chronology and biography in question and answer for the use of schools. Comprising ancient history, Greece, Rome, etc., modern history, England, France, Spain, Portugal, etc., the discovery of America, rise, progress and final independence of the United States. Boston: Richardson and Lord. 170p.

Other

Bacon, John Barnitz. (nd). The history of Lucy Temple. (np).

(nd). Memoir of Mrs. Susanna Rowson. (np).

Munsell, John. (1870). A memoir of Mrs. Susanna Rowson. Albany, NY. Nason, Elias.

(April 20, 1932). Proceedings of the American Antiquarian Society, 42, Part 1.

Vining, J.W., & Smith, B.A. (1998). Susanna Rowson: Early American Geographer. The Social Studies, 89(6), 263-270.

SAMPSON, EZRA (1749-1823)

Several publications, including a geography text, by the author were found (The American Library Association, 1977). At this time, no biographic information has been unearthed.

(1807). The youth's companion, or an historical dictionary; Consisting of articles selected chiefly from natural and civil history, geography, astronomy, zoology, botany and mineralogy; arranged in alphabetical order. Hudson, NY: Balance-Press; Printed for the author. [4], 7, [1], 456p.

(1813). As above- Albany, NY: Printed by Websters and Skinners, for themselves, and Nathan Elliot, of Catskill. 12,412p. 18cm.

(1816). As above- Hudson, NY: Printed for Nathan Elliot, by W. L. Stone and R. Corss.

(1832). As above- Carefully revised and abridged...by John B. Longgley... St. Clairsville, OH: Printed by Horton J. Howard. 300p. 19cm.

SARGENT, WINTHROP (1753-1820)

Sargent was a patriot, born in Gloucester, Massachusetts, the son of Winthrop and Judith (Saunders) Sargent (Brown 1897). He graduated from Harvard and became captain of one of his father's ships. He became associated with the Ohio Company in 1786 at which time he was appointed by congress Surveyor of the Territory Northwest of the Ohio river. Sargent subsequently became secretary of the Northwest Territory and Governor of the Mississippi Territory. He was a member of the Massachusetts Historical Society and a member of the American Philosophical Society. He died in New Orleans, Louisiana (V.6, p.613).

Pattison (1957) notes that Sargent succeeded Edward Dowse as Surveyor for New Hampshire. Hutchins had returned to Pittsburgh in June of 1786. Congress authorized resumption of the surveys of the east-west line which was begun, but not completed in 1785. Sargent, sponsored by the Secretary of War, was from Massachusetts, and was soon to be elected Secretary of the Ohio Company of Associates. Because of delays in getting the survey, under Thomas Hutchins, underway, Sargent took the opportunity to reconnoiter and survey the lower valley of the Muskingum River (p.133). The preceeding is confirmed by White (1893, V.VI. p.152).

(1812). Account of several shocks of an earthquake in the southern and western parts of the United States... Memoirs of The American Academy of Arts and Sciences, 3, pt. 2, art. 50, pp.350-360.

(1814). Communicated for the Natchez Intelligeneer, no. w. The following information from the record of meteorological observations of the Hon. Wonthrop Sargent, Esq. of Gloster Place, two miles south of Natchez. Natchez, MS: (np). 22x36cm. Broadside.

(1819). Detail of the rise and fall of the Mississippi River at Natchez, for more than 20 years... Natchez, MS: (np). 15x44cm.

(1819). From the meteorological observations at Gloster Place, two miles south of Natchez... Mean temperature of each and every month in the year, and of every year from 1810 to 1818, inclusive... Natchez, MS: (np). 21x25cm. Broadside.

(1796). A letter to Dr. B. S. Barton, accompanying drawings and some account of certain articles, which were taken out of an ancient tumulus, or grave, in the western country. Transactions of the American Philosophical Society, 5(9), pp.74-76.

(1787). List of forest and other trees, northwest of the River Ohio. Memoirs of The American Academy of Arts and Sciences, 2, pt.1, art.21, pp.156-59.

(1855). Plan of an ancient fortification at Marietta, Ohio. Cambridge: (np). 4°.

(nd). Proclamation establishing the County of Wayne in Northwest Territory. (see under Northwest Territory, U.S. Governor, Acting 1787-1798, Winthrop Sargent). (np).

(1924). Winthrop Sargent's diary while with General Arthur St. Clair's expedition against the Indians. In Ohio Archaeological and Historical Quarterly, 33, pp.237-282. Columbus, OH.

Other

Bartlow, Bert S., Todhunter, W. H., Cone, Stephen D., Pater, Joseph J., Schneider, Frederick, and others. (Eds.). (1905). Centennial history of Butler County, Ohio, to which is appended a comprehensive compendium of local biography and memoirs of representative men and women of the country... (np): B. F Bowen & Co.

(1851). Diary of Col. Winthrop Sargent, adjutant general of the United States' army, during the Campaign of MDCCXCI. Wormsloe, GA: (np). 58p. 2 plans. 33.5x27cm.

Pershing, Benjamin H. (1927). Winthrop Sargent: A builder in the old Northwest. Unpublished Ph.D dissertation. Dept. of History, University of Chicago.

Smith, Benjamin B. and Sargent, Winthrop. (1796). Papers relative to certain American antiquities. (np).

SAUTHIER, CLAUDE JOSEPH (C.1778)

Sauthier and Ratzer collaborated on publishing "A Map of the Province of New York. Which included New Jersey" (Schwartz & Ehrenberg 1980, p.187). Although Sauthier may fall outside the dates for the first generation of American geographers, his contributions likely had influence on the outcome of some Revolutionary War skirmishes and perhaps the American Revolution.

(1778). Carte des troubles de l"Amerique levee par ordre du chavalier Tryon, Capitaine General et Gouverneur de la province de New York, ensemble la province de New Jersey, par Sauthier et Ratzer. Traduit de l'anglais. Paris: Chez Le Rouge. 70.5x52cm. Scale 1:1,070,000. [Accents omitted]

(1799). A chorographical map of the province of New York in North America divided into counties, manors, patents and townships; exhibiting likewise all the private grants of land made and located in that province. London: Faden. $71^{1/4}$x$54^{1/4}$inches. Scale: none.

(1849). As above- Albany: Pease. Scale: About 11 miles to 1 inch.

(1777). A map of the inhabited part of Canada from the French surveys; with the frontiers of New York and New England from the large survey by Claude Joseph Sauthier. London: Wm. Faden. 34x23inches.

(1776). A map of the province of New York reduced from the large drawing of that province compiled from actual surveys by order of Excellency William Tryon... By Claude Joseph Sauthier. To which is added New Jersey from the topographical observations of C. J. Sauthier and B. Ratzer. Engraved by William Faden. London: William Faden. 61x76cm.

(1777). A map of the provinces of New York and New Jersey, with a part of Pennsylvania and the province of Quebec. Augsburg: Lotter. $28^{1/2}$x$21^{3/8}$inches. Scale: 16.1 miles to 1 inch.

(1778). Mappa geographica provinciae Novei Eboraci ab Anglis New York dictae ex ampliori delineatione ad exactas dimensiones concinnata in arctius spatium redacta cura Claudii Josephi Sauthier oui accedit Nova Jersey ex topographicia observartionibus. Norimbergae, sumtibus Homannianor, heredum. 71x56.5cm. Scale 1:1,060,000. [Accents omitted]

(c.1771). Plan of the camp and Battle of Alamance the 16 May 1771. Between the Provincials of North Carolina. Commanded by His Excellency Governor Tryon, and rebels who styled themselves Regulators. 33x57cm. Scale: 1 inch to 2.1 miles.

(1777). A plan of the operations of the King's Army under the command of General Sr. William Howe, K. B. in New York and East New Jersey, against the American forces commanded by General Washington, from the 12th of October, to the 28th of November 1776. Wherein is particularly distinguished the engagement on the White Plains the 28th of October. Engraved by Wm. Faden. London: William Faden. 73x49cm. Scale 1:88,000.

(1921). Plan of the town of Wilmington, in New Hanover County, surveyed and drawn in Dec. 1769. In Janet Schaw. Journal of a lady of quality. (np).

(1776). A topographical map of Hudson's River...from Sandy Hook... as high as Fort Chambly on Sorel River by Claude Joseph Sauthier.. Engraved by William Faden. London: W. Faden. 81.2x55.5cm.

(1777). A topographical map of the north part of New York Island, exhibiting the plan of Fort Washington, now Fort Knyphausen, with the rebel lines to the southward, which were forced by the troops under the command of the Rt. Hon. Earl Percy, on the 16th Nov. 1776. London: Wm. Faden. 18 1/4x10inches. Scale: 1730ft. to 1 inch.

SCHUYLER, PHILIP JOHN (1733-1804)

Schuyler was surveyor general of New York. A native of Albany, New York. He, James Clinton and Simeon DeWitt were engaged by the New York Legislature (1785) to determine the boundary between Pennsylvania and New York (Ristow 1985, p.75). The Biographical Dictionary report that he was a US general and a politician, a member of the Continental Congress (2001. Schuyler, Philip).

During the American Revolution he was commissioned one of the four major generals. Ill health appears to have brought on a series of misfortunes that included the planned invasion of Canada and the fall of Ft. Ticonderoga, NY. He was accused of incompetence and neglect of duty and was court-martialled. The court-martial acquitted him of all charges. He went on to work for ratification of the U.S. Constitution and was one of New York's first two U.S. Senators. Schuyler served as Commissioner of Indian Affairs, negotiating several treaties with the Six Nations (Encarta Encyclopedia 2001. Schuyler, Philip John).

(1807). An essay, suggesting a plan to introduce uniformity in weights and measures of the United States of America. In Transactions of the Society for the Promotion of Useful Arts, 2, pp.36-69. Albany, NY.

SCOLES, JOHN (C.1796)

Ristow (1985) includes Scoles as engraver of one of Roberts' maps in the Matthew Carey atlas (p.153). The National Union Catalog Pre-1956 Imprints also notes a John Bentley Scoles who published in 1839. Although he is likely not the same as John, one would need to closely examine the publications ((The American Library Association 1976, V.533, p.490)

(1796). The American atlas. Philadelphia: Matthew Carey.

(1825). Geological sketch of the lead mine district of Missouri, embracing the granitical tract of St. Michael. Scoles, sculp. In Henry R. Schoolcraft. Travels in the central portions of the Mississippi Valley. New York. P.275.

(nd). Georgia from the latest authorities. (np).

(1825). A sketch of the western country, designed to illustrate the present work. Scoles sculp. In Henry R. Schoolcraft. Travels in the central portions of the Mississippi Valley. New York. p.15.

SCOT, ROBERT (C.1796)

Charles Varle' conducted a survey of the city of Philadelphia in 1796 that was engraved by Robert Scot of that city (Ristow 1985, p. 250). The National Union Catalog Pre-1956 Imprints does not list any publications under the name (The American Library Association 1976).

(c.1796 & 1802). Plan of the city of Philadelphia. By P. C. Varle', geographer & engineer. Philadelphia: author.

(1794). A grand plan of the city of Philadelphia and its environs; taken from actual survey by A. P. Folie, geographer from St. Domingo. (np).

SCOTT, JOSEPH T. (C.1796)

The National Union Catalog Pre-1956 Imprints labels Scott a "geographer" (The American Library Association 1976, V.534, p.181). According to Ristow (1985) Scott was a resident of Philadelphia when he worked as an engraver with Carey (p.138). It appears that Scott went on to prepare maps and atlases (p.154). Scott's map of the Tennessee Government was included in Guthrie's Geography Improved (Schwartz & Ehrenberg 1980). It is also included in Carey's American Atlas.

(1796). An atlas of the United States. by Joseph Scott. Philadelphia: F. & R. Bailey, B. Davis, & H. & P. Rice. 19 maps. 12o.

Note.-Earliest atlas of the United States.

(1807). Elements of geography, for the use of schools, illustrated with six diagrams, and a map of the world. With an appendix containing an account of the division of time... Philadelphia: Printed by Kimbler, Conrad & Co. 165p. 17.5cm.

(1800 & 1806). A geographical description of Pennsylvania; also of the counties respectively, in the order in which they were established by the Legislature. With an alphabetical list of the townships in each county; and their population in 1800. Philadelphia: Printed by R. Robert Cochran. 147p. map. 16cm.

(1807). A geographical description of the states of Maryland and Delaware; also of the counties, towns, rivers, bays and islands... Philadelphia: Printed by Kimber, Conrad, and Co. 191p. 2 folded maps. 16cm.

(1805). A geographical dictionary of the United States of North America. Containing a general description of each state... With a succinct account of Indiana, and upper and lower Louisiana territories... Philadelphia: Printed by Archibald Bartram, for J. Johnson. 384p. folded map. 21.5cm.

(1805). As above- Philadelphia: Printed by A. Bartram for T. Armstrong.

(1811). As above- ...With an appendix, comprising a more general description of the rivers, springs, towns, counties, &c. of the western country; and the population of the United States, as ascertained by the census of 1810. Pittsburgh: Cramer, Spear & Eichbaum. 20.2cm.

(1793). A map of the Tennasee [sic] government formerly part of North Carolina taken chiefly from surveys by Genl. D. Smith & others. Engraved by Joseph T. Scott. Philadelphia: Mathew Carey. 24x52cm.

(1799). The new and universal gazetteer; or, modern geographical dictionary... To which is added, a new and easy introduction to geography and astronomy; with a nomenclature, explaining the essential terms in each science...in four volumes. Philadelphia: Printed by Francis & Robert Bailey; Patterson & Cochran. 22cm.

(1799 & 1800). The new and universal gazatteer; or, modern geographical dictionary, containing a full and authentic description of the different empires, kingdoms...in the known world...to which is added, a new and easy introduction to geography and astronomy...illustrated with twenty-five maps, and armillary sphere, and several diagrams...in four volumes. Philadelphia: Printed by Francis & Robert Bailey. (V.1, & 2 - 1799, V.3, & 4 - 1800]

(1795). The United States gazatteer: Containing an authentic description of the several states. Their situation, extent, boundaries, soil, produce, climate, population, trade and manufactures. Together with the extent, boundaries and population of their respective counties... Illustrated with nineteen maps. Philadelphia: Printed by F. & R. Bailey. 292p. 19 folded maps. 18cm.

SCULL, WILLIAM (C.1775)

Scull worked under Samuel Holland to survey the British American colonies. Whether he became a United States citizen is in question.

(1775). A map of Pennsylvania, exhibiting not only the improved parts of that province, but also its extensive frontiers; laid down from actual surveys and chiefly from the late map of W. Scull...1770 and humbly inscribed to the Hon. Thomas Penn and william Penn... London: Bennett, John and Robert Sayer. 134x70cm.

(1770). To the honorable Thomas Penn and Richard Penn, esq., true and absolute proprietors and governors of the province of Pennsylvania... Philadelphia: James Nevil for the author. . 82.5x57cm

(1770). To the honorable Thomas Penn and Richard Penn esquires true absolute proprietaries and governors of the province of Pennsylvania and the territories thereunto belonging ...this map of the province of Pennsylvania is...dedicated by...W. Scull. Philadelphia: James Nevil. . Henry Dawkins Sculpt. 21.5x32in.

Other

(1930). The family of scull... Philadelphia: The John C. Winston Company. 1p. 1.,27, 25p., illus (incl. maps, facsims.) plates (1 col.) ports. (1 col.) 25cm.

SCULPSIT, ALLARDICE (1757-1837)

Sculpsit and Robert Scot are credited with engraving Folie's plan of Philadelphia (Ristow 1985, p. 250). The National Union Catalog Pre-1956 Imprints (The American Library Association 1976) provided no publications by Sculpsit.

(1794). A grand plan of the city of Philadelphia and its environs; taken from actual survey by A. P. Folie, geographer from St. Domingo. (np).

SEYMOUR, JOSEPH H. (C.1791-1822)

Guthorn (1984) reports that Seymour worked as an engraver in Worcester, Massachusetts prior to moving to Boston and then to Philadelphia. He is responsible for engraving six of the plates in Matthew Clark's A Complete Set of Charts of the Coast of America, extending from Cape Breton on the East, into the Gulph [sic] of Mexico, with Nautical Directions... The National Union Catalog Pre-1956 Imprints (The American Library Association 1976) revealed no additional publications..

(1790). Chart of the coast of America from New York to Rhode Island. Boston: Matthew Clark.

(nd). Vermont, from actual survey... (see under Mathew Carey).

SHALLUS, FRANCIS (C.1801)

Shallus was the engraver for two of Zebulon Pike's maps published, in 1810 by Fielding Lucas, Jr. (Schwartz & Ehrenberg 1980, p. 225). The First Part of Captn. Pike's "Chart of the Internal Part of Louisiana" and "A Chart of the Internal Part of Louisiana, including all the Hitherto Unexplored Countries, Lying between the River LaPlatte of the Missouri on the N: and the Red River on the S: the Mississippi East and the Mountains of Mexico West: With a part of New Mexico and the Province of Texas," were compiled by Antoine Nau, a draftsman serving under General James Wilkinson in St. Louis. According to Schwartz and Ehrenberg, "They are the first published maps to depict graphically the geographical knowledge of the entire southwest derived directly from exploring parties" (p. 225). Ristow (1985) notes that Shallus based his map of Baltimore on Folie's map, but he included only the city proper.

(1817). Chronological tables, for every day in the year. Compiled from the most authentic documents. Philadelphia: Merritt, printer. 2v. 18.5cm.

(nd). Plan of the investment and attack of York in Virginia. Engraved by Francis Shallus. Philadelphia: C. P. Wayne. 20.6x32.3cm. (see Phillips, Maps of America, p.1135)

(1801). Plan of the city and environs of Baltimore. (np): Warner & Hanna.

SHERMAN, ISAAC (C.1785)

Sherman was one of eight surveyors who reported for duty in 1785 to the newly appointed Geographer of the United States, Thomas Hutchins beginning work on the surveying of America. He, like many of his fellow geographer/surveyors, was a replacement for the man selected by the Connecticut delegates for the job of surveying. Sherman was a colonel in the Continental Army who was also the son of an influential man and a skilled politician successful in Connecticut's claim to the Western Reserve (Pattison, 1957). No entry is found in The National Union Catalog Pre-1956 Imprints (The American Library Association 1976).

SIBLEY, JOHN DR. (1757-1837)

One of those engaged in exploring the Missouri, Red River and Washita, according to a Message from the President of the United States, Thomas Jefferson. One of the Lewis and Clark expedition (Meisel 1926, v.2, p.86-87).

(1803). An account of Louisiana. Being an abstract of documents, in the offices of the departments of state, and of the Treasury. Washington, DC: Duane, printer.

(1804). An account of Louisiana. Being an abstract of documents delivered in, or transmitted to, Mr. Jefferson, President of the United States of America; and by him laid before Congress, and published by their order. Printed at Washington and reprinted at Philadelphia, and all the other states of the Union. London: Reprinted by J. Hatchard.

(1902). ...An account of Louisiana, 1803. Being an abstract of documents in the offices of the Departments of State and of the Treasury. Boston: Directors of the Old South Work.

(1803). An account of Louisiana, laid before Congress by direction of the President of the United States, November 14, 1803; comprising an account of its boundaries, history, cities, towns and settlements... Providence: Printed by Heaton & Williams.

(nd). An account of the Red River and country adjacent. (np).
(1806). Historical sketches of the Indian tribes in Louisiana, south of the Arkansas River... (np).
(1803). Letter [on Louisiana] to J. Gales. (np).
Lewis and Clark expedition, pt.3: Historical sketch of Indian tribes in Louisiana and ethnological observations, by John Sibley, p. 721, American State Papers: Indian Affairs, V.VII (Doc. 113).
Lewis and Clark expedition, pt. 4: Geographical description of Red River and adjacent country, by John Sibley, p. 725, American State Papers: Indian Affairs, V.VII (Doc. 113).
(1806). Message from the President of the United States, communicating discoveries made in exploring the Missouri, Red River and Washita, by Captains Lewis and Clark, Doctor Sibley, and Mr. Dunbar; with a statistical account of the countries adjacent. Washington: By order of the Senate, A. & G. Way, printers.
(1803). A topographical and statistical account of the province of Louisiana, containing, a description of its soil, climate, trade, and produce...&c. ...together with new and interesting particulars, relative to the Indian tribes: to which is annexed, a copious preface; and, the recent conventions, between the United States, and the French Republic. Comp. by different individuals, possessed of the best information, and from the documents communicated to Congress, by the President... Baltimore: Printed for J. Rice, S. Butler by Martin & Pratt.
(1807). Travels in the interior parts of America: communicating discoveries made in exploring the Missouri, Red River, and Washita, by Captains Lewis and Clark, Doctor Sibley, and Mr. Dunbar; with a statistical account of the countries adjacent. As laid before the Senate, by the President of the United States. London: R. Phillips.

Other

Abel, Annie Heloise. (Introduction). John Sibley (Ed.). (1922). ...A report from Natchitoches in 1807, by Dr. John Sibley, ed., with an introduction by Annie Heloise Abel. New York: Museum of the American Indian, Heye Foundation. 102p.

SIMPSON, JAMES (C.1785)

Simpson was one of eight surveyors who reported for duty in 1785 to the newly appointed Geographer of the United States, Thomas Hutchins beginning work on the surveying of America. Maryland gave its appointment to Simpson of York County, Pennsylvania. Pattison (1957) notes that, "Through Simpson, the party of surveyors encamped at the mouth of Little Beaver Creek, made their only known contact with the boundary commissioners who established the beginning point for federal surveying" (p.127). No entry is found in The National Union Catalog pre-1956 Imprints (The American Library Association 1976). The only possible person dates, 1781-1853.

SMITH, ANTHONY (C.1796)

Smith is credited with preparing a chart of the Chesapeake and Delaware Bays, "Pilot of St. Marys" (Ristow 1985, p. 227). The National Union Catalog Pre-1956 Imprints (The American Library Association 1976) provided no additional publications.

(1796). The American pilot. By William Norman. Boston: W. Norman.
(nd). A new and accurate chart of the bay of Chesapeake... (np).

SMITH, DANIEL (1748-1818)

According to Durham (1976), Smith is an important American figure: "ranking General of the Mero District, Secretary of the Territory Southwest of the River Ohio, and United States Senator" (p.xi). He is also credited as builder of civilization on the frontier."

Smith engaged in surveying boundaries, maintaining public records, fought hostile Indians, protected friendly Indians, negotiated treaties, served on college boards, collected books, and aided in the organization of local and state governments. There is no surviving image of Smith, but a good bit of information about the character of the man is known from correspondence.

On June 20, 1773, he was married to Sarah Michie Smith who, during the first 36 years of their marriage, performed the duties of mother and father since he was absent much of the time attending to business and public service. He and a twin sister, Sarah, were born on October 17 near Aquia Creek in Stafford County, Virginia (p.264).

Brown (1897) notes that Smith was born in Fauquier county, Virginia in 1740 (V.7, p.112). Durham (1976) goes on to report that following his early childhood in a rural setting he was sent to Baltimore for an education. Prior to 1768 Smith left the Tidewater to go to Albemarle County Virginia to study medicine with a Dr. Thomas Walker. While there he tutored Dr. Walker's son. Smith meanwhile exhibited little interest in medicine, so the Dr. proceeded to train him in surveying, mathematics, and accounting.

His work as a surveyor began by the time he was twenty years old, much of that in the Albemarle area. Many of his surveys noted natural features of the places being surveyed. Daniel and Sarah moved, in November of 1773, to land on the upper Clinch River near the present Mt. Airy in Wythe County, Virginia. While there he continued his surveying activities which often took him away from home. During 1774 Smith was made captain in the Fincastle Militia. His responsibilities included protecting settlers and training the militia. In 1775, on the eve of the Declaration of Independence, Smith became a member of the Fincastle Committee. Its members were somewhat divided on the matter of independence from Britain, but the Committee declared that, "submitting our liberty or property to the power of a venal British Parliament or to the will of a corrupt ministry" would not be considered (Durham 1976, p.41).

In 1776 the Continental Congress passed a resolution that Virginia, North Carolina, and Georgia were to cooperate with South Carolina to make war against the Cherokee. Smith and some of his men proceeded to Fort Patrick Henry, near the present Kingsport, Tennessee. There he became associated with James Robertson, father of Middle Tennessee, and with John Sevier, first governor of Tennessee.

From 1776 Smith was active in the development of what was to become known as Tennessee. When, in 1776, Fincastle County was broken into three counties, Smith was appointed a major in the Washington County Militia. That was followed by military and treaty activity. During that period Smith was appointed Justice of the Peace and on March 17, 1779 was elected Commissioner of the Tax for Washington County. The Virginia General Assembly and the General Assembly of North Carolina authorized the extension of their state line to the Mississippi. Smith and Anthony Bledsoe did the job.

During that survey he began to keep a detailed journal:

> The journal was also to be a great value in later Kentucky-Tennessee map making as it listed and described topographic features from east to west along much of the line. In it is found convincing descriptions of the unusually cold winter of 1779-1780 and candid documentation of the hardships travelers would probably encounter in passing through the wilderness to the Cumberland settlements. (Durham 1976, p.66)

Though some activities took Smith away from surveying, his interests remained very much there. During the following years Smith was intimately involved in the formation of the state of Tennessee. He became U. S. Senator for Tennessee in 1798. His demise came after a long and rewarding life. Smith died at Rock Castle, near Hendersonville, Tennessee. He was buried next to his wife near his Rock Castle home.

Known by the presidents during his lifetime, Jefferson praised him, "For intelligence, well cultivated talents, for integrity, and usefulness, in soundness of judgement, in the practice of virtue and in shunning vice he was equaled by few men, and in purity of motive, excelled by none" (Durham 1976, p.264). The Nashville Clarion noted that, "To his industry and talents we are indebted for the only map ever published of this state, and for the geography of the country printed in a small volume by Carey" (p.264).

(1799). A geography of the state of Tennessee; with map. (np).

(1915). The journal of Daniel Smith, with introduction and notes by St. George L. Siousatt... Nashville: (np). 66p. foulded [sic] map. 23.5cm.

Note. Caption title: The journal of General Daniel Smith, one of the commissioners to extend the boundary line between... Virginia and North Carolina, August, 1779, to July, 1780.

(nd). A map of the Tennassee [sic] Government, formerly part of North Carolina. Taken chiefly from surveys by Genel. D. Smith & others. J. T. Scott, Sculp. (np). 20.5x9.5inches.

Note. Engraved for Careys American edition of Guthries geography improved.

(1797). As above- ...Engraved for Imlay's American Topography. T. Conder Sculpt. London: J. Debrett. map. 17x41cm.

Note: In Gilbert Imlay, A topographical description of the western territory of North America. 3d ed. (opposite p.512). London: (np).

(1793). A short description of the state of Tennessee, lately called the Territory of the United States, south of the river Ohio. To which is prefixed, the Constitution of that state. Philadelphia: Printed for Mathew Carey.

(1796). As above- Printed for Mathew Carey by Lang and Ustick, Sept. 20, 1796. 44p. 16.5cm.

(1797). As above- March 9, 1796. Note. In Gilbert Imlay, A topographical description of the Western Territory of North America... 3d ed. (pp.512-543). London: (np).

(1797). As above- New York: Printed by J. S. Mott. 47p. 17.5cm.

(1793). A short description of the state of Tennassee [sic], lately called the Territory of the United States, south of the river Ohio; to accompany and explain a map of that country. Philadelphia: Printed by Mathew Carey. 20p. 23.5cm.

(1936). As above- Boston: Photostat Americana. 36p. 22cm.

Other

Bryant, Gladys Eugenia. (1960). Daniel Smith, citizen of the Tennessee frontier. Nashville: Tennessee State Library & Archives. 178p. 28cm.

Durham, Walter T. (1976). Daniel Smith: Frontier Statesman. Unpublished masters thesis, Vanderbilt University, Nashville, TN. 318p. 8 plates. 24cm.

SMITH, SAMUEL STANHOPE (1750-1819)

Smith's contribution to the knowledge of geography was through two publications relating to humans and their distribution. Brown (1897) notes that Smith was an educator born in Pequea, Pennsylvania the son of Robert and Elizabeth (Blair) Smith. Smith graduated from the College of New Jersey then assisted at his father's school and was a tutor at the college. He became a Presbyterian minister. Smith married Ann Witherspoon, the daughter of Dr. John. He was awarded an honorary D.D. by Yale in 1783 and an L.L.D. by Harvard in 1810. He participated in preparing the Form of Presbyterial Government. Smith was president of the university and died at Princeton, New Jersey (V.7, pp.137-138; The Biographical Dictionary 2001. Smith, Samuel Stanhope).

(1787). Essay on the causes of the variety of complexion and figure of the human species. (np).

(1787). On the causes of diversity of figure and complexion in the human species. Proceedings of the American Philosophical Society, 22, appx. p.148.

Other

Ramsay, David. (1813, 1816, 1817 & 1818). History of the United States, 1607-1808; or, the thirty-third of their sovereignty and independence, by David Ramsay. Continued to the Treaty of Ghent by S. S. Smith and other literary gentlemen. Second ed. rev. and corrected. Philadelphia: M. Carey. 3v.

SMITH, THOMAS (1775/6-1830)

Brown (1897) reports that a Thomas Smith (1745-1811) was a deputy surveyor of a frontier district (V.7, p.138). Beyond the listed publications, we have no other knowledge about Smith.

(1799). ...Geological remarks on some parts of New Jersey on a small tour through the lower parts of New Jersey... The Medical Repository, New York City, 3(2), pp.151-54.

(1818 & 1821). Le cabinet du jeune naturaliste, ou, Tableaux interessans de l'historie des animaux; offrant la description de la nature, des moeurs et habitudes des quadrupeds, oiseux, poissons, amphibies, reptiles, etc., les plus remarquables du monde connu, et classes dans und ordre systematique. Ouvrage enrichi de soixante-cinq belles gravures; traduit de l'anglais, de M. Thomas Smith... Paris: Ledoux et Tenre. 16.5cm. [Accents omitted]

(1815). A compendious system of astronomy. Enlarged and improved by Donald M'Donald. New York: M'Donald & Gillet. 151p. 32°.

(1818). A compendious system of astronomy. Rev. and improved. New York: S. Wood & Sons; Baltimore: S. S. Wood & Co. 134p. 12cm.

(1815). Elements of geography... By the Rev. Thomas Smith... Enlarged and improved by Donald M'Donald. New York: Printed by Largin & Thompson, for D. Macdonald & J. Gillet. 152p. 11.5cm.

(1818). As above- New York: S. Wood & Sons. 142p. 11.5cm..

(1838 & 1839). History and progress of the missionary societies; containing faithful accounts of the voyages, travels, labours [sic], and happy results of the various missionaries... in different parts of the habitable globe: including the latest discoveries... The whole forming a complete missionary repository... Compiled and arranged from authentic documents, by the Rev. Thomas Smith... With a continued history to the present time. To which is also added, an interesting narrative of the great success attending the judicious exertions of... The rev. John Williams in the South-Sea Islands. London: T. Kelley. 2v. 22cm.

(1806 & 1807). The naturalist's cabinet: containing interesting sketches of animal history; illustrative of the natures, dispositions, manners, and habitats, of all the most remarkable quadrupeds, birds, fishes, amphibia, reptiles, &c. in the known world... London: J. Cundee. 6v. 21.5cm.

(1832 & 1833). The origin and history of missions... compiled from authentic documents; forming a complete missionary repository; illustrated by numerous engravings, from original drawings made expressly for this work... By the Rev. Thomas Smith and Rev. John O. Choules... Boston: S. Walker, and Lincoln & Edmands. 2v. maps. 28.5x23cm.

(1837). As above- Boston: Gould, Kendall and Lincoln.

(1838). The origin and history of missions; a record of the voyages travels, labors, and successes of the various missionaries, who have been sent forth by Protestant societies and churches... compiled from authentic documents; forming a complete missionary repository; illustrated by numerous engravings, from original drawings made expressly for this work. By the Rev. O. Choules... and the Rev. Thomas Smith... Fifth edition. Boston: Bould, Kendall and Lincoln, and Crocker and Browster. 2v. plates. maps. 29.5cm.

(1815). The scientific library; or, Repository of useful and polite literature; comprising astronomy, geography, mythology, ancient history, modern history, and chronology... Enlarged and improved by Donald M'Donald & J. Gillet. New York: Printed by Largin & Thompson for D. M'Donald & J. Gillet. 6v. 32°.

(1816). As above- New York: S. Wood & Sons.

(1802). The universal atlas, and introduction to modern geography: in which are described, the most celebrated empires, states and kingdoms of the world... also... a new geographical clock. The whole illustrated with thirty-one maps and plate accurately delineated by an eminent geographer, engraved by John Cooke. The introduction, and geographical descriptions by the Rev. Thomas Smith. London: Printed for J. Harris, successor to E. Newbery, and J. Cooke, by H. Bryer. [vii]-xxiv,87p. 24cm.

(c.1802). The wonders of nature and art; or, a concise account of whatever is most curious and remarkable in the world. London: J. Harris. 14cm.

(1803 & 1804). As above- ... compiled from historical and geographical works of established celebrity, and illustrated with the discoveries of modern travellers. London: Printed for J. Walker. London: J. Harris. 12v.

(1806 & 1807). As above- Revised, corrected, and improved by James Mease... Philadelphia: Printed by Robert Carr, for Birch & Small: sold by M. Carey, Kimber, Conrad & Co. and Jacob Johnson. 14v. Illustrated. 15.5cm.

(1838 & 1839). The wonders of nature and art, comprising upwards of three hundred of the most remarkable curiosities and phenomena in the known world, with an appendix of interesting experiments, in different arts and sciences... by J. Taylor... 8th ed. London: J. Chidley. vi[2],568p. plates.

SMITH, WILLIAM (1727-1803)

Barton (1813) notes, "On the first day of January, 1790, Dr. Rittenhouse was elected one of the Vice-presidents of the American Philosophical Society; his colleagues, in this office, being the Rev. William Smith, D. D. and John Ewing, D. D. both of them able and distinguished astronomers" (pp.342-343). Smith earlier (May, 1754) had held the post of Provost of The College,

Academy and Charitable School, of Philadelphia, in Pennsylvania (p.367).

Brown (1897) recorded information about seventeen William Smiths (V.7, pp.140-1145). The above information, however did not appear in any of those biographies. William Smith, born in Aberdeen, Scotland in 1727 may be the best candidate for the person we are seeking (V.7, p.140).

Guthron (1984) also lists a W. Smith who was very active in the production of coastal charts. Smith is listed among others in the annual report of 1855 as one of those who produced, "Fine engraving of lettering, topography, and hydrography on charts..." (p.20). He is not the Smith identified by Barton.

(1797). An abstract of the account of Col. Bouquet's expedition against the Ohio Indians, in 1764... In John Knox, of London Publisher. A new collection of voyages... 2, pp.186-210. London.

(1778). [Astronomical notes including] Astronomical observations, 1776; eclipse of the sun January 9, 1777; an eclipse of he sun June 24, 1778. In J. T. Mayer. Tabulae motuum solis et lunae... (np).

(1756). Etat present de la Pensilvanie, ou l'on trouve le detail de ce qui s'y est passe dupuis le defaite du General Braddock jusqu'a la prise d'Oswego, avec une carte particuliere de cette colonie. Paris: (np). 128p. map. 17cm. [Accents omitted]

(1774). An examination of the Connecticut claim to lands in Pennsylvania. With an appendix, containing extracts and copies of original papers. Philadelphia: Printed by Joseph Crukshank. 2,94p. map. 23cm.

(1868). Historical account of Bouquet's Expedition against the Ohio Indians, in 1764. With preface by Francis Parkman... and a translation of Dumas' biographical sketch of General Bouquet. Cincinnati, OH: R. Clarke & Co. xxiii,162p. map. 24cm.

(1907). As above- (Ohio Valley Historical Series [1]).

(1765). An historical account of the expedition against the Ohio Indians, in the year 1764 under the command of Henry Bouquet, Esq: Colonel of Foot, and now brigadier general in America. Including his transactions with the Indians, relative to the delivery of their prisoners, and the preliminaries of peace. With an introductory account of the preceding campaign, and battle at Bushy-Run. To which are annexed military papers, containing reflections on the war with the savages; a method of forming frontier settlements; some account of the Indian country, with a list of nations, fighting men, towns, distances and different routs. The whole illustrated with a map and copper-plates. Pub. from authentic documents, by a lover of his country. Philadelphia: Printed and sold by W. Bradford. 12,xiii,71p. map. 24x19cm.

(1766). As above- Philadelphia & London: Reprinted for T. Jefferies.

(1769). As above- Dublin: Printed for John Miliken. xx,99p. map. 17cm.

(1795). As above- Philadelphia.

(nd). The history of the province of New York, from the first discovery to the year 1732. (np).

(see under William Smith, 1728-1893.)
(1769). Relation historique de l'expedition contre les Indiens de l'Ohio en MDCCLXIV. Commandee par le chevalier Henry Bouquet... contenant ses transactions avec les Indiens, relativement a la delivrance des prisonniers & aux preliminaires de la paix; avec un recit introductoire de la campagne precedente de l'an 1763; & de la bataille de Bushy-Run. On y a joint des memoires militaires contenant des reflexions sur la guerre avec les sauvages; un methode de former des etablissemens sur la frontiere: quelques details concernant la contree des Indiens: avec une liste de nations, combattans, villes, distances, & diverses routes. Le tout enrichi de cartes & tailes-douces. Tr. de l'anglois, par C.G.F. Dumas. Amsterdam: M. M. Rey. [v]-xvi,147,[10]p. maps. plans. 20.5cm. [Accents omitted]
(1778). Voyage historique et politique, dans l'Amerique, du Chevalier Henry Bouquet... On y a joint des memoires militaires contenant des reflecxions sur la guerre avec les sauvages: une methode de former des estblissemens sur la frontiere: quelaues details concernant la contree des Indiens: avec une liste de nations... Traduit de l'anglois, second edition. Paris: Chaz Merlin. xvi,147(1)p. maps. plans. 8°. [Accents omitted]

Other

Adams, Thomas R. (Ed.). (1951). Account of the College, Academy, and Charitable School of Philadelphia in Pennsylvania; with a commentary by Thomas Woody. Philadelphia: University of Pennsylvania Library. 43p. 22cm.
Barton, Thomas. (1755). Unanimity and public spirit; a sermon preached at Carlisle and some other Episcopal churches, in the counties of York and Cumberland, soon after General Braddock's defeat. To which is prefixed, a letter from the Reverend Mr. Smith, concerning the office and duties of a Protestant ministry, especially in times of public calamity and danger. Philadelphia: Printed and sold by B. Franklin and D. Hall.
Smith, Horace Wemyss. (nd). Life and correspondence of William Smith. (np).

SPAFFORD, HORATIO GATES (1778-1832)

Boyd (1942) described Spafford as a person who was greatly concerned with learning, but who was not in complete control of his personal finances, "Courage born of desperation and courage born of an ineradicable love of learning were alike characteristic of his entire career" (p.3). He and some better known contemporaries, during the period of 1809 through 1825, were actively engaged in promoting the arts and the sciences, including geography. The freedom, probably spawned by Americans of common origin who experienced a liberty previously unknown, embraced that freedom with their own and original opinions flowing from the free exercise of thought and the lack of restraints from religious or political orthodoxy.

Spafford, struggled with poverty though some of his products were then considered outstanding, "Your Gazetteer is by all considered as the very best work of the kind which has ever appeared, but, you know Sir, that in Vermont, we are not all reading animals" (p.7). He was born in Dorset, Vermont on February 18, 1778 and died in 1832 in Lansingburgh, New York. He was a Jeffersonian Republican intimate with Jefferson, Webster, Franklin, Madison, and others of note. To Jefferson he addressed his reason for publishing a general geography, "The writings of Dr. Morse are not of the right character for this nation" (p.14). He was no doubt referring to the fact that Morse was a Federalist, so he could not to present Jeffersonian Republicanism to his readers.

(1813). A gazatteer of the state of New York; carefully written from original and authentic materials, arranged on a new plan in three parts... Albany: H. C. Southwick., 334p. fold map, 22cm Contents.. Part 1. A comprehensive geographical and statistical view of the whole state, conveniently disposed under separate heads. Part 2. An ample general view of each county...with topographical and statistical tables, showing the civil and political divisions, population, post-offices, &c. Part 3. A very full and minute topographical description of each town or township, city, borough, village...as also its lakes, rivers, creeks...forming a complete gazatteer or geographical dictionary of the state of New York. Albany: H. C. Southwick. 334,iip. map. 22cm.

(c.1810). As above- Albany, NY: author.

(1824). As above {possibly] - A Gazatteer of the state of New York: embracing an ample survey and description of its counties, towns, cities, villages, canals, mountains, lakes, rivers, creeks, and natural topography...with an appendix... Albany: B. D. Packard and Troy: the author. 620p. folded map. 21.5cm.

(1809). General geography, and rudiments of useful knowledge. In nine sections...illustrated with am elegant improved plate of the solar system...a map of the world...of the United States...and several engravings on wood. Digested on a newspaper and designed for the use of schools. Hudson, NY: Crosswell & Frary. 393p. 2 folded maps. diagrams. 18cm.

(1808). Map of the world from the latest discoveries engraved for Spafford's geography. (np).

(1824). A pocket guide for the tourist and traveller, along the line of the canals, and the interior commerce of the state of New York. New York: Printed by T. & J. Swords. 72p. 14.5cm.

(1825). As above- 2d ed., with additions and corrections. Troy: W.S. Parker. 88p. 16cm.

(1825). The school boy's introduction to the geography and statistics of the state of New York: designed for the use of common schools, under the patronage of the Legislature. Troy: E. Platt & Co. 46p. 17.5cm

Other

Boyd, J. P. (1942). Horatio Gates Spafford, inventor, author, promoter of democracy. Worcester, MA: The Society.

Jones, S. (1821). Communications from the late Honourable Samuel Jones... to John Pintart, esq... then recording secretary of the Historical Society. New York: (np).

SPROULE, GEORGE (C.1784)

Sproule is named by Samuel Holland as one of his deputies helping to survey New Hampshire. No other information is to be found about the man.

(1784). Topographical map of the state of New Hampshire, surveyed under the direction of S. Holland, Surveyor General of Lands for the Northern District of North America; by the following gentlemen, his deputies: T. Wright, G. Sproule, J. Grant, T. Wheeler, & C. Blaskowitz. London: Printed for W. Faden, geographer to the King. In New Hampshire. Geological Survey, 1868-1878. Geology of New Hampshire, Atlas. Scale: 8 miles to an inch.

(1781). A plan of the environs of Brooklyn showing the position of the rebel lines and defences on the 27th of August 1776 drawn by Geo. Sproule...from a survey made by him in September 1776. Drawn in March 1781. (np). map. 47x59cm. Scale 1000 feet to 1 inch.

(1780). A sketch of environs of Charlestown in South Carolina. London: (np). 14 1/2 x 20 3/8 inches. Scale (computed),. 2.1 miles to 1 inch.

SPROUT, EBENEZER (C.1785)

Sprout was the surveyor for Rhode Island during the 1786 survey (Pattison 1957). Thomas Hutchins had returned to Pittsburgh in June of 1786 and prevailed on Congress to authorized resumption of the surveys of the east-west line which were begun, but not completed in 1785.

Colonel Sprout was one of the several surveyors that were identified with the Ohio Company, therefore interested in speculation. He was among the group of surveyors who landed at the mouth of the Muskingum River, founding Marietta in 1788 (p. 134). According to Pattison (1957), Sprout was one of three men who were engaged in surveying the Seven Ranges, inaugurating the subdivision of the Ohio Company's lands. He later returned to federal surveying after the Land Act of 1796 was passed by Congress (p. 179). No entry was found in The National Union Catalog Pre-1956 Imprints (The American Library Association 1976).

STANSBURY, ARTHUR JOSEPH (1781-1865)

Ritow (1985) reveals that Stansbury worked jointly with Amos Lay to produce a map of New York (p.99). No other information about the person or his products has been located.

(1801). Map of the north part of the state of New York. New York: (np).

STEUBEN, FREDERICK WILHELM LUDOLF GERHARD AUGUSTIN, BARON VON (1730-1794)

Harley et al. (1978) relate that several European people joined the Continental Army. Among those were de Kalb, Pulaski, Lafatette, and von Steuben. One may raise the argument that these persons were not Americans, but their contributions most certainly added to the creation of the United States of America. Their contribution to American professionalism in the military was great, "One small aspect of which was their understanding of the value, use, and specific roles of maps in warfare" (p.48). Washington respected Steuben's roots, having hung a portrait of Frederick the Great in his headquarters.

While in the service of the Continental Army, Steuben rose to the position of inspector general. Duportail and von Steuben drew up, in 1783, the plans for the United States Military Academy at West Point. Steuben's Regulations for the Order and Discipline of the Troops of the United States became the "blue book" for the Continental Army. He became an American citizen in 1783 and in 1790 Congress provided him a pension (The Biographical Dictionary 2001. Steuben, Friedrich; Encarta Encyclopedia 2001. Steuben, Friedrich; Encyclopedia Britannica 2001. Steuben, Frederick).

Other

U. S. War Department, Inspector General's Office. (1810). The military assistant: being a collection of company discipline, founded on the Regulations of Baron Steuben, late Major-General and Inspector-General of the Army of the United States. Containing, the manual exercise, facings, steps, turnings, wheelings, miscellaneous evolutions, and firings. Together with the duty of officers and privates. Ornamented with handson [sic] copper plates of company evolutions. Designed for the use of the militia. 3d ed., improved. Newburyport: Published by Thomas & Whipple.

U. S. War Department, Inspector General's Office. (1809). Regulations for the order and discipline of the troops of the United States. Philadelphia: M. Carey.

STEVENSON, ROGER (C.1770)

Stevenson published the first book dedicated to George Washington: Military Instructions for Officers Detached in the Field (Harley, et al. 1978). That publication provided instruction regarding the use of a "whipcord," instead of the surveyor's usual instruments. That instrument aided in the measurement and recording of geographic features found in the United States.

(1770). Military instructions for officers detached in the field: containing, a scheme for forming a corps of a partisan. Illustrated with plans of the maneuvers necessary in carrying on the petite guerre. Philadelphia: Printed and sold by R. Aitken. vii,252p. 12 plates. 18cm.

(1770). As above- London: Cadell, Wilson, Payne. (4)vii,252p. plans. 16°.

(1774). As above- London: (np). 2d ed. 252p. 16°.

(1775). As above- Philadelphia: Printed and sold by R. Aitken. vii,232[4]p. 12 plates. 15.5cm.

Other

Hoyt, Epaphras. (1811). Practical instructions for military officers: comprehending a concise system of military geometry, field fortification and tactics of riflemen and light infantry. Also the scheme for forming a corps of a partisan, and carrying on the petite guerre. To which is annexed, a new military dictionary; containing the French words, and other technical terms now used in the art of war; with other matter connected with military operations. Greenfield, MA: Printed by J. Denio.

STODDARD, AMOS (1762-1813)

White (1906) reports that Stoddard was a soldier born in Woodbury, Connecticut on October 26. He was the son of James Stoddard and a descendant of Anthony Stoddard who came to Boston from England in 1630. He served in the Revolutionary War, afterward becoming Clerk of the Supreme Court of Massachusetts. Stoddard practiced law in Hallowell, Maine. In 1812 he was appointed captain of Artillery, then Major, and ultimately Quartermaster of the Regular Army.

The governments of France and the United States designated Stoddard their representatives in the ceding of Louisiana. Thereafter he was appointed Governor of the new territory while he served as Commander at St. Louis. His participation in the siege of Fort Meigs, Ohio resulted in a wound that brought an end to his life on May 11, 1813 (V.XIII, p.477).

(1812). Sketches, historical and descriptive, of Louisiana [sic]. Philadelphia: Published by Mathew Carey: A. Small, printer. viii,488p.

STROTHER, JOHN (1782-1862)

A North Carolina map compiled by John Strother and Jonathan Price in 1789 was prepared and engraved by Harrison (Schwartz & Ehrenberg 1980, p.224). The absence of one degree of longitude caused distortions that were ultimately corrected by Henry S. Tanner in his New American Atlas of 1823. According to Guthron (1984), "Price and Strother produced the first survey of the State of North Carolina which was engraved in Philadelphia by C. P. Harrison in 1808," (p.8). There is some mistake or two different maps, based on the one entry we have thus far identified (see 1808).

(1808). First actual survey of the state of North Carolina. Engraved by William Harrison Jr. Philadelphia: Printed by C. P. Harrison. 74x152cm. Scale 1:500,000.

(1808). A chart of the coast of North Carolina between Cape Hatteras & Cape Fear from a survey taken in the year 1806 by Thomas Coles and Jonathan Price pursuant to an Act of Congress. Engraved by C. P. Harrison. (np).

(c.1800). A map of Cape Fear River and its vicinity from The Frying Pan Shoals to Wilmington by actual survey addressed to the commissioners of navigation of Port Wilmington by Price & Strother. Engraved by W. Barker. Philadelphia: Price & Strother.

(1796). To navigators this chart being an actual survey of the sea coast and inland navigation from Cape Henry to Cape Roman is most respectfully inscribed by Price & Strother. Engraved by W. Johnston. New Bern, NC: Price & Strother. 38x13 3/4inches. Scale 1:490,00.

STURGES, DANIEL (C.1797)

Sturges was named surveyor general of Georgia in 1797 and was reappointed in 1817, 1823, and 1823 (Ristow 1985). He prepared a plan for Milledgeville, Georgia and assembled surveys and data for a map of the state (p.128). The National Union Catalog Pre-1956 Imprints provided no publications by Sturges (The American Library Association 1976).

(1818). Map of the state of Georgia prepared from actual surveys and other documents for Eleazer Early by Daniel Sturges. Engraved by Samuel Harrison. Published and sold by Eleazer Early of Savannay Georgia and John Melish and Samuel Harrison of Philadelphia. 112x140cm. Scale 1:500,000.

SULLIVAN, JAMES (1744-1808)

Sullivan was Governor of Massachusetts, born in Berwick, Maine, the son of Owen Sullivan (Brown 1897). He died in Boston. Sullivan studied law with his brother John in Durham, New Hampshire. He was appointed by the President to settle a dispute regarding the line between the United States and British North America (V.7, pp. 254-255). The Biographical Dictionary (2001. Sullivan, James) confirms the above.

(1898). Letter from Judge Sullivan concerning the eastern boundary of Maine. Copied from the original, in the Emet Collection of Manuscripts belonging to the New York Public Library... Communicated by Hon. Joseph Williamson. (In Collections and Proceedings of the Maine Historical Society, 9, pp.207-212, Portland.

(1801). History of land titles in Massachusetts. Boston: I. Thomas and E. T. Andrews. x,1,[13]-392p. 24cm.

(1795). History of the district of Maine. Boston: I. Thomas and E. T. Andrews. vii,iv,[5]-421p. map. 21cm.

Other

Amory, Thomas C. (1859). Life of James Sullivan. 2v. (np).

(1808). Interesting correspondence between His Excellency Governour [sic] Sullivan and Col. Pickering; in which the latter vindicates himself against the groundless charges made against him by the Governour [sic] and others. Boston: Printed by Greenough and Stebbins. 32p. 21.5cm.

(1808). As above- Newburyport: From the Herald-office; sold at the bookstore of Thomas & Whipple.

SYMMES, JOHN CLEVES (1742-1814)

Symmes took leadership for the Ohio Company by opening a second tract on the Ohio (Pattison 1957). Following Congressional approval, in the year 1788, he set off from Limestone, Kentucky to explore and lay plans for settlements in the new Miami Purchase (p.180). Symmes, principal surveyor Israel Ludlow, and others set out in that year to establish the boundaries of the Purchase. By the spring of 1794 the boundaries were established.

Brown (1897) reports that Symmes was born on Long Island New York and died in Cincinnati, Ohio. He served in the revolutionary army, was promoted to colonel, and became Chief Justice in the New Jersey Supreme Court. He and some associates purchased land along the Ohio and Miami rivers known as the Miami Purchase, later founding North Bend and Cincinnati. He was one of the three judges of the Northwest Territory appointed by Congress (V.7, p.272).

White (1909) goes on to report that Symmes married Anna, the daughter of Henry Tuthill of Southold, Long Island. Their daughter Anna, became the wife of President William Henry Harrison. Symmes' second wife was the daughter of Governor William Livingston of New Jersey (V.XI, p.452).

Claimants under John Cleves Symmes, American State Papers: Public Lands, V.XXVIII (Doc. 55).

Claimants under John Cleves Symmes for land in Dayton and Livingston, Ohio, American State Papers: Public Lands, V.XXVIII (Doc. 34).

Contract with John Cleves Symmes, American State Papers: Public Lands, V.XXVIII (Doc. 23, 33, 70, & 72).

Land for support of colleges in Ohio, American State Papers: Public Lands, V.XXXII (Doc. 637).

Land for support of religion in Ohio, American State Papers: Public Lands, V.XXIX (Doc. 187).

(1789). Letter on the Miami (Ohio) Purchase. May 18, 1789. In Carolos Cist. Cincinnati in 1841. p.196-231.

Petition of trustees of Woodward High School of Cincinnati, Ohio, for grant of land, American State Papers: Public Lands, V.XXXV (Doc. 1478).

Preemption rights to claimants under John Cleves Symmes, American State Papers: Public Lands, V.XXVIII (Doc. 62).

U.S. House of Representatives. 18th Congress, 1st Session. Presidential message on execution of act of 1803 respecting township of land lying within patent of John Cleves Symmes, (H.doc.135). Washington: Government Printing Office, 1823-1824. (Serial Set 102).

Reservations in Ohio, American State Papers: Public Lands, V.XXVIII (Doc. 22).

U.S. House of Representatives. 18th Congress, 1st Session. Symmes, John Cleves, right of United States to township of land reserved in contract entered into with, (H.rp.88). Washington: Government Printing Office, 1823-1824. (Serial Set 106).

U.S. House of Representatives. 32nd Congress, 1st Session. Symmes land purchase unsold, (H.exdoc.73). Washington: Government Printing Office, 1851-1852. (Serial Set 641).

Title of United States to township of land in purchase of John Cleves Symmes, American State Papers: Public Lands, V.XXX (Doc. 405).

U.S. House of Representatives. 33rd Congress, 1st Session. Unsold lands in Symmes's purchase, Ohio, (H.rp.243). Washington: Government Printing Office, 1853-1854. (Serial Set 743).

Other

Bond, Beverly W., Jr. (Ed.). (1926). The Correspondence of John Cleves Symmes, Founder of the Miami Purchase, chiefly from the collection of Peter G. Thomson. New York: The Macmillan Company. xiip,2,312p. 2port. 22.5cm.

U. S. Attorney General. (nd). Letter from the Attorney General... Relative to the contract between... the United States and John Cleves Symmes. (np).

TANNER, BENJAMIN (C.1796)

Ristow (1985) identifies Tanner as "one of the foremost American engravers of the early nineteenth century" (p.110). He also includes Tanner as engraver for nine of Roberts' maps in the Matthew Carey atlas (p.153).

(1796). The American atlas. Philadelphia: Matthew Carey.

(c.1796). Georgia, from the latest authorities. New York: Reid, Waland & Smith. 7x15inches. Scale: $29^{1/3}$miles to 1 inch.

(1822). Map of Pennsylvania, constructed from the county surveys authorized by the state; and other original documents, by John Melish. (np).

(1795). A map of the Tennasee [sic] Government, formerly part of North Carolina, from the latest surveys. New York: Reid, Waland & Smith. 7x15inches. Scale 407 yards to 1 inch.

(1796). A plan of the investment of York and Gloucester, Virginia. New York: C. Smith. $17x15^{1/8}$inches. Scale $29^{1/3}$miles to 1 inch.

(1797). The United States of America. $6^{1/4}$ x $8^{1/4}$inches. Scale: 200 miles to 1 inch. In Smith. The American war. Between pp.182 and 183. New York: (np).

TANNER, HENRY S. (1786-1858)

Ristow (1985) lists Henry Tanner as engraver of the frontispiece map for volume one of John Melish's Travels (p.179-181). The younger brother of Benjamin Tanner, Henry was born in New York City, moved to Philadelphia where he served as Benjamin's apprentice and began an association with Melish.

(1843). An accompaniment to Tanner's universal atlas... Comprising statistics of the U.S.... and states in the world. Carey and Hart's new and improved ed. Philadelphia: Carey and Hart. 250p.

(1831). Africa. Edition materially improved. Philadelphia: (np). $47^{1/2}$x$55^{3/4}$inches. No scale.

(c.1836). An alphabetical index to the four sheet map of the United States. Philadelphia: Rackliff and Jones, printers. Vp1, 99p. 18.5cm.

(1834). The American traveller; or Guide through the United States. Containing brief notices of the several states, cities, principal towns, canals and rail roads, &c. With tables of distances, by stage, canal and steam boar routes. The whole alphabetically arranged, with direct reference to the accompanying map of the roads, canals and railways of the United States. Philadelphia: author. iv[5]-144p. map. plans. 15.5cm.

(1836). As above- 2d ed.

(1837). As above- 3d ed.

(1838). As above- 3d ed.

(1839). As above- 4th ed.

(1840). As above- 6th ed.

(1841). As above- 7th ed. New York: T. R. Tanner.

(1842). As above- 8th ed.

(1844). As above- 9th ed.

(1846). As above- 10th ed. New York: The Map establishment.

(1830). Asia. Edition improved to 1830. Philadelphia: (np). $47^{1/8}$x$55^{1/8}$inches. No scale.

(1826). An atlas of ancient geography: comprehended in sixteen maps, selected from the most approved works, to elucidate the writings of the ancient authors, both sacred and profane. Philadelphia: Tanner. 16 double col. maps.

(1829 & 1840). Atlas classica; or Select maps of ancient geography, both sacred and profane. Engraved by Thackara, Vallance, Tanner, and P. E. Hamm. Philadelphia: H. S. Tanner. 46 maps.

(nd). Atlas general... (np).

(1835). Atlas of the United States, containing separate maps of each state and territory of the North American union. Philadelphia: author.

(1834 & 1840). A brief description of the canals and railroads of Pennsylvania and New Jersey, comprehending notices of all the most important works of internal improvement, in those states. Philadelphia: author. 31p. 15cm.

(1844). As above- 2d ed. New York: T. R. Tanner. v,524p. maps. 15cm.

(1840). A description of the canals and rail roads of the United States, comprehending notices of all the works of internal improvement throughout the several states. New York: T. R. Tanner & J. Disturnell. vii[9]-272p. maps. 23cm.

(nd). General outline of the United States of North America. (np).

(1832). A geographical and statistical account of the epidemic cholera, from its commencement in India to its entrance into the United States: comprehended in a series of maps and tables, exhibiting the names of places visited by the pestilence - the time of its commencement - the number of cases, and deaths, and duration, at each place. Compiled from a great variety of printed and manuscript documents. Philadelphia: author. iv[5]-35p. maps. 18.5cm.

(1841). A geographical, historical and statistical view of the central or middle United States; containing accounts of their early settlement; natural features; progress of improvement; form of government; civil divisions and internal improvements... Philadelphia: H. Tanner, Jr.; T. R. Tanner. v[3]-524p. maps. 15.5cm.

(1841). As above- 2d ed.

(1823) Georgia and Alabama. Philadelphia: author. 53x68cm.

(c.1823). Illinois and Missouri. Philadelphia: H. S. Tanner. $27^{3/4}$x$22^{3/8}$inches. Scale: 50 American miles to $2^{3/4}$ in. 40 geographical miles to $2^{1/2}$ inches.

(1836). As above- 71.5x58.5cm.

(nd). Index to the new plan of Philadelphia & adjoining districts... (np).

(1839). Kentucky and Tennessee. Philadelphia: author. 55x73cm.

(nd). List of maps, charts & geographical works recently published... by him. (np). 16p.

(1820). Louisiana and Mississippi... Philadelphia: author. 60x79cm. Scale: $2^{3/4}$ inches equal 30 miles.

(1846). Map of Boston. 6x$4^{5/8}$inches. Scale: 2215.3feet to 1 inch. In his Massachusetts and Rhode Island. Philadelphia: author.

(1823). Map of Florida, compiled and drawn from various actual surveys and observations. Philadelphia: author.

(1841). Map of Georgia and Alabama. Philadelphia: author.

(1827). Map of New England. (np).

(1846). Map of New Jersey reduced from T. Gordon's map. Philadelphia: Williams.

(1822). A map of North America, constructed according to the latest information. Philadelphia: author. 54x73cm.

(c.1823 & 1825). Map of North and South Carolina. Philadelphia: author. 52x70cm.

(1839). As above- 55x74cm. Scale: 18 miles to 1 inch.

(1851). As above- New York: author.

(c.1823). 1829 & 1830). Map of Pennsylvania and New Jersey. Philadelphia: author. $20^{1/2}$x28inches.

(1826). Map of Philadelphia and adjacent country... Philadelphia: author. 24p.

(1830). Map of Schuylkill county, Pennsylvania. Philadelphia: author. 1p.

(1822). Map of South Carolina, constructed and drawn from the district surveys, ordered by the legislature, by John Wilson, late Civil and Military Engineer of So. Cara. Engraved by H. S. Tanner. Philadelphia: H. S. Tanner.

(1833). Map of the canals and railroads of Pennsylvania & New Jersey. Philadelphia: author. 1p.

(1834, 1937, 1839 & 1840). As above- 18 7/8x25 3/4 inches. Scale: 18 1/2 miles to 1 inch.

(1813). Map of the Seat of War in North America. Philadelphia: author. Accompanied by : Description of the Seat of War in North America.

(1813). Map of the southern section of the United States. Engraved by Benjamin Tanner. Philadelphia: author. Accompanied by : Description of the Seat of War in Southern Section.

(1831, 1838, 1845 & 1846). Map of the states of Maine, New Hampshire Vermont, Massachusetts, Connecticut, & Rhode Island. Philadelphia: author. 25 3/4x21inches. Scale: 18.18 inches to 1 mile.

(1832). Map of the United States. Third ed. Philadelphia: author.

(1841 & 1843). Map of the United States including every county enumerated in the census of 1840, with the population of the several states, rail roads, canals, &c. Philadelphia: author. 115x153cm.

(1825, 1829 & 1834). Map of the United States of America. Philadelphia: author. 39x64cm. No scale.

(1825). Map of the United States of Mexico, as organized and defined by several acts of the Congress of that republic. Philadelphia: author.

(1846). As above- 22 1/2x28 1/2inches. Scale: 85.1 miles to 1 inch.

(1846). As above- 3d ed. New York: author.

(1846). Maps of Maine, New Hampshire Vermont, Massachusetts, & Rhode Island, Connecticut, New York, and New York City. Philadelphia: Mitchell. 6 maps. 44cm.

(1833). Massachusetts and Rhode Island... Philadelphia: author. 39x31.5cm.

(1813-1814). Mayo's ancient geography and history. Henry Tanner, engraver. Philadelphia: John P. Watson.

(1829 & 1830). Memoir on the recent surveys, observations, and internal improvements, in the United States, with brief notices of the new counties, towns, villages, canals, and railroads, never before delineated. By H. S. Tanner. Intended to accompany his new map of the United States. Philadelphia: author. 1-108p. 18.5cm.

(c.1815). A military and topographical atlas of the United States; Including the British possessions & Florida. Philadelphia: G. Palmer Printer

(1823 & 1825). A new American atlas containing maps of the several states of the North American Union, Projected and drawn on a uniform scale from documents found in the public offices of the United States and state governments and other original and authentic Information. Philadelphia: H. S. Tanner.

(1837). A new American atlas, containing a series of maps of the various countries of North and South America, West Indies, &c. Together with maps of the several states and territories of the North American Union. Philadelphia. 40 col. maps. 42x34cm.

(1819). A new American atlas no. 1. Containing maps of the world, Europe, and South America in two sheets, arranged from the most authentic documents. Philadelphia: Published by Tanner, Vallance, Kearny & Co. 22 maps. 44x62cm.

(1816). A new and elegant general atlas containing maps of each of the United States. Baltimore: Fielding Lucas, Jr.

(1833). A new and elegant universal atlas. no. 5. Philadelphia: author.

(1845). A new general atlas, comprising a complete set of maps, representing the grand divisions of the globe, together with the several empires, kingdoms, and states in the world; compiled from the best authorities, and corrected by the most recent discoveries. Written and engraved by Jos. Perkins. New York: Tanner's Geographical Establishment. 61 maps. 41cm.

(1828). A new general atlas consisting of maps of the several grand divisions of the known world. Philadelphia: H. S. Tanner.

(1840). A new guide to the public buildings; places of amusement; literary institutions; banks; hotels; railroad depots, &c. &c. of the city of Philadelphia, and adjoining districts: with a brief description of a plan of the city and environs. Philadelphia: author. 36p. map. 12.5cm.

(1813-14). A new juvenile atlas and familiar introduction to the use of maps: With a comprehensive view of the present state of the earth. By John Melish, Vallance, and Henry Tanner. Philadelphia: Printed by G. Palmer.

(1839). New map of the city of New York with part of Brooklyn & Williamsburgh. Compiled by J. Calvin Smith. New York: Henry Tanner.

(1828). A new map of Colombia. Philadelphia: author.

(1833). A new map of Louisiana, with its canals, roads & distances from place to place, along the stage & steam boat routes. Philadelphia: author. 28x35cm.

(1839). As above- Philadelphia: Carey & Hart.

(1845). A new map of Maryland and Delaware, with their canals, roads and distances. Philadelphia: author.

(1836). New map of Mississippi, with its roads and distances. Philadelphia: author.

(1846). A new map of North Carolina, with its canals, roads & distances from place to place, along the stage and steam boat routes. Philadelphia: Thomas, Cowperthwait & Co. 10.75x12.5cm. Scale: 40 miles to 1 inch.

(1833 & 1836). New map of Ohio; with its canals, roads and distances. Philadelphia: author.

(1833 & 1840). New map of Pennsylvania, with its canals, rail-roads and distances from place to place along the stage roads. Philadelphia: author.

(1831). New...map of the world embracing all the recent discoveries & exhibiting particularly the nautical researches of the most distinguished circumnavigators. Philadelphia: author.

(1840). A new picture of Philadelphia, or the stranger's guide to the city and adjoining districts... With a plan of the city and map of its environs. Philadelphia: author; New York: T. R. Tanner. iv[5]-156p. maps. 15.5cm.

(1841). As above- 2d ed.
(1844). As above- 3d ed.
(c.1845). As above- 4th ed.
(1828). A new pocket atlas of the United States with the roads and distances designed for the use of travellers. Philadelphia: Printed by G. Palmer. maps. 15cm.
(1836 & 1839). A new universal atlas, containing maps of the various empires, kingdoms, states and republics of the world. With a special map of each of the United States, plans of cities &c. Comprehended in seventy sheets and forming a series of one hundred and seventeen maps, plans and sections. Philadelphia: author. colored maps. plates. 41cm.
(1843, 1844, 1845 & 1846). As above- Philadelphia: Carey & Hart.
(1840). New York. Philadelphia: author.
(1836). Philadelphia. Philadelphia: author. $14^{3/4}$x$11^{3/4}$inches. Scale: 1282 feet to 1 inch.
(1846). Sectional map of Ohio and Indiana... Philadelphia: author. 49.5x64cm.
(nd). Sketch of Trenton as it was December 26, 1776. (np).
(1821). South America, with improvements to 1821. Philadelphia: author. 8°.
(1836). Spain and Portugal. Philadelphia: author. 11x$14^{1/8}$inches. Scale: 48.2 miles to 1 inch.
(c.1811). Statistical account of the United States, with topographical tables of the counties, towns, population, &c. From the census of 1810. Philadelphia: Melish & Tanner.
(1828). The stranger's guide to the public buildings, places of amusement, streets ... steam-boat landings, stage offices, etc., etc., of the city of Philadelphia and adjoining districts... Philadelphia: author. 38p. maps. 14cm.
(1845). As above- New York: T. R. Tanner.
(nd). Tanner's travelling map of California &c. (np).
(c.1839). Tanner's travelling map of Virginia, Maryland and Delaware with the roads, canals, railroad, counties, etc. Philadelphia: author.
(1846). The traveller's and emigrant's guide through Texas; consisting of a general description of the country... with tables of distances in all directions... illus. by a colored map of Texas, showing its connexion [sic] with Mexico, the United States, California, & Oregon. New York: Tanner's Geographical Establishment. Broadside. 54x65cm. map. 24x31cm.
(1837). The traveller's guide. Philadelphia: author.
(c.1825). The traveller's guide map of the roads, canals and steam boat routes of the United States; with the distances from place to place carefully noticed, including a series of tables shewing the routes pursued by the various steam boats, and the distances between important places throughout the country. Designed for the use of travellers. Philadelphia: author. 47.5x62cm.

(1844). The travellers' hand book for the state of New York and province of Canada... With maps, &c. 2d ed. New York: T. R. Tanner. iv[5]-166p. maps. 15cm.
(1845). As above- 3d ed.
(1846). As above- 4th ed.
(1830). The travellers pocket map of Illinois with its proposed canals, roads & distances from place to place along the stage & steam boat routes. Philadelphia: author. $12^{3/4}$x$10^{3/8}$inches. Scale: 80 miles to $2^{1/2}$in.
(1830). The travellers pocket map of Tennessee, with its roads and distances. Philadelphia: author.
(1841). Travelling map of New England, with the roads, canals, railroads, countries... (np).
(c.1839). Travelling map of Pennsylvania & New Jersey with the roads, canals, railroads, counties, &c. Philadelphia: author.
(1812). Travels in the United States of America in the Years 1806 & 1807, and 1809, 1810, & 1811. Philadelphia: author.
(1829). United States of America. Engraved by H. S. Tanner, assisted by E. B. Dawson, W. Allen & J. Knight. Philadelphia: author. 116x153cm. Scale 1:2,000,000.
(1830). As above- 2d ed.
(1832, 34, 36, 38 & 1839). As above- 3d ed.
(1845). Universal atlas... of the world. Philadelphia: Carey.
(1820). Virginia, Maryland and Delaware. Philadelphia: author.
(1822). The Western Navigator: Charts of the Ohio River in its whole extent, and of the Mississippi River from the mouth of the Missouri to the Gulf of Mexico, accompanied by directions for the navigation of the Ohio and Mississippi. In two volumes. By Samuel Cumings. Philadelphia: H. S. Tanner.

Other

Allen, William. (1830). Plan of the city of Philadelphia and adjoining districts shewing the existing and contemplated improvements, compiled from original documents by William Allen. Philadelphia: H. S. Tanner.
Baird, Robert. (nd). View of the valley of the Mississippi; or the emigrant's and traveller's guide... (np).
Boye, Hermann. (nd). A map of the state of Virginia. (np).
Boykin, J. (nd). Lancaster district, South Carolina. (np).
Hough, B. (nd). A map of the state of Ohio from actual survey by B. Hough & A. Bourne. (np).
Lowery, John. (nd). Chesterfield district, South Carolina. (np).

TATHAM, WILLIAM (1752-1819)

William Tatham arrived from England at the Virginia colony "without profession, trade or employment" (Herndon 1969), p.346). He was destined to become an inventor, agriculturalist, adventurer, author and book collector.

He corresponded with Thomas Jefferson and was one of only two Americans to hold membership in the Royal Society of Arts. Tatham was born at Hutton-in-the-Forest, Cumberland County, England on April 13 to Reverend Sanford Tatham and his wife.

Because of the poor financial position of the Reverend, young Tatham was sent to his maternal grandmother in Lancaster where he lived until her death in 1760. While in Lancaster he attended Friend's School. Thereafter he moved to Liverpool (1768) for schooling and on to work with learn the tobacco trade from two American merchants, Carter and Trent. He worked at a branch on the James River in Amherst County, Virginia.

Upon the demise of the Carter and Trent organization, 1774, Tatham attempted to operate his own trading business, but it failed perhaps because of British policies and the American Non-importation associations. He proceeded to join the Patriotic forces to fight the British. Tatham was unfortunate, in his eyes, to miss a battle in which his Gatlands Volunteers was involved. So he decided to move to the western frontier and found John Carter, a relative of his former employer, already established at the Watauga-Holston settlements, located in the present East Tennessee, North Carolina at the time (1776).

There Tatham found employment as a clerk, bookkeeper, and apprentice surveyor. Thomas Jerrerson, in a letter to Nathanael Green dated December, 1780, noted of Tatham, "He is particularly acquainted with the Western Country of Virginia and Carolina beyond the mountains of which he has made a pretty good map; he is versed in writing and accounts and posses understanding" (Herndon 1969, p.352). Jefferson went on to note that Tatham had skill as a draftsman.

Meanwhile Tatham studied law under the North Carolinian, William R. Davie, and was admitted to the bar in 1784. Tatham, during the next years, appears to have maintained his interest in surveying, exploring , gathering geographical information, and working on projects relating to inland navigation. He had been engaged, while still a boy, in the improvement of navigation on the James River. Herndon (1969) notes that Tatham became acquainted with Robert Fulton in 1796 and discussed with him a mutual interest in inland coastwise canals and boats. "According to Tatham, the two of them jointly worked out plans for an inland coastal canal from Fayetteville, North Carolina to Georgetown, South Carolina, using the waters of Cape Fear, Downing Creek, and the Waggamaw and Pee Dee rivers" (p.378).

Tatham and Fulton, supposedly, drew up an agreement that they would only work together on such projects. A dispute arose when Fulton presented his steamboat. Tatham insisted that specifications for the "Fulton" boat were registered by himself in London during June 1797 and later deposited in the Patent Office in Washington.

Tatham, during the next few years, sought to achieve his greatest ambition, that of respected and honored public servant. He attempted to do so during the War of 1812 and during the presidencies of Jefferson, Madison, and Monroe, but he was only able to gain employment. Herndon observes that, "He was a failure for several reasons: he was a victim of his own versatility, he lacked discipline and perseverance, and his ideas and schemes were too grandiose and frequently too far ahead of his time" (p.390).

On February 22, 1819, near Richmond, Virginia the life of Tatham suddenly ended. He was attending a celebration of Washington's birthday. As a contingency of the Public Guard was firing a salute, Tatham stepped in front of the second cannon immediately after the order to fire had been given. He committed suicide. Herndon reports that, "the wad from the cannon which ended Tatham's life was composed of tobacco stems" (p.396). Herndon (1969) provides a bibliography of works used by him (pp.480-482). Among those noted are publications relating to geography by Marc Catesby, Richard Hakluyt, Thomas Harriot, Thomas Jefferson, Sir William Keith, Jean Baptiste Labat, Jedidiah Morse, Lionel Wafer, and Isaac Weld, Jr.

Books, drafts, and models, for War Department; collection of Colonel William Tatham, American State Papers: Public Lands, V.XXXVII (Doc. 210).

(1803). Circular architecture, being a new method of building... London: Lawrence. 24p.

(1800). Communications concerning the agriculture and commerce of America: containing observations on the commerce of Spain with her American colonies in time of war. Written by a Spanish gentleman in Philadelphia, this present year, 1800. With sundry other papers concerning the Spanish interests. Ed. in London, by William Tatham. London: Printed for J. Ridgway. viii,120p. 21cm.

(1808). A comparative view of the four projected coastwise canals, which are supposed by some to be in competition for the trade between Norfolk and North Carolina. Norfolk: Printed by J. O'connon. 27p. 21cm.

(1806). Copy of manuscript report by William Tatham on survey of the coast of North Carolina from Cape Hatteras to Cape Fear. Washington: (np). 56p.

Department of works and public economy subject to executive branch, reorganized plan of colonel William Tatham, p. 460, American State Papers: Public Lands, V.XXXVII (Doc. 210).

(1800). Communications concerning the agriculture and commerce of America. London: (np).

(1796). A few hints and remarks for the use of the officers, physicians and students of His Majesty's botanic garden at Madrid, intended to introduce a botanical correspondence with the principal botanists of the United States of N. America, thereby exchanging good offices for the benefit of mankind in the science of botany in general; and M.C. Majesty in particular the numerous productions of the several American states which appear to have been hitherto omitted. Humbly submitted to their reflection by William Tatham, a traveller from the said United States now sojourning in Madrid 10th June, 1796. North Billerica, MA: author. 8°.

(1800). An historical and practical essay on the culture and commerce of tobacco. London: Printed for Vernor & Hood. xv,330p. Plates. 21.5cm.

(1801). National irrigation, or the various methods of watering meadows; affording means to increase the population, wealth, and revenue of the kingdom, by an agricultural, commercial and general economy in the use of water. London: Printed for J. and T. Carpenter. 1p.[v]-xix,427p. plates. 21.5cm.

(1803). Navigation and conservancy of the river Thames. Report on a view and examination of certain impediments and obstructions in the navigation of the river Thames, made pursuant to a resolution of the worshipful the navigation committee, dated the 7th of December, 1802: and also pursuant to farther indulgences, communicated by their clerk on the 12th day of the same month... from a survey of the premises, made in co-operation with Samuel Miller. London: C. Whittingham. 2plates. 126p. 22.5cm.

(1799). Plan for improving the city of London by means of navigable canals & commercial basons. (np).

(1799). The political economy of inland navigation, irrigation, and drainage; with thoughts on the multiplication of commercial resources; and on means of bettering the condition of mankind, by the construction of canals, by the improvement of their various capacities for commerce, transfer, agriculture, household supplies, and mechanical power; and by the unlimited extension thereof into the remotest interior of Great Britain and of foreign parts. London: Faulder. xvi,500[7]p. maps. 27cm.

(1790). Proposals, for publishing a large and comprehensive map of the southern division of the United States of America. Richmond: author. Broadside. 52.5x42.5cm.

(1798). Remarks on inland canals, the small system of interior navigation various uses of the inclined plane, &c. &c. In a letter from William Tatham, to a proprietor in the Colebrook-Dale and Stratford canals. London: Printed for J. Taylor. 20p.

(1807). The separate report of William Tatham, one of the commissioners appointed to survey the coast of North Carolina from Cape Hatteras to Cape Fear inclusive, under the Act of Congress of 10th of April last. To the Hon. Albert Gallatin, Secretary of the Treasury. Washington: (np).

(1794). To His excellency Henry Lee, Governor, and the respective gentlemen of the council... to the share holders and others interested in the canals of Virginia; to the commercial interest and property holders thereof... Richmond: author. 16p. 23cm.

(1790 & 1791). A topographical analysis of the Commonwealth of Virginia: compiled for the years 1790-1. Shewing [sic] the extent, and relative situation of the several counties, their distance from the seat of government; population; force; county lieutenants; representatives, &c. Also the district and county courts; the civil list of the commonwealth, &c. carefully collected from public records, and other authorities. To be continued annually... Philadelphia: Printed by Charles Cist.

(1805). Traite general de l'irrigation, contenant diverses methodes d'arroser les pres et les jardins la maniere de conduire les prairies, pour les recoltes du foin; avec les mohens d'augmenter ses revenus... avec huit planches represcventantdiverses machines pour elever et conduire l'eau... tr. de l'anglais. Paris: Galland. xii,[2]309p. diagrams. 21cm. [Accents omitted]

Other

U. S. Congress. House Committee on letter of William Tatham. (1806). Report from the Committee appointed on the tenth of February last, on the letter of William Tatham to the speaker... Washington: a. & G. Way, printers.

William Tatham Papers, Archives Division, Virginia State Library, Richmond.

TAYLOR, B. (C.1797)

Taylor is identified as the compiler of James McComb, Jr.s' A New & Accurate Plan of the City of New York in the State of New York in North America, engraved by John Roberts (Ristow 1985). The map is described as:

> One of the most accurate and beautiful engraved plans of the city, and particularly interesting on account of its tiny bird's-eye views of some of the most important buildings, as well as for the clear idea which it gives of the country lying along the East River in the vicinity of Crown Point. (p.247-24)

No publications were found in The National Union Catalog, Pre-1956 Imprints (The American Library Association, 1976).

(1797). New & accurate plan of the city of New York in the state of New York in North America. By Christopher Colles. Compiled by B. Taylor and engraved by John Roberts. New York: (np).

THACKARA, JAMES (1767-1848)

Thackara worked with John Vallance preparing the engraving for Andrew Ellicott's plan of Washington (p.117). No other biographical information regarding Thackara has been located. His primary contribution to American geography was through his publications.

(1795). American atlas. Engraved by Barker, Scott, James Thackara, and John Vallance of Philadelphia, Samuel Hill of Boston, Amos Doolittle of New Haven, and Benjamin Tanner of New York. Philadelphia: Mathew Carey.

(1829). Atlas classica; or select maps of ancient geography, both sacred and profane. Engraved by Thackara, Vallance, Tanner, and P. E. Hamm. (np).

(c.1790). Early American engravings of flowering plants. Philadelphia: author. 10 plates. 29.5cm.

(1795). General atlas for the present war. Engraved by Barker, Scott, James Thackara, and John Vallance of Philadelphia, Samuel Hill of Boston, Amos Doolittle of New Haven, and Benjamin Tanner of New York. Philadelphia: Mathew Carey.

(1805). A map of Pennsylvania, Delaware, New Jersey, & Maryland, with parts adjacent. Philadelphia: author, 1797-8. In W. Blodgett, Facts and arguments respecting inland navigation. Philadelphia: (np).

THOMAS AND CARLISLE (C.1795)

Isaiah Thomas, born in Boston, Massachusetts in 1749 and died in Worcester, Massachusetts was a printer and historian of the press. He became the leading book publisher of his day (Hopkins 1964, p.1057). It is likely that he and a colleague were the publisher of the following book:

(1795). The pocket gazatteer: or, a description of all the principal cities and towns, in the United States of America. To which is prefixed a table of the divisions of the United States. Walpole, NH: author.

TIEBOUT, CORNELIUS (1777-1832)

Ristow (1961) reveals that Tiebout was the engraver of Colles' title page and first three maps of his Proposals for Publishing a Survey of the Roads. The author goes on to assert that there are a number of inconsistencies in the maps ultimately presented in Survey of the Roads of the United States of America (1789). Therefore, there are questions regarding whether or not Tiebout engraved them. Tiebout, none the less, was one of the first native born American engravers. In 1789 a map of the city of New York, engraved by him, was published. His engraving, during the period 1789 and 1793, was produced additional maps and landscapes.

During 1793 he went to England to study engraving under James Heath. Tiebout was born in New York City to Tunis and Elizabeth Lamb Tiebout. Ristow (1961) relates, "It is possible that Tiebout's mother and Christopher Colles' good friend, Col. John Lamb, were related, and that the latter was the contact between the survey's map maker and engraver" (p.48).

It is likely that Tiebout was in the military service during the American Revolution. He is identified in Simms (1845) History of Schoharie County as a member of a scouting party during April 1780. That and other military experience may have provided Colles a source of geographical data, some of which may have aided in the planning of Colles New York waterway proposal. Tiebout died in New Harmony, Indiana (Hopkins 1964, p.1068).

(1793). 1st Sheet of De Witt's State Map of New York. Engraved by C. Tiebout. New York: (np).

(nd). New map of the states of Georgia, South and North Carolina, Virginia and Maryland... (np).

(1789). Plan of the city of New York. Compiled by James McComb, Jr. Engraved by Cornelius Tiebout. New York: (np).

Other

Josephus, Flaviua. (nd). The whole genuine and complete works of Flavius Josephus... (np).

TIFFIN, EDWARD (1766-1829)

Tiffin was the first Commissioner of the General Land Office, therefore responsible for disposition of the public domain (Pattison 1957, p.206-207). He also served as the first governor of Ohio, 1803-1807 (The Biographical Dictionary 2001. Tiffin, Edward). Tiffin is credited with collecting and organizing papers concerning public lands and saved those papers when they were endangered by the British in 1814.

Brown (1897) notes that Tiffin was a statesman born in Carlisle, England who studied medicine and moved to Charlestown, Virginia in 1784. He abandoned medicine in 1790 to become a preacher in the Methodist Episcopal Church. He was a member of the House of Representatives in the first and second territorial legislatures and he was Speaker of the Ohio House (1809-1811). He exchanged his Commissioner of the General Land Office with Josiah Meigs, Surveyor-General of Public Lands Northwest of the Ohio River (V.7, pp.346-347).

President Madison made him Land Commissioner, and afterward surveyor-general of the West. In 1798 he married Mary Worthington who died in 1808. He next married Mary Porter of Delaware, who survived him. Tiffin had one son and three daughters: the son died in 1853 (White 1891, V.III, p.137).

Other

Conover, Milton. (1923). The General Land Office: Its history, activities and organization. Baltimore: The Johns Hopkins Press.

Gilmore, William E. (1897). Life of Edward Tiffin, first Governor of Ohio. Chillicothe, OH: Horney & Son.

TONGUE, JAMES (C.1801)

The following publication is the only information found about Tongue. The National Union Catalog Pre-1956 Imprints had no entries under the name (The American Library Association 1976).

(1807). A letter addressed to the people of Maryland, giving a short account of the country on the south shore of lake Erie; Including a brief description, of its situation, climate, soil, productions... Shewing [sic] the advantages it offers to the middling people of Maryland to emigrate. Washington: Printed by J. D. Westcott and R. Dinmore. 32p. 23.5cm.

TUPPER, ANSELM (C.1796)

According to Pattison (1957), Tupper was one of three men who were engaged in surveying the Seven Ranges, inaugurating the subdivision of the Ohio Company's lands. He later returned to federal surveying after the Land Act of 1796 was passed by congress (p.179). No publications were found in The National Union Catalog, Pre-1956 Imprints (The American Library Association, 1976).

TUPPER, BENJAMIN (1738-1792)

Brown (1903) identifies Tupper as a "pioneer and soldier." He was born in Stoughton, Massachusetts and died in Marietta, Ohio. His father was Thomas and his mother was a Perry. It is of interest that he was apprenticed to a tanner in Dorchester, Massachusetts. He spent time teaching at a district school in Easton, Massachusetts. He married Huldah White and moved to Chesterfield, Massachusetts. During the Revolutionary War, in addition to other duties, he served as an aid to General Washington. General Tupper was involved in the organization of the Ohio Company. He made the second survey of Ohio in 1786. He was a very active citizen (V.VII, pp.392-393).

Tupper was one of eight surveyors who reported for duty in 1785 to the newly appointed Geographer of the United States, Thomas Hutchins beginning work on the surveying of America. He was sent by his fellow Massachusetts-man, Rufus Putnam because Putnam, who was the official appointee for Massachusetts had conflicting appointments. Tupper had been a general in the Continental Army. He served as an advance scout for the Ohio Company of Associates and was among the people from Massachusetts who provided leadership in the settlement of Ohio (Pattison, 1957).

Other

(1891). Letter, Magazine of American History, 25, p.503.

VALLANCE, JOHN (1770-1823)

Vallance was a Philadelphia engraver born in Scotland. He moved from Scotland to Philadelphia in 1791. He continued engraving independently and with partners until his death (Ristow 1985). He worked with James Thackara preparing the engraving for Andrew Ellicott's plan of Washington (p.117). He collaborated with H. S. Tanner to do the engraving for Melish's Travels; Vallance completing seven of them (p.179).

(1795). The American atlas. Engraved by Barker, Scott, James Thakara and John Valance of Philadelphia, Samuel Hill of Boston, Amos Doolittle of New Haven, and Benjamin Tanner of New York. Philadelphia: Mathew Carey.

(1795). Map of the state of Maryland. Laid down from an actual survey of all the principal waters, public roads, and divisions of the counties therein. Philadelphia: Engraved and published by John Vallance.

(1811). A map of the state of Pennsylvania by Reading Howell. Engraved by John Vallance. Philadelphia: Emmor Kimber.

(1812). Travels in the United States of America in the years 1806 & 1807, and 1809, 1810, & 1811. Philadelphia: John Melish.

VARLE', CHARLES P., AKA PETER, AKA PETER C., AKA P. C., AKA CHARLES (C.1770)

Varle' conducted a survey of the city of Philadelphia in 1796, which was engraved by Robert Scot of that city. He also collaborated to prepare plans of the city of Baltimore, Maryland (Ristow 1985).

Varle' was among those French people who were forced to leave Saint-Domingue (the present Haiti) at the time of the 1791 forced exodus of whites from that island. He moved to Philadelphia where he joined the Fourth Company of the Third Regiment of the Philadelphia Militia.

(1827). Candid considerations, respecting the canal between the Chesapeake and Delaware Bays. Whereby the impracticabilities of proceeding successfully with the same agreeably to the plan now pursued, is demonstrably proved. Baltimore: author, Wm. Woody, printer. 19p. 12mo.

(1833). A complete view of Baltimore, with a statistical sketch, of all the commercial, mercantile, manufacturing, literary, scientific, and religious institutions and establishments in the same, and in its vicinity... derived from personal observation and research... To which is added, a detailed statement of an excursion on the Baltimore and Ohio rail road to the Point of rocks... and an advertising directory. Baltimore: S. Young. vi[9]-166p. plates. 14.5cm.

(1834). Map of North America. New York: author. 65x47cm. Inset in his map of the United States.

(1815). Observations on the cause, symptoms and cure, of the prevailing epidemic called typhus fever, designed to promote the cause of humanity, by saving those who might otherwise be sacrificed by self injurious treatment... Charles Varle, engineer. Frederick-Town: author. 21p. 20.5cm.

(1833). Plan of the city of Baltimore including the South Baltimore Cos. Grounds. (np).

(c.1796 & 1802). Plan of the City of Philadelphia. By P. C. Varle', geographer & engineer. Philadelphia: author.

(1810). Topographical description of the counties of Frederick, Berkeley & Jefferson, situated in the state of Virginia, in which the author has described the natural curiosities of those counties, their mineralogy and lithology: also, the quality of the soil, the manufactories, mills, &c. the number of inhabitants... &c. To which is added a beautiful map of those counties. Winchester, VA: Printed by W. Heiskell. 34p. 20cm.

VINCE, REV. S. (C.1811)

Barton (1813) names Vince author of an American astronomy (p.xlix). That is true, but The National Union Catalog, Pre-1956 Imprints (The American Library Association 1976, V.638, pp.243-246) reveals only two American editions of numerous texts, most of which were published in London or Cambridge, England. Therefore, Vince may be a questionable member of this first generation of American geographers.

(1811). The elements of astronomy: Designed for the use of students in the university. 2d ed. Philadelphia: Kimer and Conrad. 242p. diagrams. tables. 22cm.

WALKER, THOMAS (1715-1794)

Walker and Thomas Abernethie collaborated to produce one of the first road maps in the United States, 1787. The two men were natives of Scotland who settled near Charleston about 1783. Walker was from Edinburgh and was apparently trained as an architect and tombstone cutter. At one time he advertised a course in architecture to be taught by himself (Ristow 1985, p.156-157).

Hopkins (1964) notes that Thomas Walker was born in King and Queen County, Virginia and died in the present Albemarle County, Virginia. He was a soldier, physician, and explorer. He studied medicine at Williamsburg, Virginia and practiced in Fredericksburg. In 1748 he joined a company of land speculators who explored the southern part of Virginia. By 1749 he became chief agent for the Loyal Land Company and explored westward. He was the first white man to have made a recorded expedition to the Kentucky Country (Encyclopedia Britannica 2001. Walker, Thomas).

Washington made him commissary-general to the Virginia troops, but in 1755 Walker was charged with misconduct for entering partnership with Andrew Lewis. He also acted as a guardian for Thomas Jefferson (p.1129).

(1888). Journal of an exploration in the spring of the year 1750. By Dr. Thomas Walker... With a preface by William Cabell Rives... Boston: Little, Brown, and Co. 69p. 20cm.

(1787). Specimen, of an intended traveling map of roads of the state of South Carolina, from actual survey. By Thomas Abernethie and Thomas Walker. (np).

Other

Rust, Francis Marion (Arr.). (1950). Dr. Walker's diary of exploration 1750 with will subjoined. Harbourville, KY: Advocate Publishing Co.

Burns, Annie Walker. (1942). Doctor Thomas Walker; first white man of any distinction to explore Kentucky (1749/50). He named the Cumberland Mountains; Gap and River. 2d ed. Washington: author.

Johnston, J. Stoddard. (1898). Doctor Thomas Walker's journal of an exploration of Kentucky in 1750, being the first record of a white man's visit to the interior of that territory, now first published entire, with notes and biographical sketch... In authors, First explorations of Kentucky. Louisville: p.1-84. map. 32.5x35.5cm.

WARREN, MOSES JR. (C.1762)

Warren is one of those geographers who may more appropriately be assigned to the second generation of American geographers, but is included because he was active prior to 1812 (Schwartz and Ehrenberg 1980). He and George Gillet prepared a state map for Connecticut that was engraved by Abner Reed and published in Hartford by Hudson and Goodwin in 1812.

Warren was born in Hopkinton, Rhode Island, but two years later his family moved to East Lyme, Connecticut. He served in the revolutionary war and produced maps with George Gillet (Ristow 1985, p.97).

(1812). Connecticut, from actual survey, made in 1811; by and under the direction of, Moses Warren and George Gillet; and by them Compiled. Published under the authority of the General Assembly, by Hudson & Goodwin. Engraved by Abner Reed, and E. Windsor. Hartford: Hudson & Goodwin.

(1897). Traverse of the portage from Cayahoga to Tuskarawa's part of the second parallel & survey of the 10 acre lots in the town of Cleaveland [sic]. (np). 10p.

WASHINGTON, GEORGE (1732-1799)

Washington is known as the Father of the Nation. He was the first President of the United States, a planter, a slave holder, and a surveyor. As a surveyor he participated in the advancement of geographic knowledge and is appropriately included as a member of the first generation of American geographers. Brown (1959) notes that, "Washington, during the Colonial period, engaged in the distribution of land grants to "soldiers who enlisted in Virginia for protecting frontiers against Indians" (p. 120). Among his duties was the surveying of those lands, attested to by his "Plat of the Survey of the Little Kanhawa 28,400 acres made in 1773" (Plate 48).

Harley, et al. (1978) remarks about the esteem in which Washington held cartographers. His employment of surveyors, engineers, and cartographer/ engravers is acknowledged several times (p. 11, 13, 24, etc.).

Numerous entries were identified in Adler (1977), however only those regarding monument locations and representation of the states might be considered geographic in nature. For non-geographic items the reader is referred to Adler.

The National Union Catalog Pre-1956 Imprints has numerous entries for Washington (The American Library Association 1976, v.649, pp.518-562). Only those publications which have an obvious geographic connection are included in this list.

(1748). Journal of a survey. (In his writings, V.1, 1889).

(1801). Map of General Washington's farm, of Mount Vernon from a drawing transmitted by the general. London: Richardson. 37x51.5cm.

(1860). Washington's journal of a tour to the Ohio, in 1753. With notes by John G. Shea. In his Diary... from 1789 to 1791. New York: pp.203-248.

Other

Fitzpatrick, J. C. (Ed.). (1931). The diaries of George Washington, 1748-1799. Boston & New York: Houghton Mifflin Co. 4v. map. 23cm.

Fitzpatrick, J. C. (Ed.). (1931). The writings of George Washington. Washington: (np).

Freeman, Douglas Southall. (1948-1951). George Washington: A biography. 4 vols. New York: Charles Scribner's Sons.

Kent, Donald H. (Ed.). (1952). Contrecoeur's copy of George Washington's journal for 1754. Harrisburg: Pennsylvania Historical Museum Commission. 86p. maps. 28cm.

Larrabee, don Marshall. (Ed.). (1950). A reprint of the journals of George Washington and his guide, Christopher Gist, reciting their experiences on the historic mission from Governor Dinwiddie, of Virginia, to the French forts in November-December, 1753. Including excerpts from the writings of historians and statesmen as to the importance of the mission, its results and effect on the world's history. Williamsport, PA: author. xii,37p. illus. map. 27cm.

Martin, Lawrence. (Ed.). (1932). The George Washington atlas. Washington, DC: U. S. George Washington Bicentennial Commission.

Sparks, Jared. (1834-37). The writings of George Washington; being his correspondence, addresses, messages, and other papers, official and private, selected and published from the original manuscripts; with a life of the author, notes and illustrations. Boston: American stationers' company; J. B. Russel. 12v. 25 maps. 26cm.

Toner, J. M. (Ed.). (1892). Journal of my journey over the mountains, while surveying for Lord Fairfax, Baron of Cameron, in the northern neck of Virginia, beyond the Blue Ridge, in 1747-8. Albany, NY: Joel Munsell's Sons.

WATSON, WILLIAM (C.1812)

Ristow (1985) asserts that little is known about Watson. His map was not original, it was based on data previously published. None the less, his was the first published map of the state of New Jersey that included township boundaries and a fairly extensive network of existing roads. Some of his roads were not accurately placed (p. 115).

Although there were several William Watson's noted in The National Union Catalog - Pre -1956 Imprints, only the one publication noted below appears to be by our person (The American Library Association 1976, V.651, pp.155-177).

(1812). A map of the state of New Jersey. Engraved by William Harrison. Philadelphia: (np). 102x72cm. Scale 1:300,000.

WEBSTER, NOAH (1758-1843)

Webster is best known as a lexicographer, born at West Hartford, Connecticut on October 16, 1758. He served as a private under his father during the American Revolution and was present when the British Army surrendered near Saratoga. He then returned to his studies at Yale College and graduated in 1778.

Webster proceeded to study law and was admitted to practice in 1781. In Goshen, New York he taught school and prepared his Grammatical Institutes of the English Language. Following the revolution, Webster gained considerable reputation as a political writer. His Sketches of American Policy, which advocated the formation of a new constitution; and papers in defence of Washington's proclamation of neutrality, and of Jay's Treaty with England prompted considerable attention. The National Cyclopedia (NAT) notes that Webster deserves recognition as an American man of letters for his numerous books on a variety of subjects. A biography describing in some detail, Websters life, American Men of Letters, was written by Horace E. Scudder. He died in New Haven, Connecticut on May 28, 1843 (NAT, pp.394-95; White, 1899, V. 2, pp.394-395).

(1810). Elements of useful knowledge. Volume I. Containing a historical and geographical account of the United States: for the use of schools... Fifth edition. New York: E. Duyckinck.

(1812). As above- Hartford: Hudson and Goodwin, 208p. 18cm.

(1804). Elements of useful knowledge. Volume 2. Containing a historical and geographical account of the United States: for the use of schools... New Haven: Sidney's Press, 206p.

(1806). As above- 2d. ed. New Haven: From Sidney's Press, 224p. 18cm.

(1808). As above- Third edition. New London: Ebenezer P. Cady for O. D. Cooke, and New Haven: Sidney's Press for Increase Cooke & Co. 114 leaves.

(1809). As above- Middletown, Conn.: Hudson & Goodwin, by T. Dunning.

(1813). As above - Hartford: Printed by Hudson & Goodnow.
"The farewell address of George Washington"

(1806). Elements of useful knowledge. Volume III. Containing a historical and geographical account of the empires and states in Europe, Asia and Africa, with their colonies. To which is added a brief description of New Holland, and the principal islands in the Pacific and Indian Oceans. For the use of schools... New Haven: Bronson, Walter & Co. 294p. 18cm.

(1815). As above- 6th ed. Hartford, Conn.: Hudson and Goodwin. 3v. 17cm.

(1848, 49). A high school pronouncing dictionary of the English language, abridged from the American dictionary of Noah Webster, with accented vocabularies of classical, scripture and modern geographical names. By William G. Webster. New York: Huntington and Savage, 360p. 17.5cm.

(1853). As above- New York: Huntington, and Mason Brothers, 352p.

(1857). As above- New York: Mason Bros.

(1857). As above- Philadelphia: (np).

(c.1860). As above- Philadelphia: J. B. Lippincott.

(1832 & 1833). History of the United States; to which is prefixed a brief account of our English ancestors, from the dispersion at Babel, to their migration to America; and of the conquest of South America, by the Spaniards. New Haven: Durrie & Peck and Louisville, KY: Wilcox, Dickerman & co. 356p. 15cm..

(1835). As above - Cincinnati: Corey, Fairbank & Webster, 318p.

(1790). On the theory of vegetation. Memoirs of The American Academy of Arts and Sciences, V.1, pt.1, art. 26, pp.178-185.

(1855). Pronouncing vocabulary of modern geographical names. Springfield, MA: G. & C. Merriam, 1321-1376p. 26cm.

Detached from the 1855 (or 1856?) edition of Webster's American dictionary of the English language, revised by Chauncey A. Goodrich." (1904). ...Webster's Empire dictionary of the English language; being the authentic unabridged dictionary by Noah Webster, LL.D. With an exhaustive appendix, including Scripture proper names, and pronouncing vocabulary of Greek and Latin proper names prepared under the direction of Noah Porter...thoroughly revised and greatly enlarged and improver over one hundred educators, specialists, and other scholars. Under the editorial supervision of Thomas H. Russell, LL. a., Albert C. Bean, M.E.LL.D., and L. B. Vaughan, PH.D.; with several thousand illustrations, color plates, and tables. Toronto: E.N. Moyer Co., Limited.

Other

Scudder, Horace E. (nd). American men of letters. (np).

WEED, ENOS, JR. (C.1797)

Other than the following, no information about Weed has been found in biographies. Other publications by Weed, found in The National Union Catalog, Pre-1956 Imprints, did not contain geographic content (The American Library Association, 1976). His limited contributions were of such a nature that they may have aided the standardization of spelling geographic terms and names.

(nd). The American orthographer. V.1, pt.1, containing the American standard for pronunciation. Pt. 2, The American's dictionarian standard... New York: (np), 120p.

(1797 & 1798). As above- Also titled, Book II, The geographical spelling book. Danbury, Conn.: Douglas & Nichols for the author, 36p., 19cm.

WHEELER, T. (C.1784)

Wheeler is named by Samuel Holland as one of his deputies helping to survey New Hampshire. No other information is to be found about the man.

(1784). Topographical map of the state of New Hampshire, surveyed under the direction of S. Holland, Surveyor General of Lands for the Northern District of North America; by the following gentlemen, his deputies: T. Wright, G. Sproule, J. Grant, T. Wheeler, & C. Blaskowitz. London: Printed for W. Faden, geographer to the King. In New Hampshire. Geological Survey, 1868-1878. Geology of New Hampshire, Atlas. Scale: 8 miles to an inch.

WHIPPLE, LEVI (C.1767)

Whipple was one of the deputy surveyors appointed under the Land Act of 1800 to work under the direction of Rufus Putnam (Pattison 1957, p.207).

The following were found, however they were identified under the surname and did not include any given name:

(1767). Deposition that Elisha Brown took tax of South County. Providence: (np).

U.S. Senate & U.S. House of Representatives. 33rd Congress, 2nd Session. Explorations and surveys for Pacific Railroad, V.3: Whipple's route, geology, Indians, etc. (S.exdoc.78 & H.exdoc.91). Washington: Government Printing Office, 1854-1855. (Serial Sets 760 & 793).

U.S. Senate & U.S. House of Representatives. 33rd Congress, 2nd Session. Explorations and surveys for Pacific Railroad, V.4: Whipple's route, botany, zoology, etc. (S.exdoc.78 & H.exdoc.91). Washington: Government Printing Office, 1854-1855. (Serial Sets 761 & 794).

U.S. House of Representatives. 33rd Congress, 1st Session. Pacific Railroad surveys, V.2: Beckwith, Whipple, and Pope (H.exdoc.129). Washington: Government Printing Office, 1853-1854. (Serial Set 737).

WHITELAW, JAMES (C.1796)

Whitlaw served as an assistant to Ira Allen and ultimately succeeded him, in 1787, as surveyor general of Vermont. He used the deposited town surveys of Vermont to compile a reasonably accurate map of that state (Ristow 1985, p. 89).

(1796). A correct map of the state of Vermont from actual survey; exhibiting the county and town lines, rivers, lakes, ponds, mountains, meetinghouses, mills, public roads &c. By James Whitelaw Esqr.: Surveyor General 1796. With Privilege of Copy Right. Engraved by Doolittle. (np): Whitelaw.

(1810). As above- Engraved by James Wilson.

(1821). As above- Engraved by Ebenezer Hutchinson.

(1824). As above- Engraved by Ebenezer Hutchinson.

(1813). Map of the Northern Part of the United States and the Southern Part of the Canadas. Engraved by James Wilson. (np).

WILLIAMS, SAMUEL (1743-1817)

Williams was an astronomer and historian born in Waltham Massachusetts on April 23, 1743 (Whiet 1898). He was the son of Waxham Willams and the grandson of "The Redeemed Captive" John Williams. He was a student at Harvard and student of Prof. John Winghtop with whom he went to Newfoundland to observe the planet Venus in 1761. Following his graduation he taught for a while then pursued divinity as a career.

He surveyed, in 1786, the western boundary of Massachusetts and went to Penobscot Bay in a vessel to observe the total eclipse of the Sun, October 20th. He moved to Rutland, Vermont where he edited the Herald and the Rural Magazine. In 1805 Williams made an official survey of the western boundary of Vermont and lectured on astronomy and natural philosophy at the state university. There he died on January 2 leaving numerous scientific manuscripts (White 1898).

(1780). Account of a very uncommon darkness in the states of New-England, May 19, 1780. (np), p. 234-246, 26cm.
(1824). A history of the American Revolution: Intended as a reading-book for schools. By Samuel Williams, LL. D. New Haven: W. Storer, Jr., 204p., illustrated, 17.5cm.
(1826). As above- 2d ed., Stonington, Conn.: W. Storer, Jr.
(1827). As above- 3d stereotyped ed.
(1830). As above- 11th stereotyped ed. New Haven: W. Storer, Jr.
(1830). As above- 13th stereotype ed.
(1831). As above- 16.5cm.
(nd). Manuscript notes, sermons, and lectures. (np), 18v. in 17.
(1794). The natural and civil history of Vermont. By Samuel Williams, LL. D. member of the Meteorological Society in Germany, of the Philosophical Society in Philadelphia, and of the Academy of Arts and Sciences in Massachusetts. Walpole, NH: Printed by Isaiah Thomas and David Carlisle, Jr., Sold in Worcester by Isaiah Thomas, 416p., 21.5cm.
[Found in the South Caroliniana Library at the University of South Carolina, Columbia.]
(1809). As above- 2d ed., Burlington, VT: Printed by Samuel Mills; Sold in Burlington by Mills and White; Worcester: Isaiah Thomas, Jr.; Boston: Thomas and Whipple; and Newburyport: S. Sawyer and Co., 2v., 20.5cm.
(1784). Observations and conjectures on earthquakes of New England. Memoirs of The American Academy of Arts and Sciences, 1, pt. 2, art. 8, pp.260-311.

Other

Williams, F. W. (1899). The life and letters of Samuel Wells Williams. New York: (np).

WILLIAMS, WILLIAM (1787-1850)

Williams was an author and patriot. He was born in Lebanon, Connecticut on April 8, 1731 and died on August 2, 1811. He was probably the man responsible for the following publications. He was the son of Solomon Williams, pastor of the First Congregational Church and Mary Porter of Hadley, Massachusetts. He graduated from Harvard College in 1751.

Williams afterward began to study theology under the direction of his father. He participated in the French and Indian War, taking part in the operations at Lake George in 1755. Williams was active in the American struggle for independence. He helped to compose many of the Revolutionary State Papers of Governor Trumbull and published claims of the colonists. He offered specie in return for paper Continental currency, thought to be near worthless. He is credited with remarking that, "If independence were established, he would get his pay; if not, the loss would be of no account to him" (Johnson & Malone 1928, V.10, pp.293-294).

(1798). First principles of geography, selected from approved authors. Charleston: Printed by T. B. Bowen, 11p. 3pl. 26cm..

"Engraved table and 3 astronomical diagrams (numbered plates I-V) " (The American Library Association 1976)

(1804). Travels in Turkey, Asia-Minor, Syria and across the desert into Egypt, during the years 1799, 1800, and 1801, in company with the Turkish Army and the British Military mission: also through Germany, Holland, &c on the return to England: to which are affixed, observations on the plague, and on the diseases prevalent in Turkey, and a meteorological journal... Philadelphia: James Humphreys, 426p., tables, 22cm.

Note. The above publication is authored by the only W. Williams who is identified as "geographer." The publications following this note are identified as W. Williams, but they do not appear to be the same person.

Other

(1896). Report by Professor W. Williams...on cattle diseases in Jamaica... (np), 15p.

Identified in The National Union Catalog Pre-1956 Imprints (The American Library Association 1976) as "Professor."

WILLIAMSON, CHARLES (1757-1808)

Williamson, in 1798, produced a book designed to promote the settlement of western New York known as Genesee Country (Schwartz and Ehrenberg 1980, pp.212-213). He was a British officer, a secret agent and a New York legislator who was born in Balgray, Scotland and died at sea. He was the representative of a land promoter, a builder of roads, bridges, and a hotel. He was an American citizen who represented New York in that legislature, but who reverted to his British citizenship. He advised a policy to overthrow the "Frenchified" Jeffersonian administration and considered Aaron Burr as an agent in the endeavor (Hopkins, 1964, p.1216). His inclusion as an American geographer is justified by the fact that Williams was a member of the New York Legislature and he did add to the geographic knowledge of Americans.

(1798). Description of the Genesee Country, Its rapidly progressive population and improvements: In a series of letters from a gentleman to his friend. Albany, NY: Printed by L. Andrews & Co. 37p. 2 folded maps. 18x15cm.

(1804). A description of the Genesee Country in the state of New York. To which is added - An appendix, containing a description of the military lands, in which the situation, dimensions, civil divisions, soil, produce, lakes and rivers, curiosities, climate, navigation, trade and manufactures, populations, and other interesting matters, relative to that country, are impartially described. (np): Printed for the author. 24p. 16°.

(1804). As above- Baltimore: Printed for the author. 21cm.

(1804). As above- By Robert Munro. [Williamson supposed author.] New York: Printed for the author. 16p. 23cm.

(1799). Description of the settlement of the Genesee Country, in the state of New York. In a series of letters from a gentleman to his friend. New York: Printed by T. & J. Swords. 63p. map. 21.5cm.

(nd). Early settlers in the Phelps and Gorham Purchase; gleanings from an account book. (np). [see under James McCall]

(1797). The legislature at their last session, having granted considerable aid to improve the great road leading from Fort Schuyler, on the Mohawk River, to the county of Ontario, and appointed a commissioner to carry the law into effect...[form of subscription for roads in the Ontario Reservation] to be paid into the hands of Charles Williamson, Esq. or his order, for the purposes aforesaid. Albany: Printed by Loring Andrews & Co. 4to. Broadside.

(nd). Observations on the proposed state road, from Hudson's River to Lake Erie. (np). [Williamson supposed author]

(1804). A view of the present situation of the western parts of the state of New York, called the Genesee Country. Frederick-Town: Herald Press for the author. 23p.

(1804). As above- ...In which the situation, dimensions, civil divisions... and other interesting matters, relative to the country are impartially described. Frederick-town, MD: Printed at the Herald press for the author; and Rochester MY: G. P. Humphrey (1892). [3]-23p. 18.5cm.

(1939). As above- (np): Monroe County Division of Regional Planning.

WILSON, JAMES (1763- 1855)

Wilson was the first American globe maker and an engraver who worked with Whitelaw, Isaac Eddy and others. He is credited with the production of at least seventy-six globes which range in size from 8 to 51 cm. (Ristow 1985, p.89, 278).

He was born in Londonderry, New Hampshire where he worked on his father's farm and received a basic education. He moved to Bradford, New Hampshire in 1796. On a visit to Dartmouth College he saw a pair of English globes. That excited his interest and encouraged him to become more knowledgeable about geography. To accomplish that he purchased a set of Encyclopedia Britannica. Wilson proceeded to move to New Haven, Connecticut and undertake engraving under the instruction of Amos Doolittle.

(c.1820). Colored globe. Albany: Wilson & Sons. 8cm. in diameter.

(1813). Chronology delineated to illustrate the history of monarchial revolutions. Engraved by James Wilson and Isaac Eddy.

(1810). A correct map of the state of Vermont from actual survey; exhibiting the county and town lines, rivers, lakes, ponds, mountains, meetinghouses, Mills, Public Roads &c. By James Whitelaw Esqr.: Surveyor General 1796. With privilege of copy right. Engraved by James Wilson. (np): Whitelaw.

JOHN WILSON(C.1778)

Very little information is to be found about Wilson. There are two maps listed in The National Union Catalog Pre-1956 Imprints that may be his (The American Library Association 1976). The only encouragement that he was an American is the 1822 map listed under the James Wilson publications.

(1822). Map of South Carolina, constructed and drawn from the district surveys, ordered by the legislature, by John Wilson, late civil and military engineer of So. Cara. Engraved by H. S. Tanner . Philadelphia: H. S. Tanner.

(1778). Plan of the decent and action of 29th Decr. 1778, near the town of Savannah. (np).. map. 47x72cm. Scale 1 inch to 600 feet.

(1822). A map of South Carolina, constructed and drawn from the district surveys, ordered by the Legislature: by John Wilson. The astronomical observations by Professor Geo: Blackburn & I. M. Elford. Engraved by H. S. Tanner. Philadelphia: (np). 111x136cm. Inset of Charleston Harbour. Scale 6 miles to 1 inch.

For only the previous map the designation"cartographer," is found in The National Union Catalog Pre-1956 Imprints (The American Library Association 1976). Perhaps not the same John Wilson.

WINDSOR, E. (C.1812)

Windsor was an engraver who worked with Abner Reed (Ristow 1985, p.97). The National Union Catalog - Pre -1956 Imprints provided no entries under that name (The American Library Association 1976).

(1812). Connecticut, from actual survey, made in 1811; by and under the direction of, Moses Warren and George Gillet; and by them compiled. Published under the authority of the General Assembly, by Hudson & Goodwin. Engraved by Abner Reed, and E. Windsor. Hartford: Hudson & Goodwin.

WINTERBOTHAM, WILLIAM, REV. (1763-1829)

The only thing we know of Winterbotham is through his publications. Those were located in The National Union Catalog, Pre-1956 Imprints (The American Library Association, 1976).

(1796). The American atlas. Engraved by Benjamin Tanner, David Martin, D. Anderson, John Scoles, and John Roberts. Philadelphia: John Reid. 20 maps. 42.5cm.

(nd). A complete view of the Chinese Empire. (np).

(1795). A geographical, commercial, and philosophical view of the present situation of the United States of America: comprehending a description of the United States, their extent, civil divisions, chief towns, climates... agriculture, commerce, and history. Copious and interesting accounts of the late war; general particular description of the eastern territory, Kentucky, &c. To which is prefixed a general account of the discovery of

America, by Columbus; general description of the whole continent of America, and the numerous tribes of American Indians, their manners, customs, &c. &c. And to which is added, a view of the present state of the Spanish, French, English and Dutch possessions in America and the West India Islands. Also an appendix, containing various interesting and valuable papers and instructions relative to cultivating tobacco, rice, maize, sugar, maple tree &c. Illustrated with numerous large maps, charts, and plates. By the Rev. W. Winterbotham. New York: Printed by tiebout & O'Brien, for J. Reid, L. Wayland, and C. Smith. 4v. charts. maps. plates. 22cm.

(1795). An historical, geographical, and philosophical view of the Chinese empire; comprehending a description of the fifteen provinces of China, Chinese Tartary, tributary states; natural history of China; government, religion laws, manners and customs, literature, arts, sciences, manufactures, &c. To which is added, a copious account of Lord Macartney's embassy, compiled from original communications. London: Printed for the editor: J. Ridgway. 435,114p. plates. map. 21.5cm.

(1796). As above- Philadelphia: Re-print for Richard Lee. Dunning, Hyer, and Palmer-printers. 2v. 21.5cm.

(1796). An historical, geographical, commercial, and philosophical view of the United States (4v.). New York: Printed by Tiebout and O'Brien for J. Reid. 4v. maps. tables.

(1795). As above- London: J. Ridgway.

(1799 & 1819). As above- Printed for the compiler.

(1796). Plan of Franklinville, in Mason County, Kentucky. New York: (np). 28x21cm.

(1796). Plan of Lystra in Nelson County, Kentucky. New York: (np). 29x21cm.

Other

Winterbotham, Sir William Howard. (1893). The Rev. William Winterbotham, a sketch. London: Ballantyne, Hanson. 69p. 27cm.

WOLCOTT, OLIVER (1760-1833)

Wolcott was Secretary of the Treasury (1795-1800). He was responsible for drawing up the first set of general instructions for federal surveying policy, 1797 (Pattison 1957, p.206).

Brown (1897) reports that Wolcott became Governor of Connecticut, was born in Litchfield, Connecticut, the son of Gen. Oliver and Lorraine (Collins) Wolcott. He graduated from Yale in 1781 and served in the army under his father. He married Elizabeth Stuyhton. Wolcott was the Comptroller of Public Accounts, Auditor, Comptroller of the U.S. Treasury, Secretary of the Treasury prior to becoming Governor (V.7, p.635).

(1820). Remarks on the present state of currency, credit, commerce, and national industry: in reply to an address of the Tammany Society of New

York: By Oliver Wolcott, governor of the state of Connecticut. New York: G. Wiley and Co. 43p. 20.5cm.

(1819). To the Senate and House of Representatives of the state of Connecticut... I herewith transmit, for the use of both houses, a sketch of a bill providing for the assessment and collection of taxes in this state, together with an explanation of the principles on which the bill is founded. Hartford: E. Babcock & Sons, printers. 23p. 24cm.

Other

Gibbs, George. (1846). Memoirs of the administration of Washington and John Adams, edited from the papers of Oliver Wolcott, Secretary of the Treasury. New York: Printed for the subscribers, W. Van Norden, printer.

Smith, William Loughton. (1796). The pretensions of Thomas Jefferson to the Presidency examined; and the charges against John Adams refuted. addressed to the citizens of America in general; and particularly to the electors of the President. Philadelphia: (np).

WORKMAN, BENJAMIN (C.1767)

Bedini (1972) reveals that Workman was "an instructor at the University of Pennsylvania," around 1787 (p.92). He was, additionally an author of geography textbooks.

(1789, 1793 & 1796). The American accountant; or, Schoolmasters' new assistant, comprised in four books...and including all the questions in the Philadelphia edition of Gough...The whole adapted to the commerce of the United States... Philadelphia: Printed by John McCulloch, for William Young. 224p. 17cm.

(1790). Elements of geography, designed for young students in that science. In seven sections..."Sect I. Of the solar system. Sect. II. Of the earth in particular. Sect III. Of maps and globes. The three foregoing sections contain the scientific or astronomical part of geography, digested in a clear and comprehensive manner. Sect. IV. Of the different religions, governments, and languages of nations. Sect. V. Of the political divisions of the earth, into empires, kingdoms, &c. or the historical part of geography. Sect VI. Of natural philosophy; or the properties of matter, &c. Sect VII. Of chronology. The 3d ed. Philadelphia: Printed & sold by John McCulloch. 148p. maps. 14cm.

(1793). As above- 4th ed. 180p. 13.5cm.

(1795, 1796 & 1799). As above- 14cm.

(1803, 1804). As above- 15cm.

(1805). As above- 196p.

(1807). As above- The 12th ed. Philadelphia: Published by Ebenezer McCulloch. 191p. illus. maps. 15cm.

(1811). As above- The 14th ed. 208p.

(1814). As above- 15th ed. 216p. 14.5cm.

(1816). As above- Philadelphia: W. McCarty.

(1810). Epitome of Workman's geography, containing such parts only as are necessary to commit to memory. Published at the desire of sundry teachers... Philadelphia: Wm. M'Carty. 106p. map. 15cm.

(1796). Father Tammany's almanac, for the year 1796; being Bissextile, or leap year. Philadelphia, W. McCulloch. 17-18cm.

Other

(Ed.). (1788). A treatise of arithmetic in theory and practice.. Adapted to the commerce of Great Britain and Ireland. By John Gough. To which are added, many valuable additions and amendments. Philadelphia: Printed by J. M'Culloch for W. Young.

WORTHINGTON, THOMAS (1769 OR 1773 -1827)

Worthington was one of the deputy surveyors appointed under the Land Act of 1800 to work under the direction of Rufus Putnam (Pattison 1957, p.207). He became U. S. Senator from Ohio and took up a surveyorship in 1799 for the government. He became an official of the local Ohio Land Office in 1800 (p. 208). The National Union Catalog - Pre -1956 Imprints provided notes that Worthington's life span was 1773-1827 (The American Library Association 1976, V.674, p.579).

White (1891) identifies him as Governor of Ohio from 1814 to 1818. He also became Surveyor of Public Lands, member of the Constitutional Convention, and subsequently U. S. Senator from Ohio. He was responsible for the introduction of a bill by which the public lands were sold in quarter sections, instead of tracts two miles square, as at first proposed, inducing vast immigration to the West (V.III, p.138).

Brown (1897) notes that Worthington was born in Jefferson County, Virginia, the son of Robert Worthington of Berkeley County, Virginia. He went to the Ohio Territory in 1796. He was a member of the territorial legislature and served as Pro Tempore Speaker, and was Surveyor of Public Lands. From 1818-1827 he was Commissioner of Canals and became a state legislature (1821-1822 and 1823-1824). Worthington was the original Vice President of the American Bible Society. He died in Chillicothe, Ohio on June 20, 1827 (V.7, p.649).

(1802). Communication to citizens of the Northwestern Territory opposed to an alteration of boundaries of the states. Chillicothe, OH: (np). Pol. Pam., V.103,

(1813). Motion. December 20, 1813. Washington: Printed by R. C. Weightman for the U.S. Senate. 22cm.

(1814). Motion. September 21, 1814. Washington: Printed by R. C. Weightman for the U.S. Senate. 22cm.

WRIGHT, S. (C.1806)

Stephen Smith Wright was a revolutionary war participant who was captured by the British and later repatriated (Johnson & Malone 1928, V.10, pp.565-567). He is one possibility for authorship of the following text .

(1806). A short introduction to geography and astronomy, designed particularly for the young masters and misses of New England... Randolph, Vermont: Author

(1863). As above.

WRIGHT, T. (C.1784)

Wright is named by Samuel Holland as one of his deputies helping to survey New Hampshire. No other information is to be found about the man.

(1784). Topographical map of the state of New Hampshire, surveyed under the direction of S. Holland, Surveyor General of Lands for the Northern District of North America; by the following gentlemen, his deputies: T. Wright, G. Sproule, J. Grant, T. Wheeler, & C. Blaskowitz. London: Printed for W. Faden, geographer to the King. In New Hampshire. Geological Survey, 1868-1878. Geology of New Hampshire, Atlas. Scale: 8 miles to an inch.

Non-American Geographers

The following people were geographers who were probably or definately not American, but due to the presence of their books may have influenced the geographic understanding of those who were American writers and citizens. It should also be noted that the inclusion of these people is based on the discovery of their publications in American archives.

ADAMS, GEORGE (1750-1795)

Adams' succeeded his father as the Mathematical Instrument Maker to George III. He was author of a large number of basic scientific works which were intended for instruction. He also authored mathematical books. Among the publications noted in his biography is the following which might place him among those who we identify as the first generation of American geographers: Astronomical and Geographical Essays' (1790), but, it must be noted that he was a "staunch tory." The fact that he spent little if any time in the United States appears to be reason to exclude him from those we can list as among the first generation of American geographers (Stephan & Lee 1917). He is included in this volume because his works were available to numerous Americans in the United States and some copies of those are still available in special collections within the United States.

(1808). Essay III: Containing a description of the most improved planetarium, lunarium, and tellurian. Philadelphia: William W. Woodward - Printed by Abel Dickinson. Issued as the third part of Adams, George. Astronomical and Geographical essays...14 x 22.4 cm.

(1800 & 1808). Astronomical & geographical essays: Containing, I. A full and comprehensive view, on a new plan of the general principles of astronomy. II The use of the celestial and terrestrial globes, exemplified in a greater variety of problems, than are to bee found in any other work.

III The description and use of the most improved planetarium tellurian and lunarium. IV An introduction to practical astronomy. Whitehall: Printed for William Young, Bookseller and Stationer - Philadelphia. 22cm.

(1789). As above- London: R. Hindmarsh. 665p.

(1790). As above- 2nd ed. London: R. Hindmarsh. 599p.

(1795). As above- 3d ed. London: Printed by R. Hindmarsh. 579p.

(1799, 1803 & 1812). As above- English author - "Mathematical Instrument Maker to His Majesty , and Optician to the Prince of Wales." London: W. and S. Jones. 531p.

(1808) An essay on the use of the celestial and terrestrial globes; exemplified in a greater variety of problems than are to be found in any other work; exhibiting the general principles of dialing and navigation. 5th ed. Philadelphia: Woodward. 238p. 23cm.

(1800). As above- 4th ed. Philadelphia: William Young.

(1792) An essay on vision, briefly explaining the fabric of the eye, and the nature of vision; intended for the service of those whose eyes are weak or impaired... 2nd ed. London: Printed for the author by, R. Hindmarsh. 172p. 22cm [Reprinted from Yearbook of optometry. New York, 1941, 289-374p.]

(1789). As above- London: Printed by R. Hindmarsh and sold by G. Adams. 153p.

(1787). Atlas - Plates for the essays on the microscope. London: (np). 4to.

(1813) Geometrical and graphical essays, containing a general description of the mathematical instruments used in geometry, civil and military surveying, leveling, and perspective; with many new practical problems, illustrated by thirty-four copper plates. 4th ed. corr. London: Printed by C. Baldwin for, and sold by W. and S. Jones. 534p. 21.5cm.

(1791). As above- London: Printed for the author by R. Hindmarsh. 500p. 22.5cm.

(1797). As above- 2nd ed. corrected and enlarged by William Jones. London: W. & S. Jones. 518p.

(1803). As above- 3d ed. corr. London: Printed by W. Glendinning, and sold by W. and S. Jones. 518p. 20.5cm.

(1790) A short dissertation on the barometer, thermometer, and other meteorological instruments: together with an account of the prognostic signs of the weather. London: Printed for the author, by R. Hindmarsh 60p. 19cm.

Other

Jones, William. (Ed.). (1806 & 1807). Lectures on natural and experimental philosophy, considered in its present state of improvement. In 4 vols. Illustrated with 43 copperplates. This American edition, printed from the last London edition, edited by William Jones, is revised by Robert Patterson. Whitehall, Philadelphia: Printed for W. W. Woodward.

ADDINGTON, STEPHEN, D.D. (1729-1796)

Addington was an independent minister born at Northampton, England on June 9, 1729 and died in London on February 6, 1796. Although numbers of his books are to be found in American archives, he does not appear to be American (Stephen & Lee 1917).

(1811). Blank maps for the use of schools, forming a new system of modern geography; containing the projection and outlines without the names, the latter to be inserted by the pupil... Also exhibiting, at a general view, the extent, population, nature of government, religion, principal towns, latitudes and longitudes, etc., etc... Philadelphia: author.

(1809). Butler's geographical and map exercises...corrected and improved by Stephen Addington...Philadelphia: M. Carey, 51p. 18cm.

(1767). An inquiry into the reasons for and against inclosing open-fields. Humbly submitted to all who have property in them; and especially the members of the British Legislature. Coventry: Printed by and for T. Luckman; and sold by J. Johnson. 39p. 20cm.

(1772). As above- The 2d. ed. ...Coventry: Printed at Hargorough by J.W. Piercy. 60p. 18cm..

(1806). Questions relating to geography; with particular reference to Workman's system. To which is appended promiscuous questions, on the study of geography in general. By S. Addington and T. Watson. Philadelphia: Printed and sold by John McCulloch. 15cm.

(1806). A syllabus of the history of England: to which is appended, A tour through the southern parts of Great Britain. In two parts...designed for the use of schools. Philadelphia: D. Hogan. 84p.

(1770). The youth's geographical grammar; containing geographical definitions...to which is added K. An alphabetical index of kingdoms, states... II. An alphabetical index of cities... London: Printed for, and sold by the author and J. Buckland. 364 p. Folded maps. 18cm.

AIKIN, JOHN (1747-1822)

Aikin was located via an on-line search (bibliomania.com, 2001). His work appears in many American collections, but he was the son of Dr. John Aikin and studied medicine at Edinburgh and London. He settled in London where he appears to have practiced medicine until his death. Aikin was, in addition, the author of numerous books. Some of his writing presented geographic views of the world.

(1727). The American instructor: or, Young man's best companion. Containing, spelling, reading, writing, and arithmetick [sic], in an easier way than any yet published; and how to qualify any person for business, without the help of a master. Instructions to write a variety of hands, with copies both in prose and verse...Also merchants accompts [sic]...with a description of the several American Colonies. Together with the Carpenter's Plan and Exact Rule: Shewing [sic] how to measure carpenters, joyners [sic], sawyers... work... Likewise the practical-granger made easy; the art of dialing, and how to erect and fix any dial; with instructions for dying, couring [sic] and making colours. London: Printed for E. Midwinter. 424p. 16cm.

(1739). As above - London: Printed for J. Rivington and sons. 384p. 17cm.

(1740). As above - 5th ed., rev. and corrected. London: Printed for Charles Hitch. 380p.

(1742). As above - 6th ed., rev. and corrected. London: Printed for F. Hodges.

(1746). As above - 8th ed., rev. and corr. London: Printed for J. Hodges.

(1748 & 1753). As above - ...To which is added, the poor planters physician. With instructions for marking on linen; how to pickle and preserve; to make divers [sic] sorts of wine; and many excellent plaisters, and medicines, necessary in all families. And also prudent advice to young tradesmen and dealers. The whole better adapted to these American Colonies, than any other book of the like kind. Philadelphia: Printed by B. Franklin and D. Hall. 378p. 17cm.

(1755). As above - 13th ed., London: Printed for J. Hodges.

(1757). As above - 14th ed., London: Ware. 384p.

(1758). As above - ...To which is added a copper-plate copy-book...Corke: Reprinted by and for D. Roche. 13.5cm.

(1760). As above - New York: Printed and sold by H. Gaine.

(1760). As above - 25th ed. cor. and improved. London: Bunney.

(1767). As above - 21st ed. Edinburgh: Printed by A. Donaldson. 382p. 18cm.

(1770). As above - 390p. 16.2cm.

(1771). As above - 23rd ed. Edinburgh: Alston. 384p. 18cm..

(1772 & 1775). As above - 21st ed. London: Printed for J. Fuller. 17cm.

(1773). As above - 22d ed. Edinburgh: A. Donaldson. 382p. 18cm..

(1776). As above - Edinburgh: Gavin.

(1777). As above - 23d ed. London: Printed for H. Woodfall.

(1779). As above - ...To which is added, the poor planter's physician. With instruction for marking on linnen; how to pickle and preserve; to make divers sorts of wine; and many excellent plaisters and medicines, necessary in all families. A historical table of remarkable events from the creation of the world to the death of Julius Caesar. And a short abstract of the history of England. Also, prudent advice to young tradesmen and dealers. The whole better to these American states than any other book of the like kind. Boston: Printed for John Boyle and J. D. M'Dougall.

(1779). As above - 23d ed. London: Printed for W. Strahan.

(1785). As above - 13th ed. Worcester, MA: Isaiah Thomas.

(1786, & 1797). As above - Glasgow: J. Duncan.

(1787). As above - ...Likewise the practical granger made easy...A compendium of the sciences of geography and astronomy. Also some useful interest tables. With alterations and amendments adapted to the use of youth of the United States of America. Philadelphia: Printed and sold by Joseph Crukshank. 372p.

(1788). As above - London: Printed for Osborne & T. Griffin.
(1791). As above - London: Printed for J. Martin. 394p. 16.5cm.
(1792). As above - London: (np). 404p. 17.5cm.
(1794). As above - Walpole, NH: Isaiah Thomas and David Carlisle, for said Thomas in Worcester, MA.
(1797). As above - Wilmington: Printed and sold by Peter Brynberg. 300p. 17cm.
(1799). As above - London: Printed for booksellers. 360p.
(1801, 1810, & 1812). As above - Philadelphia: Printed by John Bioren for John Conrad.
(1806). As above - 29th ed. London: Printed for J. Johnson. 324p. 18cm.
(1810). As above - 30th ed. London: Printed for F. and C. Rivington. 328p.
(1822). As above - London: Printed for Dean and Hunday.
(1833). As above - Philadelphia: Thomas I. Bonsal.
(1856, & 1862). As above - London: Dean. 288p. 17.5cm.
(1785, 1785 & 1787). The calendar of nature; designed for the instruction and entertainment of young persons. Warrington, England: Printed by W. Eyres, for J. Johnson. 82p. 23cm.
(1787). As above - London: Joseph Johnson.
(1801 & 1806). As above - 72p. 17cm.
(1850). As above - London: E. Lacey, etc.
(1822). As above - 65p. 16cm.
(c.1780). As above - New York & Baltimore: (np).
(1815, 1818, &1823). As above - New York & Baltimore: Printed and sold by Samuel Wood & Sons, at the Juvenile Book-store. Illus. by Alexander Anderson. 112p. 14.5cm.
(nd). As above - see also under Aikin, Arthur, 1773-1854. The natural history of the year.
(1789). The calendar of nature: or, Youth's delightful companion. Containing, details of natural history, the narrative varigated with poetical quotations from the most celebrated authors. Embellished with engravings, representing the four seasons. Calculated principally to inspire a taste for the most curious products of art, as well as for the principal beauties of nature. London: Printed for the author by J. Johnson. 147p. plates. 15.5cm.
(1815, 1818, & 1823,). As above - New York: Printed and sold by Samuel Wood & Sons.
(c.1850). As above - London: E. Lacey, etc.
(1795). A description of the country from thirty to forty miles round Manchester: Containing its geography, natural and civil; principal productions; river and canal navigations...buildings, government, &c. London: Printed for J. Stockdale. 624p. 42 plates. 16 maps. 4 plans. 31cm.
(1788 & 1790). England delineated; or, A geographical description of every county in England and Wales, with a concise account of its most important products, natural and artificial, for the use of young persons. London: Printed by T. Bensley for J. Johnson. 425p. maps.

(1795, 1800, & 1803). As above - 396p.
(1804). As above - 2v. London: Lackington, Allen & Co.
(1818). As above - London: Baldwin, Cradock and Joy. 499p.
(1777). An essay on the application of natural history to poetry. Warrington, Eng.: J. Johnson. 156p. 18cm.
(1814). An essay on the plan and character of Thomson's Seasons. In Thomson, James. The seasons. pp.33-57. Georgetown: (np).
(1817 & 1822,). The farm-yard journal. Also the history of the mouse and marten... New Haven: Sidney's Press. 31p. illus. 10.5cm.
(1806 & 1807). Geographical delineations; or, A compendious view of the natural and political state of all parts of the globe. Philadelphia: F. Nichols. 416p.
(1806). As above - Philadelphia: Printed for F. Nichols, by Kimber, Conrad & co. 22.5cm.
(1806). As above - London: J. Johnson. 2v. 20cm.
(1811). The history of the environs of London, containing an historical and topographical account of every thing that is interesting and curious in all the towns, villages, and country, within twenty-five miles of the metropolis. By John Aikin, M. D. and other gentlemen...London: Stockdale. 811p. plates. maps. 30.5cm.
(1819). Questions adapted to Aikin's Geography. Salem: (np). 52p. 12mo.
(1797). A survey of the counties of Lancashire, Chesshire, Deryshire, West Riding of Yorkshire, and the northern part of Staffordshire, describing the rivers, lakes, soil, manure, climate, productions, minerals, property, and civil and ecclesiastical divisions; with a general account of the river and canal navigations within those districts. London: J. Stockdale. 216p. 23cm.
(1773). The works, in natural history, of the late Gilbert White... Comprising the natural history of Selborne; The naturalist's calendar; and miscellaneous observations, extracted from his papers. To which are added, a calendar and observations, by W. Markwick. London: J. Johnson. 2v. 19cm.
(1802). As above - London: Printed for J. White by T. Bensley.

Other

Aikin, Lucy. (1823). Memoir of John Aikin, M.D. By Lucy Aikin. With a selection of his miscellaneous pieces, biographical, moral and critical. London: Baldwin, Cradock, and Joy. 2v. 275p. 22cm.
Aikin, Lucy. (1823 & 1824). Miscellaneous pieces, biographical, moral and critical. (With Lucy Aikin's Mem. of him). Philadelphia and London: (np). 2v. 22cm.
Hale, Sarah. (1855). Juvenile budget opened being selections from the writings of Dr. John Aikin with a sketch of his life. New York: Harper & Bros.
Howitt, Mrs. Mary (Botham). (Ed.). (1862) Pictorial calendar of the seasons, exhibiting the pleasures, pursuits and characteristics of country life for

every month in the year, and embodying the whole of Aikin's Calendar of nature. Edited by Mary Botham Howitt. London: H. G. Bohn.

Pennant, Thomas. (1790). Indian zoology. 2d ed. London: Printed by H. Hughs, for R. Faulder.

BARTHELEMY, JEAN JACQUES (1716-1795)

Perusal of a variety of biographical collections turned up nothing about Barthelemy until his name appeared in The Encyclopedia Americana: International Edition (1988). Although numerous geography books by Barthelemy are to be found in American collections, it appears that he was a Frenchman (V.3, p.278).

(1821). Abrege du voyage d'Anacharsis; ou, Le Barthelemy de la jeunesse. Paris: Delalain. 2v. [Accent omitted]

(1831). Abrege du voyage der Jeune Anacharsis en Grece, dans le milieu du quatrieme siecle avant l'ere vulgaire. Ouvrage... arrange a l'usage des ecoles par Jean Henri Meynier. 5th ed. Halle: Renger. xvi, 656p. Colored map. 18cm. [Accent omitted]

(1794). Abrege du voyage du jeune Anacharsis en Grece, dans le milieu du quatrieme siecle avant l'ere vulgaire. A l'usage des ecoles. Nuremberg: Grattenauer. 637p. 18.5cm. [Accent omitted]

(1804). As above. arrange a l'usage des ecoles par J. H. Meynier. 2d ed. 656p.

(1838). As above. Limoges: 2v.

(1801). Abrege du voyage du jeune Anacharsis en Grece, dans le milieu du quatrieme siecle avant l'ere chretienne. De l'abbe Barthelemy, a l'usage de la jeunesse. Avec la vie de l'auteur, par M le duc de Nivernois. 2 ed. La Haye: Van Cleef. 2v. [Accent omitted]

(1894). Athenes et les Atheniens; voyages dans la Grece ancienne. Notes et introd. par Eugene Muller. Paris: Delagrave. 234p. [Accent omitted]

(1821). Atlas du myage du jeune Anacharsis. Paris: Menard & Desenne. 3p. 3 diagrams. 28 maps. [Accent omitted]

(1782). Voyage du jeune Anacharsis en Gr'ece, Vers le milieu du quatri'eme si'ecle avant l' 'ere vulgaire. (np). [Accent omitted]

(1789 &1792). As above. Paris: De Bure l'aine. 9v. 17cm.

(1789). As above. 2 ed. 7v. 20cm.

(1790). As above. 3 ed.

(1792). As above. 8 ed. Paris: Chez De Bure l'aine.

(1793). As above. Deux-Points: Sanson & Cie. 9v. & atlas.

(1795). As above. Paris: Chez De Bure l'aine, libraire de monsieur frere.

(1796). As above. A Londres: Charles Dilly.

(1798). As above. Londres: de l'Imp. de Baylis.

(1799). As above. Paris: De l'imprimerie de Didot jeune. L'an septi 'eme, 7v., 21cm. Note. Memoires sur la vie et sur quelques-uns des ouvrages de J. J. Barthelemy, becrits par lui-m^eme en 1792 et 1793.

(1801). As above. Paris:chez de Bure l'aine. 7v.

(1806). As above. Londres: Cox, fils, et Baylis. 21.5cm.
(1809). As above. Paris: L. Fantin.
(1810). As above. Paris: Chez les librarires associes.
(1813). As above. A Belin: (np).
(1815). As above. Paris: Nicolle.
(1817). As above. Paris: A. Egron. 14cm.
(1817). As above. Paris: Chez Pigoreau.
(1821). As above. Paris: E. Ledoux. New ed. 7v. & atlas.
(1822 & 1824) As above. Paris: Mme veuve Dabo, a la librairie stereotype.
(1823). As above. Avignon: Chez H. Offray.
(1824). As above. Londres: Booey et fils. viii, 481p.
(1824). As above. Paris: Janet.
(1825). As above. Paris: Berquet.
(1825). As above. Paris: Mme. Dabo-Butschert.
(1825 &1826). As above. Paris: E. Ledoux.
(1827). As above. Paris: Sanson. 810p.
(1829). As above. Paris: Bossange pere. 23cm.
(1830). As above. Paris: Rolland. 7v. and atlas. 22.5cm.
(1881). As above. Paris: Hachette. 3v. 18.5cm.
(1834). As above. Paris: Lebigre. 5 v. illus. maps. plans. 21cm.
(1836). As above. Paris: A. Payen. 8v. 14cm.
(1838). As above. 4v. 17.5cm.
(1839). As above. Paris: A. Hiard. 561p. 27 maps. 4 plans.
(1939). As above. Paris: Didot. 765p.
(1843). As above. Paris: Didier. 2v. 18cm.
(1853). As above. ... Abrege... par M. le Duc de Nivernois. Revue et soigneusement corrigee par Nicholas Wanostrocht. 11 ed. Londres: (np). 471p. [Accent omitted]
(1853). As above. Paris: Chez Firmin Didot freres, libraires, imprimeurs de l'Institut de France. 765p. 26.5cm.
(1856). As above. Paris: Hiard. 8v.
(1859). As above. Paris: Didot.
(1860 &1861). As above. Paris: Hachette. 3v. atlas (34 plates). 22cm.
(1876). As above.
(1880). As above. Paris: Firmin-Didot. xxviii, 765p. 26cm.
(1868). Voyage du jeune Anacharsis en Grece. Par Barthelemy. Schulausgabe mit worterbuch von A. de Saules. Leipzig: G. Brauns. 351p. 17cm. [Accent omitted]
(1817). Recueil de cartes geographiques, plans vues et medailles de l'ancienne Grece, relatifs au voyage du jeune Anacharsis, precede d'une analyse critique des cartes. Paris: Garney. xxxix p.32 plates. [Accent omitted]
(1819). As above.
(1790 &1791). Travels of Anacharsis the younger in Greece, during the middle of the fourth century before the Christian era by the Abbe Barthelemi...

Tr. from the French... In seven volumes and an eighth in quarto, containing maps, plans, etc. London: G.G.J. & J. Robinson. 7v. 22cm.

(1800). As above. London: Printed by J.D. Dewick for Vernor and Hood. viii, 562p. map. plates. 21.5cm.

(1801). As above. 1st American edition. Philadelphia: Jacob Johnson & co. 4v. 22cm.

(1806). As above. London: J. Johnson. 7v. 24.5cm.

(1810). As above. 4th ed. London: Vernor, Hood and Sharpe. viii, 468p.

(1816 &1817). As above. 7th ed. London: W. Green, printer for the Proprietors, Military classics' office.

(1817). As above. London: J. Mawman.

(1825). As above. London: Printed for C. and J. Rivington. 6v. tables. 22cm.

(1829). As above. Baltimore: F. Lucas, jr. ix, 396p. 19cm.

(1891). As above. London: (np).

(1802). Travels in Italy, by the late Abbe Varthelemy, author of the Travels of Anacharsis the Younger; in a series of letters written to the celebrated Count Caylus. With an appendix, containing several pieces never before published... Translated from the French. London: Robinson. xxiii, 108p. 22cm.

(1811 &1812). Viage de Anacarsis el joven por la Grecia, a mediados del siglo quarto antes de la era vulgar. Por mr. el abate Barthelemy... Tr. del frances al castellano. Per don Ignacio Pablo Sandino de Castro... Mallorca: Impr. de M. Domingo. 9v. 15cm. [Accent omitted]

(1813 &1814). Viage del joven Anacharsis a la Grecia a mediados del sigio quarto antes de le era vulgar. Traducido del frances al castellano. Madrid: Impr. de Collado. [Accent omitted]

(1835). As above. Paris: Lib. de Rosa.

(1790 &1791). Viaggio del giovine Anacarsi in Grecia nella meta' del quarto secolo avanti l'era volgare. Tradotto dal franchese... Napoli: L. Coltellini. 9v. 22cm. [Accent omitted]

(1791 &1793). As above. Venezia: Antonio Zatta e figli. 12v. maps. plans. and plates.

(1793). Voyage en Italie... (np). 111p.

(1801). Voyage en Italie de M.L'Abbe Barthelemy imprime sur ses Lettres originales ecrites au comte de Caylus, avec un Appendice ou se trouvent des morceaux inedits de Winckelmaim du P. Jacquir... et d'autres savans. pub. par A. Serieys... Paris: F. Buisson. xxxiv, 432p. 21cm. [Accent omitted]

(1802). As above. ...imprime sur ses lettres originales ecrites au comte de Caylus. Avec un appendice, ou se trouvent des morceaux inedits de Winckelmann, du P. Jacquier, de l'abbe Zarillo. et d'autres savans: pub. par A Serreys... 2ed. augm. d'une notice sur Madame de Choiseul. Paris: Chez F. Buisson, an X. xxxiv, 448p. 22cm. [Accent omitted]

BENNETT, JOHN (C.1770)

Bennet and Robert Sayer were partners in the publishing of numerous books. Those relating to American information are to be found under Sayer.

BIGLAND, JOHN (1750-1832)

Bigland was a school-master and author born in 1750 at Skirlaugh in Yorkshire, England and died at Finningley, England on February 22, 1832. Many of his textbooks are in archival collections in the United States, but his residence and work were in the United Kingdom (Stephen & Lee 1917).

(1811). A geographical and historical view of the world: Exhibiting a complete delineation of the natural and artificial features of each country; and a succinct narrative of the origin of the different nations, their political revolutions, and progress in arts sciences, literature, commerce, &c. The whole comprising all that is important in the geography of the globe, and the history of mankind. By John Bigland, Author of "Letters on ancient and modern history," "Essays on various subjects,", &c. &c. with notes, correcting and improving the part which relates to the American continent and islands. By Jedidiah Morse, Author of the American Universal Geography, &c. In five volumes. Boston: Printed by Thomas B. Wait and co. Sold by them and Philadelphia: Mathew Carey; and Richmond: Samuel Pleasants.

(1806 & 1814). Letters. On the study and use of ancient and modern history; containing observations and reflections on the causes and consequences of those events which have produced conspicuous changes in the aspect of the world, and the general state of human affairs. Philadelphia: W. E. Woodward. 586p. 21.5cm.

(1814). As above- 342p. 18.5cm.

(1806). As above- London: J. Cundee. 610p.

(1810). As above- 546p.

(1808). As above- 2d ed. London: (np). 615p.

(1828, 1831, 1832 & 1839). A natural history of animals. Philadelphia: J. Grigg. 189p. 19cm.

(1844). As above- Philadelphia: Grigg & Elliott.179p. plates.

(1845 & 1849). As above- 190p.

(1865). As above- Philadelphia: J. B. Lippincott & Co.

(1828, 1832, 1834 & 1839). A natural history of birds, fishes, reptiles, and insects. Philadelphia: J. Grigg. 179p. 11 colored plates. 19cm.

(1844 & 1845). As above- Philadelphia: Grigg & Elliot.

(1851, 1854 & 1865). As above- Philadelphia: Lippincott, Grambo, & Co.

(1819). Precis de l'histoire politique et militaire de l'Durope, depuis l'annee 1783 jusqu'a l'annee 1814...Traduit de l'anglais, augm., quant a la partie militaire, et continue jusqu'a l'annee 1819, par J. MacCarthy. Paris: P. Mongie l'Aine. 3v. 21cm. [Accent omitted]

Translation of: A sketch of the history of Europe from the peace of 1783..."

(1815). A sketch of the history of Europe from the 1783 to the general peace in 1814. 2d ed. London: Longman, Hurst, Rees, Orme, and Brown. 2v.

(1810). A sketch of the history of Europe, from the peace of 1783 to the present time, exhibiting a view of the commotions in Holland and Brabaut; the wars between Russia, Austria, the Ottoman Porte, and Sweden; the annihilation of the kingdom of Poland; the revolution of France, and the wars which have proceeded from that extrodinaty event, with the recent revolution in Spain, Portugal, and Sweden. London: Longman, Hurst, Rees and Orme. 2v. 21.5cm.

(c.1819). A topographical and historical description of the county of York; containing an account of towns, cathedrals, castles, antiquities, churches, monuments, public edifices, picturesque scenery, the residences of the nobility, gentry, &c. accompanied with biographical notices of eminent and learned men to whom this county has given birth. London: Sherwood, Neely, and Jones. 938p. 25 plates. 3 folded maps. 21.5cm .

(1815). Yorkshire, or original delineations, topographical, historical, and descriptive, of that county; the result of personal survey. London: J. Harris. 2v. illus.

BLANCHARD, JOSEPH, COL. (1704-1758)

Blanchard's map of New Hampshire was the first printed map of New Hampshire. It was revised by Samuel Langdon and published in Boston in 1784 (Schwartz and Ehrenberg 1980). Blanchard predates the establishment of the United States, therefore he was a British citizen and did not become an American citizen.

(1784). An accurate map of the state and province of New Hampshire in New England... by Col. Blanchard, and the Revd. Mr. Langdon. Engraved by Thomas Jefferys... Boston, April 21st, 1784. with many additions by Abel Sawyer Junr.

BOWLES, CARINGTON (1724-1793)

Bowles appears to have remained an British subject; before, during, and after the American Revolution. His publications were edited both in England and in the United States. None the less, several of his geographic and other publications are to be found in American archives. The following is representative of his publications:

(1771). A general map of the middle British colonies in America, viz Virginia, Maryland, Delaware, Pensylvania, New Jersey, New York, Connecticut & Rhode Island. Wherein is also shewn [sic] the antient and present seats of the Indian nations, carefully copied from the original published at Philadelphia by Lewis Evans. London: (np). 19x24$^{3/4}$inches. Scale: 36 miles to an inch.

(1776). Bowles's map of the Seat of War in New England. Comprehending the provinces of Massachusetts Bay and New Hampshire; with the colonies of Connecticut and Rhode Island. London: author, $24^{1/2}$x$19^{3/4}$inches. Scale: $19^{3/4}$miles to 1 inch.

Note. Submap, plan of Boston with its harbour and environs, $4^{5/8}$x$6^{1/2}$inches. Scale: 1.6 miles to 1 inch.

(1786). Carington Bowles's New and enlarged catalogue of useful and accurate maps, charts, and plans; curious and entertaining engraved and mezzotinto prints, single or in sets: writing books... London: (np). 280p. 22.5cm.

(1785). New medium English atlas; or complete set of maps of the counties of England and Wales: divided into their respective hundreds, wapontakes, wards, rapes, lathes, &c., exhibiting the roads, rivers, cities; borough, market, and sea-port towne; principal villages and places of note... and a variety of historical extracts relative to natural produce, trade, manufactures and commerce. Compiled and laid down from the latest and most approved authorities. London: author. 44 colored maps. 27cm.

BROOKES, RICHARD (C.1750)

Brookes was a physician and author who lived and worked as a rural practitioner in Surrey, England. He traveled in America and Africa prior to 1762. Although he was an industrious compiler producing a General Gazetteer, he was not an American (Stephen & Lee 1917).

(1806). Brookes's general gazetteer improved; or, a new and compendious geographical dictionary...illustrated by eight maps. Philadelphia: Jacob Johnson. "first American Edition." [1750]

(1762). The general gazetteer. London: (np).

BUTLER, SAMUEL, DR. (1774-1839)

Butler was the bishop of Lichfield and Coventry, born at Kenilworth, England on January 30, 1774 and died and was buried at St. Mary's Church in Shrewsbury on December 4, 1839. He wrote a number of geography books that can be found in American archives, but he was not an American (Stephen & Lee 1917).

(1831 & 1832). An atlas of ancient geography. Philadelphia: Carey & Lea, 34p. [26 folded colored maps], 24cm

(1834, 36, 38, 39 & 1841). As above- Carey, Lea, & Blanchard.

(1843, 44, 47, 49, 50, 51, 54, 55 &1857). As above- Philadelphia: Lea & Blanchard. 34p. 21 color plates.

(1871 & 1888). As above- Edited by the author's son (George Butler 1819-1890). London: Longmans, Green & Co. 39p. 24 maps. 23cm.

(1888). As above- London: Green.

(1842, 1844 & 1853). An atlas of modern geography. London: Longman Rees, Orme, Brown, Green & Longmans. 23 folded maps. 22cm.

(1822, 1825 & c.1834). As above- London: Longman. 32p. 20 folded & colored maps. 22x15cm.

(1829). As above- London: Longman, Rees, Orme, Brown & Green. 19 colored maps. 50p. 21.5cm.

(1871). As above- ...New ed., with additional maps and with corrections from the government surveys and the most recent sources of information. Edited by the author's son, T. Butler. London: Longmans, Green. 32 folded & colored maps. 60p. 24cm.

(1809). Butler's geographical and map exercises; designed for the use of young ladies and gentlemen. Corrected and improved by Stephen Addington. Philadelphia: Matthew Corey. 51p. 22.5cm.

(1820 & 1821). Geographica classica, or the application of ancient geography to the classics. For the use of students. [1821 ed. "From the 4th ed.] New York: F. Nichols. 137p. 19.5cm.

(1831 & 1835). As above- 2d & 3d American, from the 9th London ed., with questions on the maps, by John Frost. Philadelphia: Carey and Lea. 262p. 21.5cm.

(1840). As above- 4th Amer., from the last Lond. ed., with questions on the maps by John Frost, Philadelphia: (np). 262p.

(1843 & 1847). As above- 5th & 6th Amer. Philadelphia: Lea & Blanchard. 264p. 23.5cm.

(1813). A sketch of modern and ancient geography. London: (np).

(1812). As above- ...For use of schools. Shrewsbury: W. Eddowes. 246p. 23cm.

(1818). As above- Ed. 4 enl. London: Longman. 260p.

(1825). As above- 7th ed. London: (np).

(1828, 1830 & 1833). As above- London: Printed for Longman, Rees, Orme, Brown, and Green. 345p. 23cm.

(1838 & 1849). As above- London: Longman. 406p.

CARY, JOHN (C.1754-1835)

Carey was a London map seller. Although many of his geography related publications are to be found in American archives, he was not an American. Virtually all of his publications were in London. A sampling of those publications follows:

(1783). An accurate map of the United States of America with part of the surrounding provinces agreeable to the Treaty of Peace of 1783. London: author. map. 73.8x62.5cm.

(1800). Carys actual survey of the country fifteen miles round London... Wherein the roads, rivers, woods and commons; as well as every market town, village &c. are distinguished; and every seat shewn [sic] with the name of the possessor... London: author. 51 maps (1folded). 18.5cm.

(1821). Catalogue of maps, atlases, globes, geological and astronomical works published by John Cary. London: author. 39cm.

(1795). Inland navigation; or select plans of the several navigable canals, throughout Great Britain: accompanied with abstracts of the different acts of Parliament relative to them; likewise the width, depth, length, and number of locks on each: With the principal articles of carriage, &c. &c. London: author. 132p. 16 folded maps. 35cm.

(1816). The names and situations of the constellations: to which are added the names of the principal stars therein, as laid down on Cary's new celestial globes; also, an explanation of the equation table, given on his terrestrial globe. Stereotyped ed. London: Stereotyped and printed by A. Wilson. 21p. 17.5cm.

(1818). Cary's new and correct English atlas: Being a new set of county maps from actual surveys. Exhibiting all the direct & principal cross roads, cities, towns, and most considerable villages, parks, rivers, navigable canals &c. Preceded by a general map of south Britain... also a general description of each county and directions for the junction of the roads from one to another. London: author. 96p. 47 colored maps. 31x25.5cm.

(1826). Cary's new itinerary: Or, an accurate delineation of the great roads, both direct and cross, throughout England and Wales; with many of the principal roads in Scotland. From an actual measurement by John Cary... under the direction and inspection of Thomas Hasker... 10th ed. with improvements. London: G. & J. Cary. 602p. 7 maps 19.5cm.

Other

Cary, H. G. (1908). Ancestry of John Cary, 1755-1823, & his descendants, compiled by H. G. Cary. Boston: (np). 48p. coats of arms. maps. 22cm.

L'ENFANT, CHARLES (C.1800)

Ristow (1985) lists L'Enfant as the Chief Engineer to the plan for Washington, D.C. He along with Andrew Ellicott, the surveyor, developed a preliminary plan for the city. L'Enfant disagreed with government officials about the plan and withdrew from the project returning to France (p.246).

FISHER, JOSHUA (1707-1783)

According to Klemp (1976), Fisher produced the first accurate map of the water link between the Atlantic Ocean and Philadelphia. The date of his demise along with the lack of information about his life may exclude him from membership in the first generation of American geographers. Although we have attempted to be inclusive for those who contributed to knowledge of geography in America, this person may be erroneously placed in this section or in this reference.

(1777). Baye de la Delaware...d'apres la carte de Joshua Fisher publiee a Philadelphie. Paris: Le Rouge. map. 46.5 x 64.5cm. [Accents omitted]

(1778). Carte de la baye et riviere de Delaware... par Josue Fischer...1778. Carte de la entree de la riviere d'Hudson. (np). [Accents omitted]

(1756). Chart of Delaware Bay and River. (np).

(1776). A chart of Delaware Bay and River containing a full and exact description of the shore...from the original chart published at Philadelphia by Joshua Fisher. Engraved and published by William Faden, March 12, 1776. London: Drury. $17^{5/8}$x$26^{7/8}$inches. Scale: 4.4 miles to 1 inch.

(1776). As above- London: Sayer & Bennett.
(1776). As above- London: W. Faden.
(1783). As above- ...corrected from the best surveys. (np).
(1904). As above- Philadelphia: (np).
(c.1801). A map of the state of Delaware and eastern shore of Maryland with the soundings of the bay of Delaware from actual surveys and soundings made in 1799, 1800, & 1801 by the author. Philadelphia: (np). 15x20 3/4inches. [interesting, at least one Joshua Fisher appears to have died in 1783]
(nd). This chart of Delaware Bay and River...from the capes to Philadelphia and the set of the tide on the several quarters of the flood and ebb. (Manuscript in Historical Society of Pennsylvania.) 16 3/4x26 1/2inches.
(1756). To the merchants and insurers of the city of Philadelphia this chart of Delaware Bay and River...is dedicated... February 28, 1756. (Manuscript in Historical Society of Pennsylvania.) 25x46inches.
(1775). As above- Philadelphia: (np). 48x74cm.

GUTHRIE, WILLIAM ESQ. (1708-1770)

Gutherie was an Englishman and author who was educated at Aberdeen. He published the Gentleman's Magazine, c.1730; A General History of the World, 1764-1767; and Geographical, Historical and Commercial Grammar, 1770 (Stephen & Lee 1917, V.4, p.547). His inclusion in this reference is based on publications that fell within our period of study.

(1794). A new system of modern geography: or, A geographical, historical, and commercial grammar; and present state of the several nations of the world. Philadelphia: Printed by Matthew Carey. "The First American Edition"
(1809). A new geographical, historical and commercial grammar. (np).
(1794 & 1795). A new system of modern geography. (np).
(1820). A universal geography. (np).
(1809). A new geographical, historical, and commercial grammar; and present state of the several kingdoms of the world.
"The First American Edition Improved." [probably English] Philadelphia: Johnson & Warner V.I "Arrowsmith" world map - foldout - facing the title page p.556
W = 14 cm H = 22.5 cm

KEITH, THOMAS (1759-1824)

Keith was baptized at Brandesburton, United Kingdom, on September 22, 1759. He, in 1810, was made Professor of Geography to the Princess Charlotte of Wales. In 1814 he was appointed accountant to the British Museum. He died on June 29, 1824 in Marylebone, United Kingdom (DNB, p.1219). Although there are numerous works of Keith in American archives and those works may have influenced numerous Americans, he was never an American citizen.

(1833). The complete measurer; or, The whole art of measuring: being a plain and comprehensive treatise on practical geometry and mensuration. For the use of schools, and persons concerned in measuring, gauging, surveying, &c. 22d ed., Corrected and improved, by Samuel Maynard. London: Printed for Longman & co., 320p., tables, diagrams, 18cm..

(1864). Guy's elements of astronomy, and an abridgment of Keith's New treatise on the use of the globes. New American ed., with additions and improvements, and an explanation of the astronomical part of the American almanac. 30th ed. Philadelphia: C. Desilver.

(1801 & 1810). An introduction to the theory and practice of plane and spherical trigonometry, and the orthographic and stereographic projections of the sphere; including the theory of navigation... London: Printed by S. Straban for the author, 414p., 21cm.

(1816). As above - London: Printed for Longman, Hurst, Rees, Orme, and Brown, 436p., 21.5cm.

(1839, 45). As above - 442p., 22cm.

(1805). A new treatise on the use of globes; or a philosophical view of the earth and heavens...preceded by an extensive selection of astronomical and other definitions...designed for the instruction of youth..."London: Longman, Hurst, Rees, and Orme, etc., 355p., illus., 17cm.

(1808). As above - 360p., 5 fold. plates, tables, 19cm.

(1811). As above - 3d ed., 356p., 18cm..

(1811). As above - 1st American, from the last improved English ed. New York: Published by Samuel Whiting & Co., and Samuel Wood; S. Wood printer, 346p., 20.5cm.

(1815). As above - 2d American ed. New York: Whiting & Watson, 344p.

(1819). As above - New York: S. Wood & Sons and Baltimore: S. S. Wood & co. 352p., 23cm. (1826, 32). As above - 4th American ed. from the last London ed., rev. and corr. by Robert Adrian. New York: S. Wood, 354p.

(1814). As above - 4th ed. London: Printed for Longman, 359p., 17.5cm.

(1830, 1831). As above - London: Longman, Rees, Orme, Brown, and Green, 18cm..

(1853). As above - 423p., 19cm.

(1855). As above - 9 fold. plates, tables, diagrams, 18cm..

Revised and corrected, by Robert Adrain - Professor of Mathematics in Rutger's College, New Brunswick, N.J.

Other

(1810). The story of Mr. Keith's captivity among the American Indians... (In Mssrs. John. The Post-Captain. pp. 54-107. London: (np). 15.5 x 9cm.

PHILLIPS, SIR RICHARD (1767-1840)

Phillips was probably British, but it should be noted that his writing may have influenced one of the most prolific of early American writers, Samuel G. Goodrich (1796-1860).

(1819). Analyses of new works of voyages and travels, published during the last six months in Great Britain. London: Printed for Sir Richard Phillips & co., 100p., 23cm. In New voyages and travels.

(nd). Blair's outlines of ancient history. (np). See Goodrich, Samuel Griswold.

(1806). Boston and its environs. Map 8x12 5/8 inches. Scale, 4400 feet = 1 inch. London: (np).

(1826). The British empire in 1826; being a popular grammar of British geography in the four quarters of the world, with seven maps and one hundred views. 15th ed., London: Whittaker, 308p., illustrations, maps.

(1805). A collection of modern and contemporary voyages & travels: containing, I. Translations from foreign languages, of voyages and travels never before published. II. Original voyages and travels never before published. III. Analyses of new voyages and travels published in England... London: printed for R. Phillips, plates, folded maps, 22cm.

(1810). A concise account of the character, manners, and customs, of he inhabitants of Japan, Kamtschatka, Cochin China, Siam, Pegu, Pelew Islands, and New Holland, in which the remarkable curiosities of those nations are particularly noticed. By Rev. J. Goldsmith [pseud]. Philadelphia: Johnson and Warner, 54p., 18cm.

(1810). A description of the character, manners and customs of the inhabitants of China, including a particular account of their government, marriage ceremonies, treatment of their women, amusements, dress, funeral rites, temples, and other curiosities. Illustrated with plates. By the rev. J. Goldsmith [pseud.]. Philadelphia: Johnson & Warner, 51p., illustrated, 18cm.

(1804). An easy grammar of geography, for the use of schools: with maps. By the Rev. J. Goldsmith [pseud.]. Philadelphia: Printed for Benjamin Johnson, 108p., maps, 15cm..

(1807). As above - 144p.

(1810). As above - Philadelphia: Johnson & Warner, 144p., 14.5cm.

(1813). As above - 3d improved edition, carefully revised with considerable additions, and a map of the United States. Boston: Printed by Manning & Loring and sold by J. Norman, 119p., maps, 18cm.

(1818). As above - ...With ten engravings. By David Blair [pseud.], 4th ed., Philadelphia: S. W. Conrad, 160p., 2 folded plates, 15cm..

(1821, 1824). As above - 5th ed., 14.5cm.

(1829). As above - revised and enlarged by Benjamin Hallowell, Philadelphia: Kimber, 230p.

(1832). As above - 2d ed., rev. and enl., Philadelphia: Kimber & Sharpless, 249p., folded plates, 15cm..

(1836). As above - 3d ed., 250p.

(1806, 1811). An easy grammar of geography... companion and introduction to the "Geography on a popular plan for schools and young persons." 31st ed., London: R. Phillips, 144p., folded maps, 14.5cm..

(1813). As above - 45th ed., London: Longman, Hurst, Rees, Orme and Brown.
(1815). As above - 47th ed.
(1812). As above - Philadelphia: Johnson and Warner.
(1817). As above - By the Rev. J. Goldsmith [pseud.], a new edition, improved, by a Citizen of Philadelphia, 179p., 15cm.
(1807). Elements of geography, principally compiled with a view to teach children, at an early age, the geography of the United States. Philadelphia: Kimber & Conrad, 36p., 15cm.
(1832). Essay on the physico-astronomical causes of the geological changes on the earth's surface, and of he changes in terrestrial temperature, with notes. Republished with a preface by William Devonshire Saull... London: Sherwood, Gilbert, and Piper, 72p., 21.5cm.
(1818). Essays on the proximate mechanical causes of he general phenomena of the universe. London: Printed by Jas. Adlard & sons., and sold by J. Souter, 96p., illustrations, diagrams, 17cm.
(1822). A general view of the Earth. New Haven: (np), 2v.
(1810 & 1813). A general view of the manners, customs and curiosities of nations; including a geographical description of the earth. The whole illustrated by fifty-four maps,. and other engravings. In two volumes. By the Rev. J. Goldsmith [pseud.]. Philadelphia, Richmond, VA, and Lexington, KY: Johnson & Warner, 2v. plates, 18cm.
(1817). As above - Philadelphia: B. Warner.
(1818). As above - 252p., 18.5cm.
(1822). As above - New Haven: John Babcock and son, 17.5cm.
(1825). As above - New Haven: John Babcock and son; and Charlestown: S. Babcock and co.
(1823). A geographical and astronomical atlas, containing ancient and modern maps, with a solar system and two hemisphere, prefaced by new problems on maps, and a copious table of latitudes and longitudes, serving also as a companion to the geographical copy books. By the Rev. J. Goldsmith [pseud.]. London: G. and W. B. Whittaker, 22 plates (part double), 13 colored maps, 23.5cm.
(1835). As above - London: Sherwood, Gilbert, & Piper, 16p., 18 maps, 21cm.
(1826) A geographical view of the world, embracing the manners, customs, and pursuits, of every nation; founded on the best authorities. By Rev. J. Goldsmith [pseud.]. New York: E. Hopkins and W. Reed, 406p., plates, 19cm.
(1827). As above - 2d & 3d editions. Improved by James G. Percival... Boston: Printed for Reed and Jewett.
(1827). As above - 7th ed.
(1827). As above - 8th & 9th editions. Boston & New York: Printed for Reed and Jewett.
(1828). As above - 30th ed. Boston.
(1829). As above - 2d American ed., rev., corr., and improved. New York: D. M. Jewett.

(1831). As above - 6th American ed. Hartford: D. F. Robinson.
(1833). As above - 2d American ed. Illustrated with eight copperplate views. 19.5cm.
(1835). As above - 2d American ed. Hartford: Robinson & Pratt.
(1836). As above - 2d ed. New York: Robinson.
(1838). As above - 3d American ed. New York: Robinson, Pratt & co.
(1841). As above - 6th American ed.
(1843). As above - 7th American ed., New York: Robinson.
(1845). As above - 8th American ed., New York: Robinson, Pratt & co., 20cm.
(1849). As above - New York: Pratt, Woodford & co.
(1851). As above - 18 plates.
(1803). Geography for the use of schools, and young persons in general. By the Rev. J. Goldsmith [pseud.]. London: Printed for author by T. Gillet, 580p., chart, maps, plates, 17cm.
(c.1800). A grammar of general geography for the use of schools and young persons with maps & engravings. A new edition, corrected and modernized by Rev. J. Goldsmith [pseud.]. London: (np), 14.5cm. Note. First edition was 1803.
(1822). As above - London: Longman, Hurst, Rees, Orme, and Brown, 188p., plates, folding map, 14cm.
(1823). As above - London: Longman, Hurst, Rees, Orme, Brown, and Green.
(1852). As above - Revised, corrected, & greatly enlarged by Edward Hughes, 335p., 17cm.
(1841). As above - Improved and enlarged by the Rev. G. N. Wright... London: William Tegg & co., 246p., plates, 7 folded maps, 15.5cm.
(1814). A grammar of natural and experimental philosophy; a new edition greatly enlarged. London: Printed for Richard Phillips.
(1822). As above - ... with one hundred engravings on wood... From the 12th London ed., New Haven: J. Babcock and son; and Charleston, SC: S. Babcock and co. 216p., illustrations & diagrams, 14.5cm.
(1822). As above - Hartford: Huntington and Hopkins.
(1823). As above - ... including physics, dynamics, mechanics, hydrostatics, hydraulics, pneumatics, acoustics, astronomy, chemistry, electricity, magnetism, geology, minerology, meteorology, including the latest discoveries. With numerous engravings. By David Blair [pseud.]. 15th ed., considerably enlarged, with five hundred questions annexed. London: Printed for Sir Richard Phillips & co., by J. G. Barnard, 398p., plates, 18cm.
(1824). As above - 20th ed., from the 12th London ed., Hartford:O. D. Cooke & co.
(1826). As above - 21st ed., from the 12th London ed., 214p., illustrated, diagrams, 15cm..
(1837). The hundred wonders of the modern world and of three kingdoms and nature, described according to the best and latest authorities and illustrated

by engravings. By C. C. Clarke [pseud.]. London: Sherwood, Gilbert, 668p.

(1821). As above - 1st American, from the 10th London ed., New Haven: J. Babcock and son; and Charleston, SC: S. and W. R. Babcock, 660p., plates, 17.5cm.

(1822). As above - 15th ed., London: Sir. R. Phillips and co., 688p., 18cm.

(1824). As above - New Haven: (np).

(c.1798). The London, Edinburgh and Dublin Philosophical Magazine and Journal of Science. V.1-68, June 1798-Dec. 1826; new ser., V.1-11, Jan. 1827-June 1832; 3d ser., V.1-37, July 1832-Dec. 1850; 4th ser., V.1-50, Jan. 1851-Dec. 1875; 5th ser., V.1-50, Jan. 1876-Dec. 1900; 6th ser., V.1-50, Jan. 1901-Dec. 1925; 7th ser., V.1-50 Jan. 1926-Dec. London: Taylor & Francis.

(1835). A million of facts, of correct data, and elementary constants, in the entire circle of the sciences, and on all subjects of speculation and practice; adapted to the closet and the active world. A new ed., much enl., and carefully revised. London: Sherwood, Gilbert, and Piper.

(1839). As above - London: (np).

(1848 & 1851). As above - Stereotyped ed., much enlarged, and carefully revised and improved. London: Darton & co., 548p., 12o.

(1871). As above - London: Ward, Lock, & Tyler, 1215p., 20cm.

(1804). Modern London; being the history and present state of the British metropolis... London: R. Phillips, 31 col. plates, 27cm. Description of the plates, representing the itinerant traders of London in their ordinary costume; with notices of the remarkable places given in the background.

(1805). As above - 571p., 28cm.

Note. Appendix contains lists of the churches, public buildings, public halls and companies, public institutions, hospitals, courts, etc.

(1817). A morning's walk from London to Kew. London: Printed by J. Adlard, 393p., map, 20cm.

(1825). The natural and artificial wonders of the United Kingdom... By the Rev. J. Goldsmith [pseud.]. London: (np), 3v. illustrations, 17cm.

(1819 & 1823). New voyages and travels: consisting of originals, translations, and abridgments... With index and historical preface. London: Printed for Sir Richard Phillips and co., 65 nos. in 9v. illustrations, plates, maps, plans, 23.5cm.

(1828). A personal tour through the United Kingdom; describing living objects, and contemporaneous interests... London: Printed for H. Phillips.

(1798 & 1926). The Philosophical Magazine; A Journal of Theoretical, Experimental and Applied Physics. V.1-68, June 1798-Dec. 1826; new der., V.1-11, Jan. 1827-June 1832; 3d ser., V.1-37, July 1832-Dec. 1850; 4th ser., V.1-50, Jan. 1851-Dec. 1875; 5th ser., V.1-50, Jan. 1876-Dec. 1900; 6th ser., V.1-50, Jan. 1901-Dec. 1925; 7th ser., V.1- Jan. 1926- London: Taylor & Francis, ltd.

(1827). A popular dictionary of facts and knowledge, for the use of schools and students, with several hundred engravings in wood. By the Rev. S. Barrow [pseud.]. London: Printed for Poole and Edwards, 232p., illustrations, 18.5cm.

(1829 & 1830). As above - 2d ed., 228p.

(c.1830). A popular display of the wonders of nature, exclusively selected from the Transactions of the Royal Society of London, by the Rev. C. C. Clarke [pseud.]. London: W. Tweedie, 700p., illustrations, 19cm.

(1827). Populare astronomie, ohne hulfe der mathematik in zwanzig vorlesungen erlautert. Nach der 13. englischen und 3. frazosischen ausg. frei bearb. und mit vielen zusatzen, erlauterungen und verbasserungen versehen von M. L. Frankenheim... Braunschweig: F. Vieweg, 474p., 6 plates including maps, diagrams, 17cm. [Accents omitted]

(1823). Russian missions into the interior of Asia... Translated from the German. London: Printed for Sir R. Phillips and co., 112p., 23cm.

Contents: I. Nazaroff's expedition to Kokand. II. Eversmann and Jakovlew's account of Bucharia. III. Mcuraview's embassy to Turkomania and Chiva.

(c.1800). The schools atlas; or, Key to Goldsmith's geographical copy-books... London: (np), 15 maps (2 folded, 13 doubled), 21.5cm.

(1824). The sweeping of my study, by the compiler of "The hundred wonders of the world." Edinburgh: Oliver & Boyd, 332p.

(1820). The universal preceptor, being a general grammar of arts, sciences and useful knowledge, with... questions and exercises. By the rev. David Blair [pseud.]. London: Printed by J. McGowan for Sir Richard Phillips & co., 364p., illustrated, 12th ed.

(1822). As above - 4th American ed., with additions and improvements. Philadelphia: E. Parker, 316p., 14.5cm.

(1825). As above - 20th ed., London: Printed for Geo. B. Whittaker, 323p., 27 engravings.

(1826). As above - 13th English ed., Greenfield, Mass.: A. Phelps, 312p., 16cm.

(1827). As above - London: (np).

(1828). As above - 21st ed., London: Printed for R. Phillips, 323p., 15cm..

(1831). As above - 13th English ed., Greenfield, Mas.: A. Phelps, 294p., 1 folded map.

(1837). As above - 69th ed., London: Sherwood, Gilbert, and Piper, 355p.

(1810). A view of the character, manners, and customs of the North Americans, comprehending an account of the northern Indians; of the inhabitants of Oonalashka and Nootka Sound; of the five Indian nations of Canada; of the inhabitants of the United States, &c... By the Rev. J. Goldsmith [pseud.]. Philadelphia: Johnson and Warner, 39p., 17.5cm.

(1826). The wonders of the heavens displayed, in twenty lectures: by the author of the hundred wonders of the world. With numerous engravings... 9th ed., London: Printed for Sir R. Phillips and co. 305p., plates, folded maps, 18.5cm.

(1822). The wonders of the world. described according to the best and latest authorities and illustrated by engravings. In numbers. No. 1. By the Rev. C. C. Clarke [pseud.]... Abridged from the 10th London ed., for young people. New Haven, Con.: J. Babcock and son, 32p.

Other

(nd). Memoirs of the public and private life of Sir Richard Phillips. (np).
Feltham, John. (c.1800). The picture of London for 18.. (np).

PINKERTON, JOHN (1758-1826)

Pinkerton was a Scottish antiquary and historian, born in Edinburgh, Scotland on February 17, 1758. On March 10, 1826 he died in Paris, France. Although many of his books were published and/or sold in the United States, Pinkerton was not an American. His biographical sketch notes that his General Collection of Voyages and Travels, published in 17 volumes (1807-1813), "was a useful compilation in its day, being the most voluminous that had hitherto appeared with the exception of the French Historie Generale des Voyages' (Paris, 1785)," (Stephen & Lee 1917, pp. 1202-1203)

(1816). An atlas for the use of schools, selected from Pinkerton's largest work and designed to accompany and illustrate either of Dr. Morse's school geographies...Boston: Thomas and Andrews, unpaged. (nd). Australasia. A general collection of the best and most interesting voyages and travels in all parts of the world.(np).

(1813). A catalogue of a valuable and curious collection of books, in geography, voyages, travels, and belled letters, &c. of John Pinkerton. London: (np), 53p.

(1787). A dissertation on the origin and progress of the Scythians or Goths. Being an introduction to the ancient and modern history of Europe. London: G. Nicol, 207p., fold. map, 20.5cm.

(1795). Dissertations sur la rarete. les differentes grandours et la contrefaction des medailles antiques, avec des tables du degre de regrete des medailles des nciens peoples, villes, rois et des empereurs romains. Le tout tr. de l'anglois de Jean Pinkerton, ...Dresde: Leu flores Walther, 84p. 21 x 17.5cm. With this volume are bound his Notitia raritatis munismatum populorum veterum graecorum [Accents omitted]

(1886). ...Early Australian voyages. Pelsart. Tasman. Dampier. New York: Cassel & Co., 192p., 14.5cm.

(nd). As above - New York: The Merabon Co., 192p.

(1789). An enquiry into the history of Scotland, preceding the reign of Malcolm III, or the year 1056. London: G. Nicol, 2v., 6 maps, 20.5cm.

(1794). As above - London: (np)

(1814). As above - Edinburgh: Bell & Bradfute.

(1808 & 1819). A general collection of the best and most interesting voyages and travels in all parts of the world; many of which are now first translated into English. Digested on a new plan. London: Longman, Hurst, Rees, and Orme, 17v., 27.5 x 21cm.

(1810 & 1812). As above - Philadelphia: Kimber and Conrad, plates, maps, 28cm.

(1814). As above - London: Longman, Hurst, Rees.

(1805 & 1806). Geografia moderna; osia, Descrixione storica, politica, civile e naturalo di tutte lo parti dells terra formata sopra un nuove piano da Giovanni Pinkerton. Roma: Stamporia Pagliarini, 3v., 20cm. [Accents omitted]

(1790). The medallic history of England. London: Edwards and sons, 112p., 33.5cm.

(1802). As above - London: E. Harding and J. Scott.

(1815). A modern atlas, from the latest and best authorities, exhibiting the various divisions of the world, with its chief empires, kingdoms and states; in 60 (ei.e. 61) maps, carefully reduced from the largest and most authentic sources. London: Printed by T. Bensley, for T. Cadell, 60 maps, 57cm.

(1818). As above - Philadelphia: T. Dobson, 7p., 61 maps, 57 x 46cm.

(1802). Modern geography. A description of the empires, kingdoms, states, and colonies; with the oceans, seas, and isles; in all parts of the world: including the most recent discoveries and political alterations. Digested on a new plan. The astronomical introd. by the Rev. S. Vince. With numerous maps drawn under the direction and with the latest improvements of Arrowsmith and engr. by Lowry. To the whole are added a catologue of the best maps, and books of travels and voyages in all languages, and an ample index. In two volumes. London: T. Cadell, 2v., maps, tables.

(1803). As above - London: Printed by A. Straban, for T. Cadell and W. Davies, 641p., 22cm.

(1807). As above - London: Printed for T. Cadell and W. Davies, Strand; and Longman, Hurst, Rees, and Orme, V.1, 739p.; V.2, 820p.; V.3, 1006p., 21x28.5cm. [Examined in UT Special Collections]
V.3

(1811). As above - 834p.

(1817). As above - London: Printed for T. Cadell and W. Davies, Strand, and Longman, Hurst, Rees, Orme, and Brown, V.1 775p., V.2 788p., 28cm.

(1804 & 1864). As above - the astronomical introduction by the Rev. S. Vince...with numerous maps, drawn under the direction of Arrowsmith...to the whole and added a catalogue of the best maps, and books of travels and voyages, in all languages; and with an ample index...Philadelphia: John Conrad & Co., and atlas of 65 maps, 28cm.

(1804). Modern geography. A description of the empires, Kingdoms, states, and colonies with the oceans, seas, and isles; in all parts of the world...theastronomical introduction by the Rev. S. Vince...with numerous maps, drawn under the direction of Arrowsmith...to the whole are added, a catalog of the best maps, and books of hamlets and voyage, in all languages; and an ample index...Philadelphia: John Conrad & Co.

(nd). The new universal traveller. see under Heron, Robert, 1764-1807.

(1812). Northern hemisphere. London: Cadell & Davies.

(1800). Petralogy. A treatise on rocks. London: White Cochrane & co., 2v.

(1811 & 1820). As above - London: Printed for White, Cochrane & co. by S. Hamilton & co., 21.5cm.

(1805) Pinkerton's geography, epitomized for the use of schools. By David Doyle . Wherein the arrangement of the original is faithfully preserved. In the description of the United States, each state is more systematically described, than in any work of the kind heretofore published. Philadelphia: Printed for Samuel F. Bradford by T.S. Manning, 348p., 18.5cm.

(1810). Pinkerton's modern atlas. Philadelphia: Dobson, 39 double-page maps, engraved by Neele, 58 x 44cm.

(1817). Pinkerton's school atlas: ...London: Cadell & Davies, 29 maps, 22cm.

(1804). Recherches sur l'origine et les divers etablissemens des Scythes ou Goths, servant d'introduction a l'historire ancienne et moderne de l'Europe; accompagnees de plusieure eclaircissemens sur la geographie ancienne de cette partie du monde; ouvrage traduit sur l'original anglois de J. Pinkerton. Paris: De l'Imprimerie de la Republique, 404p., fold. map, 21cm. [Accents omitted]

(1805 & 1806). Recollections of Paris, in the years 1802-3-4-5. London: Longman, Hurst, Rees & Orme, 2v., 21.5cm.

RENNELL, JAMES (1742-1830)

Rennell was a major in the British army during the American Revolution. Although his publications are available in American archives, he was not an American.

(1781). A Bengal atlas: containing maps of the theatre of war and commerce on that side of Hindoostan. Compiled from the original surveys; and published by order of the honourable the court of directors for the affairs of the East India Company. London: East India Co., 21 double plates (part folded) including 32 maps, illustrations, and diagrams; 49cm.

(1801). A comparison of the geographical system of Herodotus, with those of other ancient authors, and with modern geography... London: (np).

(1800). The geographical system of Herodotus, examined; and explained, by a comparison with those of other ancient authors, and with modern geography. In the course of the work are introduced, dissertations on the itinerary stade of the Greeks, the expedition of Darius Hystaspes to Scythia, the position and remains of ancient Babylon, the alluvions of the Nile, and canals of Suez; the oasis and temple of Jupiter Ammon, the ancient circumnavigation of Africa, and other subjects of history and geography. The whole explained by eleven maps, adapted to the different subjects; and accompanied with a complete index. London: Printed by W. Bulmer and Co. for the author, 766p., 11 maps (10 folded), 28x22cm.

(1792). Memoir of a map of Hindoostan; or, The Mogul empire: with an introduction, illustrative of the geography and present division of that country: and a map of the countries situated between the heads of the Indian rivers, and the Caspian Sea.... To which is added, an appendix, containing an Account of the Ganges and Burrampooter rivers. The 2d ed. with very considerable additions, and many corrections: and a supplementary map, containing the geography of the countries, contiguous to the heads of the Indus. London: Printed by W. Bulmer & co. for the author, 428p., 5 maps (4 folded), 27cm.

Other

Markham, Clements R. (1895). Major James Rennell and the rise of modern English geography. New York: Macmillan, 232p., illustrated, 19cm.

ROSS, CHARLES, LIEUT.

Samuel Holland identifies a Lt. Ross as one of those engaged in surveying the "British colonies." Because of the date, we are not sure about the nationality of the man.

(1775). Course of the River Mississippi: from Balise to Fort Chartres: taken on an expedition to the Illinois, in the latter end of the year 1765/ by Lieut. Ross of the 34th Regiment. -- Improved from surveys of that river made by the French. -- London: Printed for Robert Sayer... Published as the Act directs. 1 sheet: col. map; 117 x 54 cm.

SAYER, ROBERT (1725-1794)

Sayer and John Bennett were partners. They a published numerous maps of North American British colonies. An extensive bibliography of those maps is to be found in The National Union catalog Pre-1956 Imprints, V.522, pp.683-686. They were both British. Only those maps relating to the United States are included in this list.

(1787). Catalogue of pilots, neptunes and charis... for the navigation of all the seas and coasts of the universe. London: author. 49p

(1757-1794). General atlas containing 41 colored maps of all countries of the world. London: author. 41 maps on 84.1 fol.

(1772). The seven United Provinces, with their roads and divisions. [Map.] London: author. 18 1/8 x 22 1/8 inches. Scale (computed), 9 1/2 miles to 1 inch.

(1775). An accurate map of North and South Carolina with their Indian frontiers, shewing in a distinct manner all the mountains, rivers, swamps, marshes, bays, creeks, harbours, sandbanks and soundings on the coast; with the roads and Indian paths; as well as the boundary or provincial lines, the several townships and other divisions of the land in both the provinces; the whole from actual surveys by Henry Mouzon and others. London: author. Size when joined 39 3/8 x 55 1/8 inches. Scale (computed), 7.5 miles to 1 inch.

(1778). An American atlas: or, a geographical description of the whole continent of America...and chiefly the British colonies, composed from numerous surveys...by Major Holland, Lewis Evans, William Scull...Engraved by Thomas Jeffereys,...and others. London:Printed and sold by R. Sayer and Bennett. (3)p. 30 maps. 55.5cm.

(nd). The American military pocket atlas.

(1775). A chart of the Banks of Newfoundland, drawn from a great number of hydrographical surveys. chiefly from those of Chabert, Cook and Fleurieu. London: author. 18 3/4 x 25 1/2 inches. Scale (computed), 10 geographic miles to 1 inch.

(1776). Chart of Hudson's River, from Sandy Hook to New York, with the banks, depths of water, sailing-marks, &c. London: author. 27 3/8 x 20 1/4 inches. Scale (computed), 1.1 miles to 1 inch.

(1775). A chart of the Gulf of St. Laurence, composed from a great number of actual surveys and other materials, regulated and connected by astronomical observations. London: autor. 21 3/4 x 18 1/2 inches. Scale (computed), 20 geographical miles to 1 inch.

(1774). A compleat map of the West Indies. Containing the coasts of Florida, Louisiana, New Spain, and Terra Firma... By Samuel Dunn... (np). 44.7 x 30.2 cm.

(1776). A general map of the Middle British colonies, in America. Containing Virginia, Maryland, the Delaware counties, Pennsylvania and New Jersey. With the addition of New York and of the greatest part of New England, as also of the bordering parts of the Province of Quebec, improved from several seurveys made after the late war, and corrected from Governor Pownall's late map 1776. London: author. 18 1/4 x 25 3/8 inches. Scale (computed), 32.5 miles to 1 inch.

(1776). A general map of the southern British colonies, in America. Comprehending North and South Carolina, Georgia, East and West Florida, with the neighbouring [sic] Indian countries. From the modern surveys of Engineer de Brahm, Capt. Collet, Mouzon & others; and from the large hydrographical survey of the coasts of East and West Florida, by B. Romans. 1776. London: author. 18 /8 x 24 1/2 inches. Scale (computed), 48.8 miles to 1 inch.

SMITH, CHARLES (1768-1808)

Although Smith published numerous geography related materials, he appears to have been "A London Map Seller." Therefore, only those books that appear to have been about and/or published in the United States are included. He, in fact, is identified in Smith's classical atlas containing district maps of the countries described in ancient history: both sacred and profane, is described as Charles Smith of London, map maker. Only those maps relating to America are included in this volume. Other maps can be found in The National Union Catalog Pre-1956 Imprints (V.550, pp;.468-470.

(1797). The American gazatteer; or, Geographical companion. New York: Printed by A. Menut for C. Smith, 53p., map, 13cm.

In the author's The gentleman's political pocket almanac for the year 1797.

(1797). The American war, from 1775 to 1783, with plans. New York: Printed for C. Smith, 183p., 2 folded maps, 6 folded plans, 21.5cm.

(1897). The engagement on the White Plains the 28th of October 1776, between the American & British forces. Map 7 1/8x8 1/4 in. In his The American war. Between pp. 120-121.

(1808). A general atlas; being a collection of maps of the principal kingdoms of the world, from the best authorities, on forty-two sheets...2d ed. improved. London: Selected and sold by C. Smith, mapseller. 1p. 1., 33 maps.

(1801). The gentleman's political and commercial pocket almanac for 1801... New York: (np), 123p., 18 mo.

(1794). The gentleman's political pocket almanack, for 1795. New York: (np), 24o.

(1795). The gentleman's political pocket almanac for the year 1796, being leap year. New York: Printed by Wayland & Davis, 142p., 14cm.

(1797). The gentleman's political pocket almanac for the year 1797. New York: Printed by J. Buel for C. Smith, 59p., 13cm.

In the author's The American Gazatteer.

(1813). A map exhibiting the frontiers of Canada and the United States; intended to illustrate the operations of the British and American armies. London: author. 19 1/8 x 29 1/4 inches. Scale (computed), 25 miles to 1 inch.

(1797). Map of the city of Quebec. 7 3/8x7 3/4 inches. Scale 1515 to 1 inch. In his The American war. Between pp. 64-65.

(1815). A new map shewing the seat of war between the allied powers & France, comprehending the whole of Germany, Poland, the United Provinces, France and Italy; with part of Spain, from the best authorities. 1814. with additions to 1815. London: author. col.map. 74x76cm. Scale 32 miles to 1 inch.

(1797). Plan of the investment and attack of York in Virginia. Map 7 5/8x8 5/8 inches. Scale 800 yards to 1 inch. In his The American war. Between pp. 168-169.

(1808). Smith's New general atlas containing distinct maps of all the principal empires, kingdoms, & states throughout the world, carefully delineated from the best authorities extant. London: Printed for C. Smith. 2p.1., 45 pl. incl. 43 col. maps. 41x32cm. Maps engraved by E. Jones, B. Smith, and J. Bye.

(1813). As above - 32 fold. col. maps. 23cm.

(1816). As above - ...arranged according the the general treaty signed in Congress at Vienna, June 1815. 48 col. maps. 40x32cm.

(1820). As above - 53 pl. 48 col maps.
(1826). As above -
(1830). As above - ...with considerable additions.
(1836). As above -
(1814). Smith's new map of America with the recent explorations. London: C. Smith, col. map, 72 x 74.5cm. (nd). Smith's new map of England and Wales. See under the title.
(nd). Smith's railway map. See under the title. (np).
(1795). Universal geography made easy: Or, a new geographical pocket comparison: Comprehending, a description of the habitable world... New York: Printed by Wayland & Davis, Water Street for the author and L. Wayland, 192p., 13.5cm.
(1800). As above -...with maps - 2nd edition. New York: C. Smith. (1801). The writing desk... New York: Printed for Charles Smith, 72p.

TAYLOR, GEORGE (C.1778)

The following works of Taylor appear in American collections. It appears that he was not an American and although his works may have been used as references by Americans, they were not published in America.
(1776). Survey and maps of the roads of North Britain, or Scotland. To His Grace, John, Duke of Argyll... this work is most humbly inscribed by George Taylor and Andrew Skinner. London: authors, 32 folded maps, 24cm.
(1778). Taylor and Skinner's maps of the roads of Ireland, surveyed 1777. London: authors, 288 pages of maps, 16p., 26cm.
(1783). As above - ... and corrected down to 1783. 289 pages of maps, 27cm.
(1776). Taylor & Skinner's survey and maps of the roads of North Britain or Scotland... This work is most humbly inscribed... London: authors, 61 plates, 23x55cm.
(1776). As above - 62 folded maps, 23cm.
(1776). As above - 66p., 63 maps, 22x16cm.
(1776). Taylor and Skinner's survey of the great post roads, between London, Bath & Bristol. London: authors, 24 maps, 14cm.
(1790). Taylor & Skinner's survey of the roads of Scotland on an improved plan, to which is prefixed an accurate map of Scotland... Edinburgh: T. Brown, 1v., maps, 19cm.
(1800). As above - 178 pages of maps (2 folded), 18.5cm.

TYTLER, ALEXANDER FRASER - LORD WOODHOUSELEE (1747-1813)

Although Tytler's books are to be found in numerous American collections, he was not a citizen of the United States. His works, none the less, must have influenced some early American citizens. Therefore, they are noted on the following pages. Tytler was born at Edinburgh on October 15,

1747 and died there on January 5, 1813. He was most distinguished as Professor of History at the University of Edinburgh. Although prolific, his publications lacked distinction. One biographer noted, "There is no kindness in insinuating that he was a man of genius, and of public or even social influence, or in describing Woodhouselee as Tusculum" (Stephen & Lee 1917, p. 1378-1379).

(1790). An account of some extraordinary structures on the tops of hills in the Highlands: with remarks on the progress of the arts among the ancient inhabitants of Scotland. Edinburgh: Royal society of Edinburgh, Transactions, v.2, 3-32, 2 plates. 22x28cm.

(1801). Elements of general history, ancient and modern. To which are added a table of chronology and a comparative view of ancient and modern geography. Illustrated by maps. Edinburgh: W. Davies, and London: Murray & Cochrane, printers, 2v. 6 maps, 22cm.

(1809). As above - Philadelphia: Printed for F. Nichols, 387, 62p. 23cm.

(1812). As above - Edinburgh: Printed for W. Creech, and T. Cadell, and W. Davies, and A. Smellie, 22cm.

(1813). As above - 2d American ed., from the 5th British. Philadelphia: Printed for Francis Nichols, 394, 66p. 2 fold. maps, 22.5cm.

(1813). As above - New York: F. Nichols, 183, 203, 50p. 23cm.

(1813, 24, 25, 36). As above - 6th ed. London: Printed for T. Cadell and W. Davies; also Edinburgh W. Creech, 3 v. 22cm.

(1818). As above - 574p.

(1819). As above - From the 7th British ed., The history continued from the close of the seventeenth century to the general peace of Europe in 1815. New York: E. Duyckinck, 448p. 19cm.

(1818). As above - Hartford: Published by Samuel G. Goodrich. Simon Lincoln, printer, 384, 78p. 19cm.

(1820). As above - 484p.

(1821). As above - : Huntington & Hopkins.

(1824, 28, 30, 31). As above - ...With a continuation, terminating at the demise of King George III, 1820. To which are added a succinct history of the United States; with additions and alterations by an American gentleman...with an improved table of chronology; a comparative view of ancient and modern geography, and questions on each section. Adapted for the use of schools and academies, by an experienced teacher...Concord, NH: Horatio Hill, 525p. 19.5cm.

(1823 & 1825). As above - : Printed by Isaac Hill.

(1827). As above - Concord, NH: Manahan, Hoag & co.

(1829). As above - ...9th ed. Corrected, with considerable additions, by Edward Nares 1825. Translated into Hindoostanee under the patronage of the Bombay native education society, by Lewis Dacosta, assisted by H.M.A. Mujeed and other learned natives. Calcutta: author, 3v.

(1837, 39, 40, 44, 49). As above - Concord, NH: J.F. Brown, also Boston: Hilliard, gray & co. 527p. (1839). London: Scott, Webster, and Geary, 534p. 2 fold. maps, 15cm..

(1846). As above - Edited by the Rev. Brandon Turner...London: A. Scott, 664p.

(1848 & 1850). As above - London: Bohn, 512p.

(1851). As above - 195th ed. New York: Clard, Austin & Smith, 527p. 20cm.

(1855). As above - London: Longman, Brown, Green, and Longmans: Simpkin, Marshall, and Co.; Whittaker and Co.; and H. G. Bohn, 703p. maps. 23cm.

(1858). As above - Glasgow: Blackie, 668p.

(1854). Elements of modern history by Alexander Fraser Tytler, Lord Woodhouselee...Continued to 1850. With a chronological table... A new edition, revised, with considerable additions, copious notes, and a continuation to the present time, by Thomas Edlyne Tomlins. London: Whittaker and Co. 202p.

(1844). As above - 3d ed. Edinburgh: A. & C. Black, 235p. map.

(1782). Plan and outlines of course of lectures on universal history, ancient and modern, delivered in the University of Edinburgh, by Alexander Tytler. Illustrated with maps of ancient and modern geography, and a chronological table. Edinburgh: W. Creech, 250p. maps, tables, 22cm.

(1823). Tytler's elements of general history, ancient and modern. From the seventh British edition. The whole work revised and continued to the general peace in Europe in 1815... By Thomas Robbins...To which is now added. A view of the state of arts, sciences, religion, laws, government, &c. by the Rev. Edward Nares...With questions adapted to the use of schools and academies. Hartford: Huntington & Hopkins, 518p. 12mo.

(1875). As above - London: Bell, 616p.

(1839, 41, 42, 43-45, 54, 54-57, 57-59, 60-68). Universal history, from the creation of the world to the decease of George III, 1820. By the Hon. Alexander Fraser Tytler, and Rev. Edward Nares, D.D. Ed. by an American... In six vols. New York: Harper & brothers.

(1839). As above - London: John Murray

(1840). As above - New York: ? [In Family lib. nos. 86-91]

(1834). Universal history, from the creation of the world to the beginning of the eighteenth century. New York: W.P. Petridge & Co., 2v. 16cm.

(1835). As above - 2. ed. London: J. Murray.

(1835, 36, 37, 38, 39, 40, 41). As above - [Edited by William Fraser Tytler]. Boston: Hillard Gray & co. 23.5cm.

(1843). As above - Boston: B.B. Mussey;

(1846, 47). As above - Boston: Jordan and Wiley.

(1848). As above - Boston: T. Wiley, 24cm.

(1850). As above - Boston: W. Crosby and H.P. Nichols.

(1852, 1853). As above - Boston: Fetridge and co.

(1844). As above - Boston: (np)

(1854). As above - Boston: (np)
(1860, 64, 66). As above - Philadelphia: J.B. Lippincott;
(1872, 1875). As above - Philadelphia: Claxton, Remsen & Haffelfinger.
(1823). Tytler's elements of general history, ancient and modern. From the seventh British edition. The whole work revised and continued to the general peace in Europe in 1815...By Thomas Robbins. to which is now added, A view of the state of arts, sciences, religion, laws, government, &c. by Rev. Edward Nares. With questions adapted to the use of schools and academies. Hartford: Huntington & Hopkins.

VOLNEY, CONSTANTINE FRANCOIS CHASSEBOEUF (1757-1820)

Volney was born in Craon, France and died in Paris. He presented 18th century rationalist historical and political thought in his publication, Les Ruines, ou Meditations Sur les Revolutions des Empires. In the publication he reflected a view of revolution as the result of abandoning the principles of natural law, religion, equality, and liberty. Although he was imprisoned during the Reign of Terror, he became a Professor of History at the Ecole Normale in Paris (Encyclopedia Britannica 2001. Volney, Constantin-Francois)

He visited the United States from 1795 until 1798. Perhaps his visit and the availability of his publications led to wide distribution of his books in the U.S. The existence of so many publications in United States collections attest to the presence of those texts in numerous United States homes.

(1796). An abridgement of the law of nature; or, Catechism of French citizens. Translated from the French of C. F. Volney, author of The Ruins of Empires, &c. London: author. 16p. 23cm.
(nd). Abridged translation of the author's Les Ruines, ou, Meditation sur les revolutions des empires. [Hebrew title.] London: (np). 81p. plate. [Accents omitted]
(1818 & 1819). L'alfabet europeen applique aux langues asiatiques... Paris: Didot. 227p. 21cm. [Accents omitted]
(1826). As above- Paris: Parmantier, etc. xix,510p. map. tables. 21cm.
(nd). Analyse raisonnee d'un nouvel ouvrage intitule: Tableau du climat et du sol des Estats Unis d'Amerique; suivi d'eclaircissemens sur la Floride, sur la colonie francaise du Scioto, sur quelque colonies canadiennes & sur les sauvages. (np). [Accents omitted]
(1788). Bemerkungen uber den jetzigen Turkendrieg, aus den Franzosischen. Amsterdam: author. 137p. map. 22cm. [Accents omitted]
(1788). Considerations sur la guerre actuelle des Turcs. Londres: author. 140p. map. 20cm. [Accents omitted]
(1799, 1800 & 1822). Lecons d'histoire, prononcees a l'Ecole normale, en l'an III de la Republique francaise. Ouvrage elementaire, contenant des vues neuves sur la nature de l'histoire; est susceptible; sur l'abus de son emploi dans l'education de la jeunesse; et sur le danger de ses comparaisons et de ses imitations generalement vicieuses en matiere de gouvernement;

accompagne de notes, et de trois plan relatifs a l'art de contruire les salles d'assemblees publiques et deliberantes. Paris: J. A. Brosson. wxiv,254p. maps. 21cm. [Accents omitted]

(1825). As above- Paris: Parmantier, etc.

(1800). Lectures on history, delivered in the Normal School of Paris. Translated from the French. London: J. Ridgway. xviii,186p. 12°.

(1801). As above - Philadelphia: Printed for John Conrad & co. J. Bioren, printer. xvi,186p. 16.5cm.

(1819). New researches on ancient history... London: w. Lewis. 2v. map. tables. 21.5cm.

(1821). As above- 2d ed.

(1821). As above- : embracing an examination of the history of the Jews until the captivity of Babylon: and showing the origin of the Mosaic legends concerning the creation, the fall of man, flood, and confusion of languages. New York: C. Blanchard. 220p. 20cm.

(1874). As above- Boston: J. P. Mendum. ix,10-263p. 19.5cm.

(1802). A new translation of Volney's Ruins; or Meditations on the revolution of empires. Made under the inspection of the author... Paris: Printed for Levrault. 2v in 1. plates. 17cm.

(1811). As above- Dublin: Printed for Hood & Cuthell. viii,280p. map. 18cm.

(1817). As above- Paris: Printed for Levrault, quai Malaquais.

(1822). Oeuvres completes... Paris: Bossange freres. 8v. plates. maps. V.3, half title, Voyage en Egypte et en Syrie. V.4-5, Recherches nouvelles sur l'historie ancienne. V.6, Lecons d'historie, prononcees a l'ecole normale... V.7, Tableau du climat et du sol des Etats-Unis d'Amerique. V.8, L'alfabet europeen applique aux langues asiatiques.

(1833). As above- Paris: Lebigre freres. 585p. map. 15.5cm.

(1836). As above- 2d ed.

(1823). Oeuvres de C.-F. Volney, comte et pair de France... Nises en ordre et precedees de la vie de l'auteur... Bruxelles: A. Wahlen et cie. 4v in 2. maps.

(1826). As above- Paris: Parmentier, etc.

(1837). As above- Paris: F. Didot. 778p.

(1838). As above- Paris: Frimin Didot.

(1839). As above- Paris: A. Desrez.

(1843, 1846 & 1854). As above- Paris: F. Didot freres.

(1857, 1864 & 1876). As above- Paris: Firmin Didot.

(1814 & 1815). Recuvelles sur l'histoire ancienne... Paris: Madam veuve Courcier. 3v. map. tables. 20cm. [Accents omitted]

(1822). As above- Paris: Bossange freres. 2v. maps. 21.5cm.

(1825). As above- Paris: Parmantier.

(1788 & 1800). Reise nach Syrien und Aegypten in den jahren 1783, 1784, 1785. Aus dem franzocischen ubers... Jena: Bey J. M. Mauke. 3v in 2. maps. 21cm. [Accents omitted]

(1803). Tableau du climat et du sol des Estats-Unis D'Amerique. Suivi d'eclaircissemens sur la Floride, sur la colonie Francaise au Scioto, sur quelques colonies Canadiennes et sur les sauvages... Tome Premier[-second]. Paris: Chez Courcier, Pmprimeur-Libraire; dentu. 2 vols. paginated continuously: [4], xvi, 300 fold. maps; fold. section. [Accents omitted]

(1804). View of the climate and soil of the United States of America: To which are annexed some accounts of Florida, The French colony on the Scioto, Certain Canadian colonies, And the savages or natives: Translated from the French... with maps and plates. London: Printed for J. Johnson; By C. Mercier and Co. xxviii, [4], 503, [1]p. 2 fold. maps.

(1804). As above- ... Translated, with occasional remarks, by C.B. Brown. Philadelphia: Published by J. Conrad & Co.; Baltimore: M. & J. Conrad & Co.; Washington: Rapin, Conrad & Co. Petersburg: Somervell & Conrad; Norfolk: Bonsal, Conrad, & Co. Printed by T. & B. Palmer. xxviii, 446p. 2 fold. maps.

(1804). C.F. Vollneys reisen durch die Vereinigten Staaten von Nordamerika. Mit einer charte und kupfer. Hamburg und Mainz: G. Vollmer. 2v. maps. 16.5cm. [Accents omitted]

(1789 & 1791). Reize door Syrie en Egypte, in de jaareen 1783, 1784, en 1785 gedaan door den heer C.F. Volney. Uit het Frensch vertaald door J. D. Pasteur... Leyden: A. en J. Honkoop. 2v. maps. 22cm. [Accents omitted]

(1868). Las ruinas de Plamira; o, Meditacion sobre las revoluciones de los imperios, por M. Volney. Barcelona: Jose Codina. 450p. plates. 24cm.

(1889). As above- Paris: Garnier hermanos. 422p. 12°.

(1821). As above- Madrid: Rosa. 358p. map. plate. 16°.

(1822). As above- Bordeos: Impr. de P. Beaume. xivii,332p. 20.5cm.

(1854). As above- Madrid: Impr. del editor.

(1869). As above- Madrid: P. Gracia y Orga.

(1822). As above- Lisbon: Typ. de D. Marques Leao. xivi,225p. 15cm.

(1800). Ruinerna; eller, Reflectioner ofver rikshvalfningarne. Ufversattning. Fran 2. upl. tryckt i Paris 1792 i januarii. Uppsala: (np). [Accents omitted]

(1926). Les ruins. Paris: (np). [Accents omitted]

(nd). Les ruines on meditations sur les revolutions des empires suivies de la loi naturelle et de l'histoire de Samuel. Nouv. ed. rev. avec soin d'apres les meilleurs testes et precedee de nouveaux apercus sur la vie et les ouvarages de l'auteur extraits des Gauseries du lundi de M. Sainte-Beuve. Paris: Garnier. xi,438p. map. [Accents omitted]

(1791 & 1792). As above- Paris: Desenne, etc.

(1821 & 1822). As above- Paris: Bossange freres.

(1823). As above- Bruxelles: A. Wahlen.

(1826 & 1827). As above- Paris: Parmantier Libraire.

(1830). As above- Bruxelles: Librairie Philosophique.

(1871). As above- Paris: Garnier.

(1876). As above- Paris: Librairie de la Bibliotheque nationale.
(1795). The ruins: or, A survey of the revolution of empires. Translated from the French. 2d ed. London: Printed for J. Johnson.
(1796). As above- New York: Printed by William A. Davis, for E. Duyckink & co., T. & J. Sword, N. Judah, etc.
(1799). As above- Philadelphia: Printed by James Lyon.
(c.1800). As above- London: Printed for Miller, Law and Cater.
(c.1800). As above- New York: C. Blanchard.
(1801). As above- Dublin: J. Stockdale.
(1801). As above- London: Printed by Bunney and Gold, for J. Johnson.
(1804). As above- 12th ed. London: Printed for the booksellers.
(1804 & 1924). an additional 40 publications, with much the same title, are noted in The National Union Catlog Pre-1956 Imprints (The American Library Association 1979, V.641, pp.620-632). Other English titles follow:
(1801). Travels in Syria and Egypt, during the years 1783, 1784, & 1785. Rev. and cor. by the author, and enl. with an abstract of two Arabic Ms. containing much new and interesting information respecting the history, population, revenues, taxes, and arts of Egypt and two engravings not in any former edition, representing the pyramids and the sphinx. Translated from the French... Perth: Printed by r. Morison for will Morison.
(1798). As above- New York: J. Tiebout.
(1787 & 1788). As above- London: G. G. J. and J. Robinson.
(1793). As above- Dublin: White.
(1804). A view of the soil and climate of the United States of America: with supplementary remarks upon Florida; on the French colonies on the Mississippi and Ohio, and in Canada; and on the aboriginal tribes of America. with occasional remarks by C. B. Brown. Philadelphia: J. Conrad & co.; Baltimore: M. & J. Conrad & co.; Philadelphia: Printed by T. & G. Palmer. xxviii,446p. maps. 21.5cm.
(1804). As above- London: J. Johnson.
(1845). Volney's ruinen, oder: Betrachtangen uber die uwalzungen der reiche. Aus dem franzosischen ubersetzt. Philadelphia: C. L. Rademacher. 220p. [Accents omitted]
(1851 & 1854). As above- Philadelphia: F. W. Thomas.
(1872 & 1880). As above- Braunschweig: F. Vieweg und sohn.
(c.1880). As above- Leipzig: P. Reclam jun.

Other

Penderton, John. (1804). Modern geography. A description of the empires, kingdoms, states, and colonies; with the oceans, seas, and isles; in all parts of the world: including the most recent discoveries, and political alterations. Digested on a new plan. By John Pinderton. The astronomical introduction by the rev. S. Vince... The article America, corrected and considerably enlarged by Dr. Barton, of Philadelphia. with numerous maps, drawn under the direction and with the latest improvements, of Arrowsmith, and enlarged by the first American artists. To the whole are added, a catalogue of the best maps, and books of travels and voyages, in all languages: and an ample index... Philadelphia: J. Conrad, &c.

Benton, Joel (Intro.). (1901). French belles-lettres from 1640-1870... humor, sentiment, romance; with special introduction by Joel Benton, and biographical notes. Washington & London: M. W. Dunne.

Holbach, Paul Henrie Thiry. (nd). Common sense or, Natural ideas opposed to supernatural... (np). [Volney supposed author]

Mackenzie, Alexander. (1802). Voyages from Montreal, on the river St. Laurence, through the continent of North America, to the frozen and Pacific Oceans; in the years 1789, and 1793. With a preliminary account of the rise, progress, and present state of the fur trade of that country. With original notes by Bougainville, and Volney... Illustrated with maps. London: Printed by T. Cadell and W. Davies.; Edinburgh: W. Creech.

Marcellus, Marie Louis Jean Andre Charles Demartin du Tyrac, comte de. (1841). Rimembranze intorno all' Oriente del visconte di Marcellus... Prima versione italiana... Prate: Tip. Giachetti. [Accents omitted]

Roberts, Peter. (nd). Christianity vindicated, in a series of letters, addressed to Mr. Volney, in answer to his book called Ruins. (np).

Vince, S., & Wood, J. (1799-1810). Mathematics. Cambridge: (np). 4v.

WALKER, JOHN (1759-1830)

Numerous volumes of Walker's publications can be found in American archives. He also published a few of them in the United States. Walker appears to have been an English citizen.

(1802). Atlas to Walker's geography. London: Printed for Vernor and Hood by Darton and Harvey, 24 folded colored maps, 22cm.

(nd). As above - London: Symonds.

(1797). An atlas to Walker's geography and gazatteer... Dublin: Printed by T. M. Bates, 31 folded plates, including maps, 21.5cm.

(1795). Elements of geography, and of natural and civil history... 2d ed., London: Printed and sold by Darton and Harvey, 616p., 21.5cm.

(1796). As above - London: (np), 620p.

(1797). As above - 3d ed., Dublin: Printed and sold by T. M. Bates, 624p., illustrations. (1800). As above - London: Darton & Harvey, 620p.

(1805). As above - 4th ed. revised and enlarged by Th. Smith. London: (np), 671p.

(1831). As above - 3rd ed., Boston: Richardson.

(1788). Elements of geography, with principles of natural philosophy, and sketches of general history... Illustrated with 10 copperplates. Dublin: Printed for the author and sold by R. Jackson and J. Crukshank, Philadelphia, 167p., folded & colored maps, 21cm.

(1802). Fragments of letters & other papers, written in different parts of Europe, at sea, and on the Asiatic and African coasts or shores of the Mediterranean, at the close of the eighteenth, and beginning of the nineteenth century. London: Printed for the author by Darton and Harvey, 440p., 22.5cm.

(1830). Map of the inland navigation, canals, and rail roads with the situations of the various mineral productions throughout Great Britain, from actual surveys projected on the basis of the trigonometrical survey made by order of the Honorable the Board of ordinance accompanied by a book of reference, compiled by Joseph Priestley. Wakefield: R. Nichols, 1 folded map, 192x156cm.

(nd). Map shewing [sic] the discoveries made by British officers in the Arctic regions from the years 1818 to 1826. (np), see under Murray, John, publisher, London.

(1833). Maps illustrative of the European connection with India. and of the British administration in its several departments. London: Cox, 5 doubled maps, 35.5cm.

(1808). Universal atlas, for the use of schools. London: Printed by W. Wilson for the author, 25 double colored maps, 22cm.

(1811). As above - 26 colored maps.

(1795). The universal gazatteer; being a concise description...of the nations, kingdoms, states, towns...&c in the known world; the government, manners, and religion of the inhabitants... of the different countries... Illustrated with a complete set of maps. London: Darton and Harvey, 736p., folded maps, 22.5cm.

(1798). As above - London: Ogilvy and son, 805p., 14 folded maps, 22cm.

(1801). As above - Revised, considerably enlarged, and improved by Arthur Kershaw. 3d ed., London: Vernor and Hood, 929p.

(1807). As above - 4th ed., London: J. Swan and Son, and E. Blackader.

(1815). As above - 6th ed., London: F. C. and J. Rivington, 802p.

(1822). As above - 7th ed.. 929p?

(1810). Walker's new geographical game, exhibiting a tour through Europe. London: W. Darton, Jr., 24p., 17.5cm.

(1813). Walker's universal atlas for the use of schools. London: Printed by J. G. Barnard, 27 folded and colored maps, 23cm.

(1814, 16, 22). As above - London: Printed by J. G. Barnard for F. C. and J. Rivington, etc. etc. Note. "The whole of the maps drawn & engraved for Walker's new atlas by M. Thomson."

(1809). Walker's tour through England and Wales, a new pastime. Published for the author, January 2nd, 1809, by W. & T. Darton. 52.5x63.6cm.

(1812). Walker's tour through Ireland. A new geographical pastime. London: W. Darton, colored map, 47x38cm.

Other

Epps, John. (1832). The life of John Walker. London: Printed by J. Haddon & co. for Whittaker, Treacher, and co., 342p.

WITTMAN, WILLIAM (C.1803)

Wittman is a person about which we know very little. His publications may indicate that he was present and publishing in the United States, but he may also have been non-American.

(1804-1805). ... Reisen in der europaischen Turkey, Kleinasien, Syrien und Aegypten in den jahren 1799, 1800, 1801, und 1802. Nebst bemerkungen uber die pest und andern in der Turkey herrschenden krankheiten, wie auch einem meteorologischen tagebuche. Aus dem englischen mit ammerkungen, und mit Desgenettes... und anderer untersuchungen uber die pest ubersetzt von J. A. Bergk... Leipzig: W. Rein und comp., 2v, plates, tables, 20.5cm. [Accents omitted]

(1803). Travels in Turkey, Asia-Minor, Syria, and Across the desert of Egypt during the years of 1799, 1800, and 1801, in company with the Turkish army, and with the british military mission. To which are annexed, observations on the plague, and on the diseases prevalent in Turkey, and a meteorological journal. London: R. Phillips. 595p. colored plates. folded map. 27cm.

(1804). As above- Philadelphia: Printed and sold by James Humphries. 426p. 22.5cm.

[Found in the South Caroliniana Library at the University of South Carolina, Columbia.]

WRIGHT, THOMAS (1711-1786)

Numerous publications about architecture and astronomy were found, which leads us to think this person was British. Other items can be found listed in The National Union Catalog Pre-1956 Imprints (V.675, pp;.461-462.

(1837). The universe and the stars, being an original theory on the visible creation, founded on the laws of nature, by thomas Wright. Ist American ed. from the London ed. of 1750. with notes by C. S. Rafinesque... Philadelphia: Printed for C. Wetherill. 1837. 2p. 1., 158, [2] p. 22.5cm

ZATA, ANTONIO (1757-1797)

Zata was not an American, but his name is here noted because of his map, "La Pensilvania, La Nuova York, Il Jersey Settentrico, con La Parte Occidentalle Del Connecticut, Massachusset-sBay E L'Irochesia" (1778), which was based on John Mitchell's work, was used in the negotiations in Paris to determine boundaries of the new United States of America (Portinaro & Knirsch, 1987, p.270-271). Zata was a Venetian publisher who produced a large number of atlases and single maps in the latter part of the eighteenth century (p.318).

Unclassified - Geographic Items

During our search for books, textbooks, and maps produced by people during the period 1784-1812, we have come across a few items that have not been clearly identified as American or otherwise. That leads us to include those few items along with the brief amount of information, regarding them, that we have acquired.

(1804). Geographical, historical and political view of the island of St. Domingo. (np).

(1806). Geographical, statistical and political amusement; by which may be obtained a general and particular knowledge of the United States. In a series of interesting games on a map designed for the purpose. Philadelphia: J. Johnson. 1,43p. 18.5cm.

(c.1805). Geography an amusement: or, A complete set of geographical cards, by which the boundaries, situation, extent...and number of inhabitants of all the countries, kingdoms...may be learned by way of amusement, in a pleasing and satisfactory manner. By several persons conversant with maps... Burlington, NJ: D. Allinson. 78 cards in case. 13.5cm.

(c.1802). [A volume made up of 28 double maps.] (np). 22cm. (NUC ref. G 1019 A88, V.25, p.146).

(c.1803). [A volume made up of 18 maps, 8 double.] (np). (NUC ref. GZ At63, V.25, p.146).

(nd). [A volume made up of 39 maps.] (np). 22cm. (NUC ref NA 0481299, V.25, p.153).

Geographers' Vocation/Avocation

This categorization is intended to help those who might wish to find people among the First Generation of American Geographers according to their life activities.

ACTRESS
Rowson, Susanna

ARCHITECT
Bridges, William
Lafon, Bartholmew
Walker, Thomas
Non-US
Wright, Thomas

ASTRONOMER
Banneker, Benjamin
Burges, Bartholmew
Carleton, Osgood
Cutler, Manasseh
Ewing, John
Franklin, Benjamin
Freeman, Thomas
Lukens, John
Mansfield, Jared
Page, John
Rittenhouse, David
Smith, William
Vince, S.
Williams, Samuel
Non-US
Adams, George
Wilson, John
Wright, Thomas

AUCTIONEER
Clark, Matthew

AUTHOR
Brown, Charles
Condy, Jonathan
Cuming, Fortescue
Dean, James
Dewitt, Simeon
Drayton, John

Dwight, Timothy
Fisher, William
Flint, Abel
Freeman, James
Gass, Patrick
Goodenow, Sterling
Griscora, Jean
Harris, Thaddeus
Irving, Washington
Jones, T.
Lewis, Samuel
Lincoln, Benjamin
Lucas, Fielding
McLean, Archibald
Malhoum, John
Mitchell, Samuel
Moore, Samuel
Payne, John
Perry, William
Smith, Samuel
Tatham, William
Tongue, James
Weed, Enos
Williamson, Charles

Non-US

Adams, George
Barthelemy, Jean
Bowles, Carrington
Brooks, Richard
Phillips, Richard
Rennell, James
Taylor, George
Volney, Constantine
Walker, John
Wittman, William

AUTHOR- GEOGRAPHY TEXTBOOK

Bingham, Caleb
Branagan, Thomas
Cottineau, Dennis
Davidson, Robert
Davies, Benjamin
Doyle, David
Dwight, Nathaniel
Fisher, George
Fraser, Donald
Gaugh, John
Gleason, Benjamin
Harris, William
Henry, Robert
Hubbard, John
Iddings, James
Johnson, Jacob
Kimber, Emmor
Lathrop, John
Lewis, Samuel
Lucas, Fielding
Melish, John
Morse, Jedidiah
Nichols, Joseph
O'Neill, John
Parish, Elijah
Pattillo, Henry
Rawson, Andrew
Rowson, Susanna
Sampson, Ezra
Scott, Joseph
Smith, Thomas
Spafford, Horatio
Vince, S.
Webster, Noah
Williams, William
Winterbotham, William
Workman, Benjamin
Wright, S.

Non-US

Addington, Stephen
Aikin, John
Bigland, John
Butler, Samuel
Cottineau, de
Guthrie, William
Keith, Thomas
Phillips, Richard
Pinkerton, John
Smith, Charles
Tytler, Alexander

Walker, John

CARTOGRAPHER
Banker, Gerard
Bauman, Sebastian
Clark, William
Colles, Christopher
Cook, James
Darby, William
Eliot, J. B.
Ellicott, Andrew
Filson, John
Forsyth, G.
Graham, G.
Gray, William
Griffith, Dennis
Hills, John
King, Nicholas
Lambert, Samuel
Lay, Amos
Lewis, Samuel
Lodge, Benjamin
Lucas, Fielding
McMurray, William
Mansfield, John
Nau, Antoine
Norman, I.
Petrie, Edmund
Poppleton, Thomas
Prentis, William
Purcell, Joseph
Ratzer, Bernard
Romans, Bernard
Sauthier, Claude
Smith, Anthony
Stansbury, Arthur
Strother, John
Tatham, William
Taylor, B.
Warren, Moses
Watson, William
Wilson, John
Non-US
Blanchard, Joseph
Fisher, Joshua
Smith, Charles

CLERGY
Belknap, Jeremy
Cutler, Manasseh
Davidson, Robert
Dwight, Nathaniel
Dwight, Timothy
Ewing, John
Freeman, James
Harris, Thaddeus
Harris, William
Johnson, Jacob
McLean, Archibald
Madison, Bishop
Morse, Jedidiah
Page, John
Parish, Elijah
Pattillo, Henry
Smith, Samuel
Smith, Thomas
Smith, William
Vince, S.
Williams, Samuel
Williams, William
Winterbotham, William
Non-US
Addington, Stephen

COLLEGE PRESIDENT
Dwight, Timothy (Yale)
Madison, Bishop (William & Mary)

COMMISSIONER
Schuyler, Philip (US, Indian Affairs)
Tiffin, Edward (US, Gen. Land Office)

DIPLOMAT
Franklin, Benjamin

ECONOMIST
Carey, Mathew

EDITOR
Carey, Mathew
Williams, Samuel

EDUCATOR
Barton, Benjamin Smith
Bingham, Caleb
Buell, Abel
Carleton, Osgood
Colles, Christopher
Cutler, Manasseh
Darby, William
Davidson, Robert
Dwight, Nathaniel
Dwight, Timothy
Ewing, John
Fraser, Donald
Harris, William
Heart, Jonathan
Hubbard, John
Lambert, Samuel
Lathrop, John
Maclure, William
Madison, Bishop
Mansfield, Jared
Mitchell, Samuel
Moore, Samuel
Morse, Jedidiah
Rowson, Susanna
Smith, Samuel
Tupper, Benjamin
Webster, Noah
Workman, Benjamin
Non-US
Bigland, John
Tytler, Alexander

ENGINEER
Abert, John
Baldwin, Jeduthan (Chief Eng. Northern Army)
Beasley, A. M.
Colles, Christopher
Duportail, Louis Lebegue
Dwight, Nathaniel
Ellicott, Andrew
Erskine, Robert
Freeman, Thomas
Gaugh, John
Geddes, James
Lafon, Bartholmew
Non-US
Eliot, J.
L'Enfant, Charles

ENGRAVER
Abernethie, Thomas
Allen, Joel
Anderson, D.
Barker, Elihu
Barker, William
Bossler, Frederick
Buell, Abel
Callendar, Joseph
Doolittle, Amos
Fairman, David
Fairman, Gideon
Fitch, John
Harrison, Samuel
Harrison, William
Hill, Samuel
Hooker, William
Hutton, Isaac
Johnston, W.
Kearny, Francis
Kneass, William
McComb, James
Martin, David
Maverick, Peter
Norman, William
Peasley, A.
Poupard, James
Reed, Abner
Revere, Paul
Roberts, John
Scoles, John
Scot, Robert
Scott, Joseph

Sculpsit, Allardice
Seymour, Joseph
Shallus, Francis
Smith, William
Tanner, Benjamin
Tanner, Henry
Thackara, James
Tiebout, Cornelius
Vallance, John
Wilson, James
Windsor, E.

EXPLORER
Clark, George Rogers
Custis, Peter
Dunbar, William
Filson, John
Freeman, Thomas
Gass, Patrick
Lewis, Meriweather
Mackay, James
Pike, Zebulon
Sibley, John
Symmes, John
Walker, Thomas

FARMER [PLANTER]
Banneker, Benjamin
Darby, William
Putnam, Rufus
Tatham, William
Washington, George

GEOGRAPHER
Colles, Christopher
DeWitt, Simeon (Geog. to US)
Eddy, John
Eliot, J. B.
Erskine, Robert (Geog. to US)
Fitch, John
Folie, A.
Freeman, James
Hutchins, Thomas (Geog. to US)
McMurray, William
Melish, John
Morse, Jedidiah
Rowson, Susanna
Williams, William

Non-US

Keith, Thomas

GEOLOGIST
Forsyth, G.
Godon, Silvain
Maclure, William
Mitchell, Samuel Latham
Maclure, William
Mitchell, Samuel

GLOBE MAKER
Wilson, James

GOVERNOR
Clark, William (MO Territory)
Drayton, John (SC)
Lincoln, Benjamin (MA LT. Gov.)
Mathews, John (SC)
Page, John (VA)
Sargent, Winthrop (MS Territory & NW Territory)
Stoddard, Amos (LA Territory)
Sullivan, James (MA)
Tiffin, Edward (OH)
Wolcott, Oliver (CN)
Worthington, Thomas (OH)

HISTORIAN
Cottineau, Dennis
Filson, John
Fraser, Donald
Freeman, James
Williams, Samuel

Non-US

Pinkerton, John
Tytler, Alexander
Volney, Constantine

INSTRUMENT MAKER

Revere, Paul (Surgical)
Rittenhouse, David (Optical)

Non-US

Adams, George

INSURANCE

Petrie, Edmund

INVENTOR

Colles, Christopher
Erskine, Robert
Fitch, John
Rittenhouse, David
Spafford, Horatio
Tatham, William

JEWELER

Buell, Abel

JURIST

Bradley, Abraham Jr.
Brown, Charles
Cleveland, Moses
Cutler, Manasseh
Drayton, John
Fisher, George
Geddes, James
Hubbard, John
Jackson, John
Lathrop, John
Marshall, John (US Chief Justice)
Stoddard, Amos
Strother, John
Symmes, John
Tatham, William
Webster, Noah

LAND SPECULATOR

Hoops, Adam
Imlay, Gilbert
Ludlow, Israel
Sprout, Ebenezer
Walker, Thomas

LEGISLATOR

Armstrong, John (US)
Cleveland, Moses
Cutler, Manasseh (US)
Ellicott, Andrew (MD)
Fisher, George (US)
Geddes, James (NY)
Jackson, John (VA)
Mathews, John (SC)
Mitchell, Samuel (US)
Pickering, Octavius (US)
Ramsay, David (SC)
Schuyler, Philip (US)
Smith, Daniel (US)
Williamson, Charles (NY)
Worthington, Thomas (US)

LEXICOGRAPHER

Webster, Noah

MATHEMATICIAN

Banneker, Benjamin
Bowditch, Nathanial
Carleton, Osgood
Churchman, John
Ellicott, Andrew
Fitch, John
Flint, Abel
Freeman, James
Hassler, Ferdinand
Mansfield, Jared
Rittenhouse, David

Non-US

Adams, George

MECHANIC

Fairman, Gideon

MERCHANT

Blodget, William
Blunt, Edmund
Clark, Matthew
Dumbiban, Daniel
Dwight, Timothy

Heart, Jonathan
Lambert, Samuel
Lucas, Fielding
Melish, John
Norman, William
Pickering, Octavius
Revere, Paul

Non-US

Cary, John

METEOROLOGIST

Cutler, Manasseh
Lewis, Meriweather

MILITARY

Allen, Ira
Armstrong, John
Buell, Abel
Clark, George Rogers
Cottineau, Dennis
Duportail, Louis (Continental Army, Commandant of Eng.)
Eliot, J. B.
Erskine, Robert
Gass, Patrick
Gray, William
Heart, Jonathan
Holland, Samuel
Lear, Tobias
Lewis, Meriweather
Lincoln, Benjamin (Sec. of War)
Nichols, Joseph
Page, John
Pickering, Octavius (Quartermaster Gen.)
Pike, Zebulon
Porter, Andrew
Putnam, Rufus
Sherman, Isaac
Steuben, Frederick (Cont. Army Inspector Gen.)
Stoddard, Amos
Tupper, Benjamin
Walker, Thomas
Washington, George
Williams, William

Non-US

Ross, Charles

MILLWRIGHT

Putnam, Rufus

MINT DIRECTOR

Moore, Samuel (Philadelphia)

NATURALIST

Barton, Benjamin Smith
Bartram, William
Cutler, Manasseh
Drayton, John
Harris, Thaddeus
Lewis, Meriweather
Mitchell, Samuel
Pickering, Octavius

Non-US

Aikin, John

NAVIGATION

Bowditch, Nathanial

NAVY

Cottineau de Klogven, Dennis Louis

NOVELIST

Brown, Charles
Rowson, Susanna

PHILOSOPHER

Non-US

Volney, Constantine

PHYSICIAN

Barton, Benjamin Smith
Dwight, Nathaniel
Mitchell, Samuel
Moore, Samuel

Ramsay, David
Sibley, John
Tiffin, Edward
Walker, Thomas

Non-US

Aikin, John
Brooks, Richard
Butler, Samuel

POSTMASTER

Bradley, Abraham Jr.

U. S. PRESIDENT

Jefferson, Thomas
Washington, George

PRINTER

Adams, James
Norman, William

PUBLISHER

Adams, James
Banker, Evart
Bartram, A.
Belknap, Jeremy
Blodger, William
Blunt, Edmund
Carleton, Osgood
Carey, Mathew
Churchman, John
Clark, Matthew
Cramer, Zadok
Franklin, Benjamin
Hooker, William
Hunter & Beaumont
Kimber, Emmor
Lucas, Fielding
Melish, John
Norman, John
Ratzer, Bernard
Reid, John
Romans, Bernard
Sauthier, Claude
Stevenson, Roger
Thomas & Carlisle
Webster, Noah

Non - US

Bennett, John
Fisher, Joshua
Sayer, Robert
Zata, Antonio

RIVERMAN

Gaugh, John

SCIENTIST

Franklin, Benjamin

STATESMAN

Allen, Ira
Franklin, Benjamin

SURVEYOR

Abert, John J.
Allen, Ira (Surveyor Gen.)
Banneker, Benjamin
Barker, Elihu
Barton, Benjamin Smith
Bauman, Sebastian
Bever, John
Biggs, Zaccheus
Blaskowitz, C.
Bowditch, Nathanial
Bridges, William
Briggs, Isaac
Buckingham, Ebenezer
Burchmore, George
Cahoone, John
Carleton, Osgood
Christmas, William
Churchman, John
Clark, George Rogers
Cleveland, Moses
Coles, Thomas
Colles, Christopher
Cook, James
Cutler, Manasseh
Darby, William

Davis, William
DeWitt, Moses
Dewitt, Simeon (NY Surveyor Gen.)
Dowse, Edward
Dumain, Louis
Dumbiban, Daniel
Duportail, Louis
Dwight, Nathaniel
Ellicott, Andrew (US Surveyor Gen.)
Erskine, Robert (Cont. Army Surveyor Gen.)
Evans, Lewis
Ewing, John
Fitch, John
Flint, Abel
Folie, A.
Fosdick, Nicoll
Freeman, Thomas
Furlong, Lawrence
Garner
Geddes, James
Gibson, Robert
Gilbert, Joseph
Gillet, George (CN Surveyor Gen.)
Goerck, Casimir
Grant, J.
Hall, John
Hardenberg, Abraham
Hardenburgh, John
Harris, Caleb
Harris, Harding
Hassler, Ferdinand (Superintendent of US Coast Survey)
Heart, Jonathan
Hills, John
Holland, Samuel (Surveyor Gen.)
Hoops, Adam
Howell, Reading
Hutchins, Thomas
Imlay, Gilbert
Jackson, John
Johnston, Robert
Lafon, Bartholmew
Langham, Elias
Ludlow, Israel
Lukens, John (US Surveyor Gen.)
Maclain, Alexander
McLean, Archibald
Madison, Bishop
Mangin, Joseph
Mansfield, Jared (US Surveyor Gen.)
Martin, Absalom
Mathews, John
Morris, William
Nancarrow, John
Neville, Joseph
Parker, Alexander
Pease, Seth
Pinkham, Paul
Poppleton, Thomas
Porter, Andrew
Price, Jonathan
Putnam, Rufus (US Surveyor Gen.)
Putnam, William
Randel, John
Rittenhouse, David
Ropes, William
Sargent, Winthrop
Schuyler, Philip (NY Surveyor Gen.)
Scull, William
Sherman, Isaac
Simpson, James
Smith, Daniel
Smith, Thomas
Sproule, George
Sprout, Ebenezer
Stevenson, Roger
Strother, John
Sturges, Daniel (GA Surveyor Gen.)
Symmes, John
Tatham, William
Tupper, Anselm
Tupper, Benjamin
Varle', Charles
Warren, Moses
Washington, George
Wheeler, T.
Whipple, Levi

Whitelaw, James (VT Surveyor Gen.)
Williams, Samuel
Wolcott, Oliver
Worthington, Thomas
Wright, T.

Non-US

Fisher, Joshua
Ross, Charles

TOMBSTONE CUTTER

Walker, Thomas

TREASURY SECRETARY

Wolcott, Oliver

UNDERTAKER

Buell, Abel

Locator

The numbers that follow names indicate page location.

Non-US

Unclassified Geographic Items

References

Adler, James B. (Ed. & Pub.). (1977). CIS US serial set indes - Part I, American state papers and the 15th - 34th Congresses, 1789-1857. Washington, DC: Congressional Information Services, Inc.

The American Library Association. (1976). The national union catalog pre-1956 imprints. Chicago: author.

American Publishers Association. (1907). Herringshaw's American biography. Chicago: author.

Anderson, Frank J. (Ed.). (1970). Geography and travels. Spartanburg, SC: Wofford Library Press.

Antonelli, M.F. 1970. Nationalism in early American geographies, 1784-1845, Journal of Geography 69: 301-305.

Barnhart, Clarence L., & Halsey, William D. (Eds.). (1954). The new century cyclopedia of names. New York: Appleton-Century-Crofts. LOC# CT103.C397

Barton, William. (1813). Memoirs of the life of David Rittenhouse, LLD. F.R.S. Late President of the American Philosophical Society, &c. Interspersed with various notices of many distingushed men: with an appendix, containing sundary philosophical and other papers, most of which have not hitherto been published. Philadelphia: Edward Parker.

Bedini, Silvio A. (1972). The life of Benjamin Banneker. New York: Charles Scribner's Sons.

bibliomania.com (2001). (on-line biographical information).

The biographical dictionary. (2001). Dwight, Timothy. [On-line]. Available: s9.com/biography.

The biographical dictionary. (2001). Ellicott, Andres. [On-line].

Available: s9.com/biography.

The biographical dictionary. (2001). Filson, John. [On-line]. Available: s9.com/biography.

The biographical dictionary. (2001). Fitch, John. [On-line]. Available: s9.com/biography.

The biographical dictionary. (2001). Freeman, James. [On-line]. Available: s9.com/biography.

The biographical dictionary. (2001). Hutchins, Thomas. [On-line]. Available: s9.com/biography.

The biographical dictionary. (2001). Lewis, Meriwether. [On-line]. Available: s9.com/biography.

The biographical dictionary. (2001). Lincoln, Benjamin. [On-line]. Available: s9.com/biography.

The biographical dictionary. (2001). Morse, Jedidiah. [On-line]. Available: s9.com/biography.

The biographical dictionary. (2001). Page, John. [On-line]. Available: s9.com/biography.

The biographical dictionary. (2001). Pike, Zebulon. [On-line]. Available: s9.com/biography.

The biographical dictionary. (2001). Putnam, Rufus. [On-line]. Available: s9.com/biography.

The biographical dictionary. (2001). Schuyler, Philip. [On-line]. Available: s9.com/biography.

The biographical dictionary. (2001). Smith, Samuel . [On-line]. Available: s9.com/biography.

The biographical dictionary. (2001). Steuben, Friedrich. [On-line]. Available: s9.com/biography.

The biographical dictionary. (2001). Sullivan, James. [On-line]. Available: s9.com/biography.

The biographical dictionary. (2001). Tiffin, Edward. [On-line]. Available: s9.com/biography.

Bowman, John S. (1995). The Cambridge dictionary of American biography. Cambridge, United Kingdom: University Press.

Boyd, Julian P. (1942). Horatio Gates Spafford - Inventor, author, promoter of democracy. Worcester, MA: The American Antiquarian Society.

Brown, John Howard. (1897). The cyclopedia of American biographies. Volume I. Boston: The Cyclopedia Publishing Company. Note.

Volume II (1900): Lamb's biographical dictionary of the United States. Boston: James H. Lamb Co.

Volume III as above.

Volume IV (1901).

Volume V (1903): The cyclopedia of American biographies. Comprising the men and women of the United States who have been identified with the growth of the nation. Boston: Federal Book Company of Boston.

Volume VI as above.

Volume VII as above.

Brown, Lloyd Arnold. (1959). Early maps of the Ohio Valley - A selection of maps, plans, and views made by Indians and Colonials from 1673 to 1783. Pittsburgh, PA: University of Pittsburgh Press.

Brown, Ralph H. (1941). The American geographies of Jedidiah Morse. Annals of the Association of American

Geographers, 31(3), 145-217.

Butterfield, Consul Willshire. (1885). Journal of Capt. Jonathan Heart on the march with his company from Connecticut to fort Pitt, in Pittsburgh, Pennsylvania, from the seventh of September, to the twelfth of October, 1784-1785, inclusive; to which is added the Dickinson-Harmar correspondence of 1784-5; the whole illustrated with notes and preceded by a biographical sketch of captain Heart by Consul Wilshire Butterfield... Albany, NY: J. Munsell's sons, 94p., 22x18cm.

DeWitt, Donald. (n.d.). Pike and Pike's Peak. A brief life of Zebulon Montbomery Pike and extracts from his journal of exploration. Colorado Springs, CO: Colorado College.

Dickinson, Robert E. (1969). The makers of modern geography. New York: Frederick A. Praeger.

Durham, Walter T. (1976). Daniel Smith, frontier statesman. Gallatin, TN: Sumner County Library Board.

Dzwonkoski, Peter. (nd). Dictionary of literary biography. Vol. 49, American literary publishing houses, 1638-1899, Part 1: A-M. Detroit, Michigan: A Bruccoli Clark Book, Gale Research Company.

Encarta encyclopedia. (2001). Dwight, Timothy. [On-line]. Available: encarta.msn.com.

Encarta encyclopedia. (2001). Hassler, Ferdinand. [On-line]. Available: encarta.msn.com.

Encarta encyclopedia. (2001). Lewis, Meriwether. [On-line]. Available: encarta.msn.com.

Encarta encyclopedia. (2001). Marshall, John. [On-line]. Available: encarta.msn.com.

Encarta encyclopedia. (2001). Pike, Zebulon. [On-line]. Available: encarta.msn.com.

Encarta encyclopedia. (2001). Putnam, Rufus. [On-line]. Available: encarta.msn.com.

Encarta encyclopedia. (2001). Rittenhouse, David. [On-line]. Available: encarta.msn.com.

Encarta encyclopedia. (2001). Schuyler, Philip. [On-line]. Available: encarta.msn.com.

Encarta encyclopedia. (2001). Steuben, Friedrich. [On-line]. Available: encarta.msn.com.

The ncyclopedia Americana - International edition. (1988). Danbury, CT: Grolier Inc. V. 3.

Encyclopedia Britannica. (2001). Cutler, Manasseh. [On-line]. Available: www.britannica.com.

Encyclopedia Britannica. (2001). Dwight, Timothy. [On-line]. Available: www.britannica.com.

Encyclopedia Britannica. (2001). Fitch, John. [On-line]. Available: www.britannica.com.

Encyclopedia Britannica. (2001). Geddes, James. [On-line]. Available: www.britannica.com.

Encyclopedia Britannica. (2001). Lincoln, Benjamin. [On-line]. Available: www.britannica.com.

Encyclopedia Britannica. (2001). Maclure, William. [On-line]. Available: www.britannica.com.

Encyclopedia Britannica. (2001). Morse, Jedidiah. [On-line]. Available: www.britannica.com.

Encyclopedia Britannica. (2001). Pike, Zebulon. [On-line]. Available: www.britannica.com.

Encyclopedia Britannica. (2001). Putnam, Rufus. [On-line]. Available: www.britannica.com.

Encyclopedia Britannica. (2001). Revere, Paul. [On-line]. Available: www.britannica.com.

Encyclopedia Britannica. (2001). Rittenhouse, David. [On-line]. Available: www.britannica.com.

Encyclopedia Britannica. (2001). Rowson, Susanna. [On-line]. Available: www.britannica.com.

Encyclopedia Britannica. (2001). Schuyler, Philip. [On-line]. Available: www.britannica.com.

Encyclopedia Britannica. (2001). Steuben, Frederick. [On-line]. Available: www.britannica.com.

Encyclopedia Britannica. (2001). Volney, Constantin-Francois. [On-line]. Available: www.britannica.com.

Encyclopedia Britannica. (2001). Walker, Thomas. [On-line]. Available: www.britannica.com.

Foster, James William. (1956). Fielding Lucas, Jr., Early 19th century publisher of fine books and maps. Worcester, MA: American Antiquarian Society.

Freeman, T. W. (1961). A hundred years of geography. Chicago: Aldine Publishing Company.

Guthorn, Peter J. (1984). United States coastal charts 1783-1861. Exton, PA: Schiffer Pub. Ltd.

Hall, G. K. (1973). The U. S. Library of Congress Geography and Map Division. Boston: author.

Harley, J.B., Petchenik, Barbara Bartz, & Towner, Lawrence W. (1978). Mapping the American revolutionary war. Chicago: University of Chicago Press.

Halsey, Francis, W. (1905). Historical and biographical introduction to Susanna Haswell Rowson, Charlotte Temple: A Tale of Truth. New York: Funk and Wagnalls.

Harley, J. B., Petchenik, Barbara Barth, & Towner, Lawrence W. (1978). Mapping the American Revolutionary War. Chicago: The University of Chicago Press.

Hartshorne, Richard. (1961). The nature of geography: A critical survey of current thought in the light of the past. Lancaster, PA: The Science Press Printing Company.

Hartshorne, Richard. (1959). Perspective on the nature of geography. Chicago: Rand McNally & co. for the Association of American Geographers.

Herndon, G. Melvin. (1969). William Tatham and the culture of tobacco. Including a facsimile reprint of an historical and practical essay on the culture and commerce of tobacco by William Tatham. Coral Gables, FL: University of Miami Press.

Herringshaw, Thomas William. (Ed.). (1898). Herringshaw's encyclopedia of American biography of the nineteenth century. Accurate and succinct biographies of Famous men and women in all wolks of life who are or have been the acknowledged leaders of life and thought of the United States since its formation. Entirely covering the field of literature, science, history, politics, commerce, and the mechanical and industrial arts. Chicago: American Publishers' Association. LOC# E663.H567

Heusser, Albert H. (see Schmidt, Hubert G.).

Hicks, Frederick Charles (Ed.). (1904). Thomas Hutchins - A topographical description of Virginia, Pennsylvania, Maryland, and North Carolina, reprinted from the original of 1778. Cleveland, OH: The Burrows Brothers Company.

Holt-Jensen, Arild. (1980). Geography: Its history and concepts. Lotowa, NJ: Barnes & Noble Books.

Hopkins, J.G.E. (Ed.). (1964). Concise dictionary of American biography. New York: Charles Scribner's Sons.

Howes, Wright. (1962). U.S.iana, 1650-1950: A selective bibliography in which are described 11,620 uncommon and significant books relating... New York: Bowker for the Newberry Library. LOC# Z1236.H8

Ingpen, Roger. (1908). The love letters of Mary Wollstonecraft to Gilbert Imlay. London: Hutchinson & Co. James, P. E.

Johnson, Allen, & Malone, Dumas. (1928). Dictionary of American biography. New York: Charles Scribner, & Sons.

Johnson, Frederick C. (1911). Rev. Jacob Johnson, M.A. - Pioneer preacher of Wyoming Valley (Wilkes-Barre, Pa.) 1772-1790 - First settled pastor, First Presbyterian Church, Wilkes-Barre. Wilkes-Barre, PA: Read before the Wyoming Historical and Geological Society and reprinted from Vol. XI, of the Proceedings.

Klemp, Egon (Ed.). (1976). America in maps dating from 1500 to 1856. New York: Holmes and Meier, Publishers.

Library of Congress. (1960). A guide to the study of the U.S.A. Washington: author.

Marquis Company. (1963). Who was who in America, Historical volume 1607-1896. Chicago: The A. N. Marquis Company. Martin, G. J.

Mathews, Catharine Van Cortlandt. (1908). Andrew Ellicott, his life and letters. New York: The Grafton Press. Meisel, Max. (Vol. I,1924; Vol. II, 1926; Vol. III, 1929). A bibliography of American natural history. The pioneer century, 1769-1865. Brooklyn, NY: The Premier Publishing Company.

Meisel, Max. (1926). A bibliography of American natural history - The pioneering century, 1769-1865. New York: The Premier Publishing Co. V.II, 1926;

Meriwether, Margaret Babcock. (Ed.). (1943). The Carolinian

florist of Governor John Drayton of South Carolina, 1766-1822 - With water-color illustrations from the author's orininal manuscript and an autobiographical introduction. Columbia, SC: The South Caroliniana Library of the University of South Carolina.

Merrian, G. & C. Co., Pub. (1972). Webster's new geographical dictionary. Springfield, MA: author.

Morris, George Spencer. (1906). William Bartram. Cassinia - A Bird Annual - Proceedings of the Delaware Valley Ornithological Club, No. X, 1-9.

Nason, Elias. (1870). A memoir of Mrs. Susanna Rowson. Albany: Joel Munsell.

(NAT) (1898). National cyclopedia of American biography. New York: White.

O'Shea, M. V. (Ed.). Eyes and no eyes and other stories by Dr. Aiken, Mrs. Barbauld, Mrs. Marcet, and Jane Taylor. Boston: D. C. Heath & Co. Parker, Patricia L. 1986. Susanna Rowson. Boston: Twayne Publishers.

Pattison, William David. (1957). Beginnings of the American rectangular land survey system, 1784-1800. Chicago: The University of Chicago Press.

Portinaro, Pierluigi, & Knirsch, Franco. (1987). The cartography of North America 1500-1800. New York: Crescent Books.

Randall, John H., and Haines, George. (1946). Controlling assumptions in the practice of American historians. In theory and practice in historical study: A report of the committee on historiography. Chapter II, pp. 15-52. New York: Social Science Research Council.

Ristow, Walter W. (1985). American maps and mapmakers - Commercial Cartography in the nineteenth century. Detroit, MI: Wayne State University Press.

Ristow, Walter W. (Ed.). (1961). A survey of the roads of the United States of America 1789 by Christopher Colles. Cambridge, MA: The Belknap Press of Harvard University Press.

Rowland, Della. (1989). The story of Sacajawea: Guide to Lewis and Clark. Richard Leonard (Illustrator). New York: Bantam Doubleday Dell Books.

Sahli, J.R. (1941). An analysis of early American geography textbooks from 1784-1840. Unpublished doctoral dissertation, University of Pittsburgh, Pennsylvania.

Schmidt, H. G. (Ed.). (1966). Heusser, A. H. (1928). George Washington's map maker - A biography of Robert Erskine. New Brunswick, NJ: Rutgers University Press.

Schwartz, Seymour I., & Ehrenberg, Ralph E. (1980). The mapping of America. New York: Harry N. Abrams, Inc.

Simms, Jeptha R. (1845). History of Schoharie County and the Border Wars of New York. Albany: Munsell and Tanner.

Smith, B. A., & Vining, J. W. (1989, October). The upper-level geographies of Samuel G. Goodrich and William C. Woodbridge: A comparative study. Paper presented at the annual meeting of the National Council for Geographic Education.Hershey, Pennsylvania.

Stephen, Leslie, & Lee, Sidney. (Eds.). (1917). The dictionary of national biography, Founded in 1882 by George Smith, From the earliest times to 1900. Oxford, UK: Oxford University Press.

Stoddart, D. R. (1986). On geography and its history. New York: Basil Blackwell, Ltd.

Tooley, R. V. (1985). The mapping of America. London, England: The Holland Press Limited.

Townsend, John Wilson. (1907). Kentuckians in history and literature. New York: The Neale Publishing Company.

Tucher, Andrea J. (Comp.). (1985). Natural history in America 1609-1806. Printed works in the collections of The American Philosophical Society, The Historical Society of Pennsylvania, and the Library Company of Philadelphia. New York: Garland Publications.

Tucker, Louis Leonard. (1990). Clio's consort: Jeremy Belknap and the founding of the Massachusetts Historical Society. Boston: The Massachusetts Historical Society.

Vining, J., & Smith, B. A. (1998 October). How Twentieth-Century Geographers Have Viewed Their Pre-Civil War Predecessors. Paper presented at the annual meeting of the National Council for Geographic Education. Indianapolis, IN, October 11-14, 1998.

Webster, A. Merriam. (1943). Webster's biographical dictionary. Springfield, Mass.: G. & C. Merriam Co. LOC# CT103.W382

Weil, Dorothy. (1976). In defense of women: Susanna Rowson (1762-1824). University Park: The Pennsylvania State University Press.

White, J. T. & Co. (1897). The national cyclopedia of American biography being the history of the United States. New York: author.

Wilson, James Grant, & Fiske, John. (Eds.). (1889). Appletons' Cyclopedia of American Biography. New York: D. Appleton and Co.

Index

This "Index" contains references to map and other publication titles, additionally one could expect to find more than what is indicated by the titles of books, maps, textbooks, and atlases within the content of those publications. References within this Index may be found by consulting the Locator for the name referenced.

About the Authors

BEN A. SMITH is Professor Emeritus of the Departments of Elementary Education and Geography at Kansas State University. He has also taught at the University of Southern Mississippi and in Georgia public schools. He is the former coordinator of the Kansas Geographic Alliance, president of the Kansas Council for the Social Studies, and coordinator of the Kansas Geographic Bee. A former editor of *Geographic Insights* and *The Journal of Social Studies Research,* he has published widely.

JAMES W. VINING is Professor of Geography at Western Illinois University.